About Island Press

Island Press is the only nonprofit organization in the United States whose principal purpose is the publication of books on environmental issues and natural resource management. We provide solutions-oriented information to professionals, public officials, business and community leaders, and concerned citizens who are shaping responses to environmental problems.

In 2003, Island Press celebrates its nineteenth anniversary as the leading provider of timely and practical books that take a multidisciplinary approach to critical environmental concerns. Our growing list of titles reflects our commitment to bringing the best of an expanding body of literature to the environmental community throughout North America and the world.

Support for Island Press is provided by The Nathan Cummings Foundation, Geraldine R. Dodge Foundation, Doris Duke Charitable Foundation, Educational Foundation of America, The Charles Engelhard Foundation, The Ford Foundation, The George Gund Foundation, The Vira I. Heinz Endowment, The William and Flora Hewlett Foundation, Henry Luce Foundation, The John D. and Catherine T. MacArthur Foundation, The Andrew W. Mellon Foundation, The Moriah Fund, The Curtis and Edith Munson Foundation, National Fish and Wildlife Foundation, The New-Land Foundation, Oak Foundation, The Overbrook Foundation, The David and Lucile Packard Foundation, The Pew Charitable Trusts, The Rockefeller Foundation, The Winslow Foundation, and other generous donors.

The opinions expressed in this book are those of the author(s) and do not necessarily reflect the views of these foundations.

Monitoring Ecosystems

Monitoring Ecosystems

Interdisciplinary Approaches for
Evaluating Ecoregional Initiatives

Edited by
David E. Busch • Joel C. Trexler

Foreword by
Lance H. Gunderson

ISLAND PRESS
Washington • Covelo • London

Library of Congress Cataloging-in-Publication Data

Monitoring ecosystems : interdisciplinary approaches for evaluating ecoregional initiatives / edited by David E. Busch and Joel C. Trexler.
    p. cm.
Includes bibliographical references.
 ISBN 1-55963-850-8 (cloth : alk. paper)—ISBN 1-55963-851-6 (pbk.: alk. paper)
 1. Environmental monitoring. I. Busch, David E. II. Trexler, Joel C.
 QH541.15.M64 M658 2002
 333.95—dc21
                        2002015727

British Cataloguing-in-Publication Data available.

Book design by Artech Group, Inc.

Manufactured in the United States of America
09 08 07 06 05 04 03   8 7 6 5 4 3 2 1

Contents

Foreword xi

Preface xvii

Part I   Introduction

Chapter 1   The Importance of Monitoring in Regional
               Ecosystem Initiatives   1
               David E. Busch and Joel C. Trexler

Part II   **Principles of Ecosystem Monitoring Design**

Chapter 2   Conceptual Issues in Monitoring Ecological
               Resources   27
               Barry R. Noon

Chapter 3   Design of an Ecological Monitoring Strategy for the
               Forest Plan in the Pacific Northwest   73
               *Paul L. Ringold, Barry Mulder, Jim Alegria, Raymond L.
               Czaplewski, Tim Tolle, and Kelly Burnett*

Chapter 4   Monitoring for Adaptive Management of the Colorado
               River Ecosystem in Glen and Grand Canyons   101
               Lawrence E. Stevens and Barry D. Gold

Chapter 5   Science Strategy for a Regional Ecosystem Monitoring
               and Assessment Program: The Florida Everglades
               Example   135
               John C. Ogden, Steven M. Davis, and Laura A. Brandt

**Part III   Information Management and Modeling for
               Monitoring Programs**

Chapter 6   The Use of Models for a Multiscaled Ecological
               Monitoring System   167
               *Donald L. DeAngelis, Louis J. Gross, E. Jane Comiskey,
               Wolf M. Mooij, M. Philip Nott, and Sarah Bellmund*

Chapter 7   Role of Knowledge-Based Systems in Analysis and
               Communication of Monitoring Data   189
               Keith M. Reynolds and Gordon H. Reeves

Chapter 8   Approaches to Quality Assurance and Information
               Management for Regional Ecological Monitoring
               Programs   211
               Craig J. Palmer

Chapter 9   Estimation of Change in Populations and Communities
               from Monitoring Survey Data   227
               John R. Sauer, William A. Link, and James D. Nichols

**Part IV   Monitoring Habitats, Populations, and
               Communities**

Chapter 10   Competing Goals of Spatial and Temporal Resolution:
               Monitoring Seagrass Communities on a Regional
               Scale   257
               James W. Fourqurean and Leanne M. Rutten

Chapter 11   Late-Successional Forest Monitoring in the Pacific
               Northwest   289
               Miles A. Hemstrom

Chapter 12   Monitoring Wetland Ecosystems Using Avian
              Populations: Seventy Years of Surveys in the
              Everglades   321
              Peter Frederick and John C. Ogden

Chapter 13   Setting and Monitoring Restoration Goals in the
              Absence of Historical Data: The Case of Fishes in
              the Florida Everglades   351
              Joel C. Trexler, William F. Loftus, and John H. Chick

Chapter 14   Monitoring Biodiversity for Ecoregional
              Initiatives   377
              William L. Gaines, Richy J. Harrod, and John F. Lehmkuhl

Part V   **Summary and Synthesis**

Chapter 15   Monitoring, Assessment, and Ecoregional Initiatives:
              A Synthesis   405
              Joel C. Trexler and David E. Busch

Glossary   425

Contributors   429

Index   433

Learning to Monitor or Monitoring to Learn?

Lance H. Gunderson

When I first reviewed the manuscript for this volume, I was reminded of a set of meetings in the early 1980s in which I participated as a scientist with the South Florida Research Center at Everglades National Park. The focus of those meetings was to develop monitoring programs in order to provide information to managers on the status and trend of the resources under the management of the National Park Service. The research center was one of a number of such centers established to help understand the complexities of ecological systems and provide better information for their long-term preservation. This was part of the national trend to apply scientific and quantitative techniques to solve resource problems.

The discussions at the meetings revealed a number of uncertainties regarding how best to provide information that was both relevant to management and would increase understanding of these dynamic systems. Looking back with the wisdom of hindsight, the questions we were confronting all centered on theoretical and operational issues of monitoring that can be summarized in three questions: Why should we monitor? What should we monitor? How should we monitor? The good news is that much has been accomplished in the past twenty years toward answering these questions, and the manifestation of that achievement is embodied in this book. I'll structure the remainder of this prologue by providing my answers to the first question regarding the context and rationale for monitoring. More elaborate answers to the other questions can be found in this volume.

Monitoring is essential to managing resource systems that are characterized by crises and surprises (Gunderson et al. 1995). Walters (1997), Johnson et al. (1999), and Gunderson (1999) have stressed the practical difficulties that humans face in attempting to manage complex ecosystems. The multiple scales of variables, as well as cross-scale and nonlinear interactions, generate multi-stable behaviors in ecosystem dynamics (Holling 1995; Gunderson and Holling 2001). In turn, the surprises generated by this multi-stable behavior create a range of problems for management (Carpenter et al. 1999; Gunderson 2000). This inherent uncertainty of resource systems and the need for continuous learning are what spawned the approach to resource management called *adaptive environmental assessment and management* (Holling 1978; Walters 1986). Adaptive management was developed as an antithesis to resource management approaches in which ecosystems were considered predictable, where impacts could be determined a priori, and where management consisted of "command and control" attitudes toward nature. Hence, the short answer to "why monitor?" is because monitoring provides the basis for learning and understanding in an unpredictable world.

Adaptive management in its early form focused on confronting the uncertainty of resource dynamics through actions designed for learning (Holling 1978; Walters 1986). This evolved from a process of testing a single hypothesis about the system to sorting among multiple hypotheses, each of which may have different social and management implications (Gunderson 1999). One of the most common problems faced by resource analysts and managers is the paucity of data required to put hypotheses at risk (Walters 1997). A number of successful adaptive assessments have been made because sufficient data were available to allow for the winnowing from a set of hypotheses to a single set that could be tested by management actions (Walters 1997; Johnson et al. 1999). In the majority of these cases, those data were the result of long-term monitoring programs.

Although such positive uses of monitoring data can be documented, the use of monitoring results to resolve resource management issues has had a checkered past. Monitoring information can be used in a positive sense to help resolve issues, or it can be used to inhibit issue resolution and learning. The uncertainties associated with resource management goals and objectives (Lee 1993), organizational complexity of bureaucracies (Westley 2001), and political pathologies (Pritchard and Sanderson 2001) are all obstacles to learning. The use of monitoring data in adaptive management frameworks can be compromised by self-serving interests of organizations, career concerns and greed among scientific experts, and disinformation campaigns by opposing sides that exploit the uncertainty of resource issues to maintain the status quo. An extreme example of this is found in a quotation from a resource manager who stated, "All I want from scientists are data that I can use in court" (Gunderson et al. 1995).

In spite of these problems, the data generated from monitoring programs have proved to be most useful in the resolution of complicated resource issues. The interpretation of data was the linchpin in the recovery from the environmental "train wrecks" of the 1980s and 1990s, such as those found in the forest issues of the Pacific Northwest and water resource issues in southern Florida. The results of monitoring programs are no panacea, nor do they resolve all issues, but at times monitoring information forms the basis of new and novel approaches to management. For example, the formula by which water is delivered to Everglades National Park was derived from the analysis of long-term hydrologic data.

In case after case, monitoring has been shown to be a critical component for understanding the loss of ecological resilience and building adaptive capacity. *Ecological resilience* is defined as the amount of disturbance that a system can take before it changes key sets of controlling processes and structures. *Adaptive capacity* is the ability to deal with unpredictable change in these dynamic systems. The key ingredients for building adaptive capacity in ecosystems involve: confronting uncertainty, monitoring key ecosystem variables (especially slow-changing ones), and developing learning institutions (Walters 1986; Gunderson 2000). Uncertainty is confronted through the articulation of a set of competing hypotheses that describe what led to the loss of resilience and what is needed to restore those lost ecosystem functions and services. Those hypotheses are being put at risk through a structured set of management actions designed to test and sort among the alternative explanations and a comprehensive monitoring plan established through decades of research. The slow response variables in the Everglades system include nutrients in sediments, and decadal hydrologic cycles. Monitoring of these variables is critical as they are the key indicators of ecosystem resilience.

Yet, though managers are increasingly aware that monitoring is important, it still seems to be the first victim of budget cuts. As contributors to this volume point out, cases when monitoring was eliminated because of budget restrictions have proved to be ecologically critical years. It was during these critical periods when the system underwent a major transformation, yet those years became missing points on time-series plots. Managers must sometimes make hard decisions about what to cut and what to keep, as well as when to begin and when to discontinue monitoring programs. In part, these decisions are made because monitoring programs are expensive and their long-term benefits are often difficult to discern. The demonstration of such benefits is one of the hard-won lessons documented in this volume. Those benefits will likely contribute to future discussions on how to balance monitoring with other programs and how to provide sustained support in uncertain and fluctuating fiscal environments.

Over the past two decades, the context of monitoring has changed. Monitoring used to be a means to its own end. That is, monitoring was done for the sake of monitoring. At that time all variables were deemed important and, hence, everything should be monitored. The results of attempts to monitor everything led to frustration and failures. Monitoring the processes and structures of large-scale ecosystems has progressed tremendously in the past two decades. That progress is demonstrated by this volume's authors, who show the value of monitoring programs in contributing to resolution of contentious and difficult resource issues. The authors also demonstrate the progress made in learning how to monitor key ecosystem variables and the appropriate spatial and temporal dimensions for those variables. Most important, the authors lay the foundation for the next step: the transition from learning to monitor to monitoring to learn. Helping to build the adaptive capacity of these systems is the legacy of this volume.

LITERATURE CITED

Carpenter, S., W. Brock, and P. Hanson. 1999. Ecological and social dynamics in simple models of ecosystem management. *Conservation Ecology* 3 (2):4. [Online at www.consecol.org/vol3/iss2/art4]

Gunderson, L. H. 1999. Resilience, flexibility and adaptive management: Antidotes for spurious certitude. *Conservation Ecology* 3(1):7. [Online at www.consecol.org/vol3/iss1/art7]

———. 2000. Ecological resilience: In theory and application. *Annual Review of Ecology and Systematics* 3:425–439.

Gunderson, L. H., and C. S. Holling, eds. 2001. *Panarchy: Understanding transformations in systems of humans and nature.* Washington, D.C.: Island Press.

Gunderson, L. H., C. S. Holling, and S. Light, eds. 1995. Barriers and bridges to the renewal of ecosystems and institutions. New York: Columbia University Press.

Holling, C. S. 1978. *Adaptive management.* New York: John Wiley and Sons.

———. 1995. What barriers? What bridges? Pp. 3–34 in *Barriers and bridges to the renewal of ecosystems and institutions,* edited by L. H. Gunderson, C. S. Holling, and S. S. Light. New York: Columbia University Press.

Johnson, N. K., F. Swanson, M. Herring, and S. Greene, eds. 1999. *Bioregional assessments: Science at the crossroads of management and policy.* Washington, D.C.: Island Press

Lee, K. 1993. *Compass and gyroscope*. Washington, D.C.: Island Press.

Pritchard, L., Jr., and S. Sanderson. 2001. The dynamics of political discourse in seeking sustainability . Pp. 147–169 in *Panarchy: Understanding transformations in systems of humans and nature*, edited by L. H. Gunderson and C. S. Holling. Washington, D.C.: Island Press.

Walters, C. J. 1986. *Adaptive management of renewable resources*. New York: McGraw Hill.

———. 1997. Challenges for adaptive management. *Conservation Ecology* 1 (2):1. [Online at www.consecol.org/vol1/iss2/art1]

Westley, F. 2001. The devil is in the dynamic. Pp 333–360 in *Panarchy: Understanding transformations in systems of humans and nature*, edited by L. H. Gunderson and C. S. Holling. Washington, D.C.: Island Press.

Preface

This book was developed partly from information presented at a symposium entitled "Interdisciplinary Approaches to Ecological Monitoring of Major Ecosystem Restoration Initiatives," held at the 1999 annual meeting of the Ecological Society of America in Spokane, Wash. However, the project's true origins stem from common interests in ecosystem monitoring identified in 1995, when we were involved in establishing monitoring plans for various aspects of a new water management regime for parts of the Florida Everglades. Dave was employed by the National Park Service in Everglades National Park, where his responsibilities included establishing a multidisciplinary monitoring plan for water management to restore ecosystem function, while Joel was on the faculty at Florida International University and assisted the Park in its efforts to monitor fishes in the Everglades. In 1997, Dave moved to the U.S. Geological Survey, where his experience in developing regional monitoring plans has proven useful in interagency monitoring of forest ecosystem management in the Pacific Northwest. With this new perspective, Dave identified parallels of regional ecosystem management initiatives in the Pacific Northwest and in South Florida and the unique challenges of monitoring at the ecoregional scale. From this was born our plan to hold a symposium bringing together scientists from these two major ecoregional initiatives to permit comparisons and contrasts of the lessons being learned.

The symposium made clear that these two initiatives were not proceeding down identical paths, and that each initiative could benefit if lessons could be transferred between them. For example, scientists working in the Pacific Northwest had made extensive progress in the development of conceptual frameworks and organizations for monitoring complex regional landscapes. On the other hand, scientists in South Florida were at the leading edge in developing and applying simulation models as tools for selecting among alternative management scenarios. In this book, we bring together these areas of expertise, along with some insights from other key ecoregional initiatives. Since we began this project, several new ecoregional initiative plans have been conceived and begun. In response to the increasing need to manage large ecosystems adaptively, we believe that additional initiatives will be developed. We hope that this book will help those responsible for new and ongoing initiatives avoid the development of new monitoring programs independent of the scientific and experiential knowledge established in previous efforts.

This project would not have been possible without financial and logistical support from our home institutions. Financial support was provided by the Forest and Rangeland Ecosystem Science Center of the U.S. Geological Survey in Corvallis, Oregon, under its Internal Competition grant program. Joel Trexler was supported by National Science Foundation grant no. 9910514 for Florida Coastal Everglades Long-Term Ecological Research. Additional funding came from the Everglades National Park (especially cooperative agreement CA-5280-6-9011) and the College of Arts and Sciences, Florida International University. The Ecological Society of America Applied Ecology Section sponsored the 1999 symposium and was helpful in providing ideas about its content.

We have been gratified by the response of potential contributors to this book and wish to thank them for their efforts. Support among the many people that we contacted about this project was unanimous, and we feel fortunate to be able to present some of the finest thinking about monitoring ecosystems in the chapters that follow. In addition to the esteemed set of authors whose work is presented in this book, a number of others helped us conduct the symposium and bring this book to completion. This includes Michael Gaines and the Department of Biology at the University of Miami, who provided us with space where we could work uninterrupted as we compiled chapters into a common manuscript format. Our thinking about this project benefited in innumerable ways from discussions about a vast array of topics relevant to ecosystem monitoring with Bob Alverts, Gary Benson, Doug Buffington, Mike Collopy, Mike Crouse, Nancy Diaz, Mike Furniss, Dave Hohler, Al Horton, Michael

Huston, Anne Kinsinger, Ron Kirby, Don Knowles, Kim Kratz, Phil Larsen, Dan McKenzie, Jon Martin, Loyal Mehrhoff, Jim Milestone, Tom Mills, Steve Odell, Tony Olsen, Tom Philippi, Stuart Pimm, Karl Stein, Michael Tehan, and many, many others. We especially thank Barbara Dean at Island Press, who gave us some key ideas about how to improve the book's impact and was especially helpful in completing the final production of the manuscript.

DAVE BUSCH
JOEL TREXLER

PART I

Introduction

CHAPTER 1

The Importance of Monitoring in Regional Ecosystem Initiatives

David E. Busch and Joel C. Trexler

Over the past two decades, an international commitment to managing ecosystems at a regional-scale has arisen. This has led to the development of programs devoted to the restoration and conservation of landscape-scale environments throughout the United States and beyond (Yaffee et al. 1996). Increasingly, this commitment is linked both conceptually and legally with requirements for ecological monitoring as a means to evaluate the efficacy of the ecosystem management actions taken. The vision for long-term research and monitoring undertaken in support of such initiatives is for ecologically relevant findings to catalyze an adaptive ecosystem management cycle resulting in improved natural resource management and sharpened focus on pertinent research questions (Holling 1978). This book integrates the thinking of leading ecologists who have worked on the development and implementation of comprehensive ecosystem monitoring programs to provide insight as to how well this vision is being achieved.

There are a number of notable North American ecosystem initiatives that have been active over the past decade (fig. 1-1). Among these, a few stand out and are the focus of this book. Despite obvious ecological and institutional differences among systems, the objectives and unprecedented need for monitoring information related to these initiatives are similar. Cadres of ecologists have been responsible for devising complex systems of integrated monitoring protocols and for putting scientifically credible monitoring programs into practice. Chapters in this volume draw upon scientists responsible for the development of monitoring programs in these disparate, but highly visible, ecosystems to share information about those factors contributing to their success.

Ecoregional Initiatives, Adaptive Management, and Monitoring

The idea of studying ecoregions, and adaptively managing resources within regional ecosystems as we learn more about them, has elevated the importance of monitoring in ecosystem conservation and restoration initiatives. We refer to the efforts to evaluate ecosystem structure and function (including human dimensions), and to alter land and water management to preserve and restore species and environments across broad landscapes, as *ecoregional initiatives*. Bailey (1989) used *ecoregion* to describe ecosystems of regional extent. At the global scale, these are similar to biomes, but finer delineations of ecoregions have been made at continental and subregional scales (Bailey et al. 1994). Ecoregions have been defined as relatively large areas of land or water containing geographically distinct assemblages of communities that share a majority of their species dynamics and environmental conditions and that function as conservation units at global and continental scales (Ricketts et al. 1999). No rigidity should be inferred by the use of the term *ecoregion*. Despite inherent ecological complexities, ecoregions will have a degree of homogeneity when compared to adjacent areas. In geographic extent, ecoregions are clearly smaller than continents, yet they are larger than most land or water management units, higher-order stream drainages, or vegetation associations. The term *initiative* is also intentionally broad and dynamic. Suffice it to say that concern for natural resources at the regional scale has catalyzed the formation of programs to preserve and restore ecosystems across broad, ecologically similar areas; these efforts comprise the "initiative" part of the ecoregional initiative concept.

The expressions *monitoring* and *status and trend detection* are often used interchangeably, and we see no reason to deviate from this convention. Assigning a value to any indicator of ecosystem condition at a single point in time provides evidence of that indicator's *status*. Because we generally refer to the measurement of ecological status over broad spatial scales, status determination and another frequently used term, *inventory*, also are synonymous in our view. A closely related concept is *baseline monitoring*,

Figure 1-1. Some major regional ecosystem initiatives within the United States. Examples depicted are the Hawaiian Archipelago, Northwest Forest Plan, San Francisco Bay-Delta, Mojave Desert, Salton Sea, Greater Yellowstone, Colorado Plateau, Southwest Strategy, Rio Grande Borderlands, Platte River, Northern Plains, Great Lakes, Mississippi-Missouri Rivers, Chesapeake Bay, South Florida-Everglades, and Coral Reef initiatives.

3

the determination of the initial status of the system to be evaluated at a given point in time. Where *status* and *inventory* imply a reporting of the current condition, *baseline* is often used to reference times other than the present. Measurement or estimation of change in an indicator's status over time is the basis for *trend detection* or monitoring. Although these definitions are simple and straightforward, this does not imply that there is not potentially substantial complexity associated with a variety of aspects of ecological trend detection and description (Dixon et al. 1998). Terminology associated with monitoring and ecoregional initiatives has been summarized in this volume's glossary.

Monitoring and Ecosystem Science

Although some organizations take pains to classify their scientifically oriented activities as either "monitoring" or "research," erecting such barriers appears to be counterproductive in most cases. Research and monitoring exist in a continuum of scientific endeavor. Within this continuum, differences do exist, but these differences tend to be gradational and not absolute. Thus, research tends to more narrowly focus on planned experimentation, while the scientific value of monitoring often comes via its temporal and spatial breadth. Well-conceived research and monitoring both purport to test hypotheses, though this often occurs within a more rigid sampling framework in research settings. Determining causation is central to the purpose of much research. Although monitoring tends to be more descriptive, it should not be divorced from research. Rather, in ecosystem initiatives with solid scientific support, monitoring and research exist within a positive feedback loop in which descriptive monitoring analyses generate hypotheses that catalyze research inquiry directed at causation and vice versa. Where framing sound research questions often dictates a degree of disciplinary, temporal, and spatial limitation, monitoring is frequently designed to be interdisciplinary, long term, and geographically broad. Over time, the development of large monitoring databases offers a realm of possibilities for ecological analyses that rarely can be achieved by more focused research programs. Thus, monitoring programs offer opportunities for experimental ecology and ecological modeling to remain relevant over time and across landscapes.

Given the shades of difference between research and monitoring, the most productive applications of science in ecosystem initiatives are those in which these pursuits are well integrated. In such situations, we envision researchers helping to develop hypotheses and protocols for monitoring but also benefiting from experimental and analytical opportunities that long-term, geographically expansive monitoring data sets tend to provide.

Monitoring programs often yield key hypotheses for research testing and such programs benefit from allegiances with researchers who may offer the capability to analyze and interpret data sets from monitoring programs. Viewed in this manner, the monitoring of ecosystem status and trends is a valid and important endeavor within the realm of ecosystem science. Monitoring of ecosystem initiatives offers key opportunities for developing the type of comprehensive inferences capable of influencing natural resource management across broad landscapes. In spite of the opportunities that a unified approach can bring to ecoregional monitoring and research, it should be noted that science is most appropriate when limited to improving our understanding of systems being managed. Scientists who go beyond this to advocate positions on initiative management can risk credibility for themselves or for the use of valid monitoring and research findings in decision making (Mills 2000).

Monitoring Concepts

The lexicon associated with monitoring topics varies somewhat among the ecoregions that are focused upon in this volume. In many ways, this variation highlights subtle differences in thought processes and contributes to the richness of the dialogue about improving the way scientifically based monitoring is conducted. Because of this, we have made no effort to force chapter authors to adhere to a common standard for monitoring terminology, but we highlight some of the more important terms here. A brief glossary also provides simple definitions and cross-references to which readers may refer.

Several authors make reference to a three-phased approach utilizing implementation, effectiveness, and validation monitoring to evaluate ecoregional initiatives. The three phases cover monitoring of how the initiative is being implemented (i.e., *implementation monitoring*), the ecosystem (including the economic and social system) effects of implementing the initiative (i.e., *effectiveness monitoring*), and an ongoing program to validate the assumptions linking implementation (causes) and effects (i.e., *validation monitoring*). This phased approach to monitoring is relatively well known in some federal land management agencies (Noss and Cooperrider 1994; Tuchman et al. 1996) but is far from universally applied. When operating in an adaptive management mode, program designers often focus much of their attention on monitoring the effects of the ecosystem initiative as implemented. However, monitoring of program implementation itself is often less well addressed, or completely ignored, leading to consternation when causes are sought in ecosystem effects analyses. Similarly, research into assumed linkages between causes and effects is often ad hoc from a monitoring program perspective and not structured well to test those hypotheses most pertinent to ecoregional initiative performance.

The goals and objectives for ecoregional initiatives are frequently categorized as sets of ecosystem *targets* or *endpoints*. Both terms are used in this volume, and most readers can consider these terms synonymous. Similarly, there are often multiple *indicators* that are used to evaluate initiative performance in achieving targets in a quantitative or semiquantitative fashion. Some initiatives use the term *performance measure* in a close approximation of the manner that others use indicator. Regardless of terminology, the best ecoregional monitoring programs have linked indicators with targets using well-thought-out conceptual models or, in a few cases, have advanced to the state where quantitative models can be used to help make this linkage. Conceptual models provide a working theory of ecosystem structure and function at a regional scale. From such models, information on ecosystem processes and stressors can be used to develop targets or endpoints. Although the use of the best scientific information available should be a part of this process, it is worth noting that the setting of targets ultimately involves assessments of values and risks on the part of decision makers.

As an example, the conceptual model in figure 1-2 could be used as the basis for a decision that the energy exchange process is key to determining habitat suitability for aquatic organisms. That might lead scientists to advise, and managers to decide, to set endpoints or targets in terms of minimum, maximum, or optimal stream temperatures. One might correctly assume that the measurement of stream temperature would then be a logical indicator of this target. Alternatively, because there are a number of logistical considerations to extrapolating point measurements of temperature across time and space, it could be more efficient and informative to use proxy indicators for this target. Other indicators that might be considered could include remote sensing of stream or riparian corridor radiation flux or the structure of riparian vegetation. While the use of direct indicators like stream temperature may provide a sense of satisfaction due to their familiarity or common usage, indirect indicators such as riparian vegetation canopy cover would provide information about energy exchange as well as causal factors important in the riparian environment.

For certain initiatives, a medical analogy employing concepts like ecosystem "health" or "vital signs" has proven useful in communicating with natural resource managers about ecological monitoring (Davis 1993). Criticisms that such analogies are an invalid means of characterizing ecosystems have been put forth (Suter 1993; Wicklum and Davies 1995; Woodward et al. 1999). From our viewpoint, medical analogies can

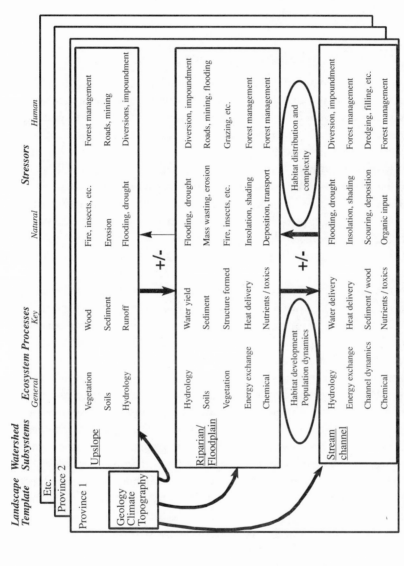

Figure 1-2. A simplified conceptual model for monitoring the effectiveness of the Northwest Forest Plan in restoring aquatic and riparian ecosystems. Arrows indicate the direction and strength of interaction among subsystems.

perpetuate misconceptions about our ability to specify an ideal ecosystem state, or lead toward thinking that normal perturbation of ecosystems is somehow undesirable. We, therefore, strive in this volume to use more value-neutral terminology such as ecosystem *condition* in discussing the status and trends of ecosystems.

Why a Book on Monitoring Ecoregional Initiatives?

As more regional ecosystem programs are started, efficiency demands that we seek ways to share valuable information among the many agencies and organizations involved. Until now, this has proven difficult, and, typically, procedures for new monitoring programs were crafted independently of one another. Now, however, there is a growing interest in what has and has not worked for others and decision makers are increasingly calling for the implementation of monitoring programs as part of the movement to manage ecosystems adaptively. As a result, resource managers and technical specialists at all levels are in need of information about ecosystem monitoring at a regional scale.

In the past, scientists developed monitoring protocols specific to the systems affected by regional ecosystem initiatives without much consideration for how these systems were linked, what the conceptual underpinnings for monitoring were, or what sort of support system was needed to produce useful monitoring information. Now, by developing a consensus about the system to be monitored and adhering to a stepwise program to develop monitoring strategies, those responsible for designing monitoring programs can more easily make use of valuable information provided by status and trend programs. As scientists are drawn to monitoring projects, attention can increasingly be paid to lessons already learned in monitoring programs elsewhere. This will enable scientists to focus more closely on the ecological intricacies of the systems they study. By describing lessons learned from a variety of recent, "cutting-edge" projects involving ecosystem monitoring, this volume will help scientists direct their productivity toward the system-specific questions with which they must contend.

As the pressure to monitor ecosystems increases in the institutional world, the subject of ecological monitoring is gaining importance in the environmental science curricula of colleges and universities. For some time, skills pertinent to environmental monitoring were addressed in individual courses on statistics, database development, and laboratory and field sampling and analysis. However, the importance of consolidating piecemeal treatment of the topic into a single, unified discipline is now recognized to be critical to the training of capable scientists and managers. It is our belief that providing educators, scientists, and managers with a solid foundation in ecosystem monitoring will ultimately result in substantially improved quality of restoration and management programs.

The literature on environmental monitoring reveals surprisingly little about the ecological aspects of monitoring. Recently published works mostly address air- and water-quality (pollutant) monitoring (McConnell and Abel 2002). A few titles address data management for environmental monitoring (Spellerberg and Holdgate 1991) but tend to focus on the statistical detail and mechanics of database development rather than the utility of data or analyses in natural resources management. Another topic at least partially addressed concerns the use of remote-sensing data in monitoring projects. Inspection of ecological monitoring titles reveals that several are directed at the basic questions of why monitoring is conducted and what sorts of resources are monitored. However, the monitoring literature offers little on the conceptual basis for monitoring programs or how the elements within such programs can be prioritized and linked. Few titles touch on the subject of integrating vegetation or fish and wildlife monitoring with important issues driving ecosystem initiatives, such as the recovery of species or habitats that are in jeopardy. Little has been written about the promise of modeling and decision support systems as means of analyzing and communicating trends detected through monitoring. A recent contribution on bioregional assessments (Johnson et al. 1999) provides a preview to initiatives operating at the regional geographic scale, as well as to the roles of science, management, and policy considerations in these initiatives. However, it touches only lightly on the subject of ecological monitoring. In this volume, we address perhaps the single biggest issue confronting those involved with moving these sorts of regional ecosystem initiatives forward: the development of viable monitoring programs.

Goals and Objectives of This Volume

The goal of this book is to set a paradigm for regional ecosystem monitoring that recognizes

- A strong emphasis on setting objectives for monitoring in conjunction with goal setting for ecoregional initiatives themselves
- The basis for most ecosystem initiatives is early establishment of linkages to natural resource management
- The development of program direction from conceptual or simulation models of the ecosystems monitored
- The establishment of information management programs, mechanisms for quality control and assurance of data, and means for communicating results to stakeholders

- The evaluation of both the physical environment and key ecological parameters based on monitoring objectives and predicted ecosystem outcomes
- A lack of boundaries between "monitoring" and "research" recognizing the interdependence of these aspects of ecosystem science
- Temporal and spatial monitoring frameworks that respond to issues of scale (i.e., How does monitoring of regional initiatives fit into national perspectives? What is its value to local projects? How can information about ecosystem initiatives be reconciled with that from outside the region?)
- Intricacies associated with the use of monitoring information in adaptive ecosystem management (i.e., monitoring as a means for affecting change where appropriate and not just an excuse for "business as usual")

The focus of this book is to critique and derive synthetic conclusions about major ecosystem initiative monitoring programs that can be extracted for generalized use. To this end, chapter authors address one or more of five basic themes, listed below, and include as part of the discussion their experiences sharing data and information between regions, and the lessons they learned as a result. Questions addressed by chapter authors include the following:

- To be successful, a monitoring program for an ecoregional initiative must be designed with clearly stated management and restoration goals in mind. How can success be evaluated by identifying ecosystem objectives and success criteria at the outset of initiative monitoring?
- Monitoring programs must include means of elucidating ecosystem processes to yield some explanation of why restoration or management efforts have succeeded or failed. How can such evaluation techniques be incorporated into ecosystem monitoring programs?
- Monitoring must precede, coincide with, and continue after management and restoration regimes are initiated. In light of this, how can sampling designs incorporate temporal or spatial controls as a reference?
- Monitoring is an empty exercise, no matter how well conceived, if it does not tie into the policy making or management process. What are the best ways to incorporate monitoring results into an ongoing adaptive ecosystem management process?
- In any large project, lessons about what works and what does not are discovered. What are the key attributes of monitoring that have lead to

success or, alternatively, have hampered progress toward project goals? To what extent are these lessons transferable to other types of restoration efforts, and to restoration efforts in other types of ecosystems?

A Comparison of Ecoregional Initiatives

Although the monitoring principles discussed here can be applied to ecosystem initiatives conducted at nearly any scale and in any geographic area, this volume focuses on the design, development, and implementation of monitoring programs in three key ecoregions. Organizations overseeing ecosystem initiatives in each of these ecoregions are expected, and in many cases mandated by policy and law, to provide status and trend information describing how well initiative objectives were met. The ecosystem initiatives on which most of the chapters in this volume are based are (1) the Northwest Forest Plan, which comprises the strategy by which management of late successional forest and associated aquatic ecosystems has been altered in the Pacific Northwest region of the United States, (2) an initiative to restore habitats and populations in the lower Colorado River Basin of the southwestern United States, and (3) restoration of marine, wetland, and terrestrial components of south Florida's Greater Everglades ecosystem. A brief introduction to each follows.

Stemming largely from the impasse between timber production on federal lands in the Pacific Northwest and the attainment of the goals of the Endangered Species Act, the Northwest Forest Plan covers the geographical range of the federally threatened northern spotted owl (*Strix occidentalis caurina*). The Plan area includes over 9 million hectares (22 million acres) of mostly forested public land encompassing nineteen national forests, five Bureau of Land Management districts, and thirteen national park units, and it extends from the San Francisco Bay in California to the U.S.-Canadian border, and from the Pacific coast to east of the crest of the Cascade Range.

Flowing from Glen Canyon Dam to the Gulf of California, the lower Colorado River is the principal drainage in the interior southwestern United States and extreme northwestern Mexico. Contrasting sharply with its surrounding desert landscape, undeveloped portions of the lower Colorado floodplain are covered by riparian vegetation including gallery forests dominated by cottonwood (*Populus fremontii*), willow (*Salix gooddingii*), and mesquite (*Prosopis* spp.), which have, over the last eighty to one hundred years, been replaced in many places by exotic *Tamarix chinensis* scrub. Embedded in this landscape are globally recognized lands and waters, such as Grand Canyon National Park, as well as a number of major water resource developments at impoundments, such as those forming Lakes Powell, Mead, and Havasu.

South Florida ecosystem initiatives are focused on the Kissimmee River drainage, Lake Okeechobee, and the area between the Miami Rock Ridge and Florida's southwest coast, including Florida Bay. The lower mainland part of this drainage network (south of Lake Okeechobee) comprises the Florida Everglades. A unique mixture of tropical and temperate ecosystem elements led to the recognition of the importance of key landscape elements such as Everglades National Park, the Florida Keys, Big Cypress Swamp, and south Florida's Water Conservation Areas.

These ecoregional initiatives benefit from a high level of commitment on the part of organizations ranging from the top tiers of federal agencies to pluralities of the public in the affected regions. Such interest has been associated with a level of budgetary support that has enabled restoration and management efforts to proceed. However, it must be noted that support for long-term monitoring of ecoregional initiatives has generally proved uncertain (Tolle et al. 1999).

Although vertebrate population monitoring is typically only one element of comprehensive monitoring programs, these initiatives share a genesis rooted in crises surrounding the status of individual wildlife or fish and species assemblages. For the Northwest Forest Plan, the species of greatest concern was the northern spotted owl (Thomas et al. 1990), with concerns about the marbled murrelet (*Brachyramphus marmoratus*) and anadromous fish populations also of high or increasing importance. Concern for imperiled avian populations in southwestern riparian ecosystems and for declines in native fish populations in regulated rivers has driven much of the effort to preserve and restore flows and habitats along the lower Colorado River and its tributaries (Ohmart et al. 1988). Conservation of the Everglades ecosystem has long been tied to the status of wading bird populations, but organisms such as the Florida panther (*Felis concolor coryii*) are also linked to the restoration of habitats in south Florida (Smith and Bass 1994).

Given their commonalities, initiatives and associated monitoring programs in the three ecoregions are compared along three dimensions using:

- the institutional drivers that mandate ecological monitoring programs
- the environmental drivers that influence ecosystem processes and structure
- the conceptual factors that serve as a basis for monitoring

Institutional Drivers for Ecoregional Initiatives

The three regional ecosystem initiatives share similar levels of public support stemming in large part from the solutions they offer for the preservation of at-risk vertebrate populations. Although they address focal species thought to be imperiled, all three initiatives have gone beyond restoration and conservation of fish and wildlife habitat to encompass concerns about preserving and reestablishing overall ecosystem function.

Each of the regional ecosystems has been exposed to intense scrutiny through environmental evaluations associated with the National Environmental Policy Act (NEPA), the Endangered Species Act (ESA), and other environmental legislation and regulations. A series of such assessments has been conducted along the lower Colorado River in association with the Glen Canyon Dam Environmental Studies, the Colorado River Basin Salinity Control Project, and similar programs. Although projects based on older legislation did not typically require ecological monitoring, an important component of the Glen Canyon Environmental Studies is the development of a long-term monitoring plan for the Grand Canyon area (National Research Council 1994; Marzolf et al. 1998). Although this plan does not encompass the entire lower Colorado River basin, it has been supplemented by a number of long-term studies of ecological status and trends in riparian habitats below the Grand Canyon (e.g., Ohmart et al. 1988).

Restoration of the Florida Everglades is rooted in legislation and intergovernmental agreements at both the federal and state level. A 1993 Interagency Agreement was the basis for unifying the comprehensive federal restoration effort under an interagency South Florida Ecosystem Restoration Task Force. More recently, commitments strengthening this effort were made in the form of the Department of the Interior's Critical Ecosystem Studies Initiative and a new strategic science planning effort aimed at integrating agency research and monitoring programs. The need to establish ecological endpoints and success criteria rooted in ecological management principles is recognized as critical to the restoration of south Florida ecosystems (Harwell et al. 1996). However, it has not been easy to define these endpoints and the manner in which progress toward such endpoints can be monitored. Although ecological monitoring is recognized for its importance in determining the scientific credibility of restoration efforts being planned and conducted, as in many complex ecosystems, there remains some uncertainty about the relative value of potential indicators in the context of overall restoration objectives.

Of the ecosystem initiatives considered, monitoring for the Northwest

Forest Plan has perhaps the strongest basis in the decisions of the courts. During the early 1990s, conflicting legal claims brought timber programs to a near standstill on federal lands in western Washington and Oregon, and in northwestern California. President Clinton's declaration of a "Forest Plan for a Sustainable Economy and a Sustainable Environment" catalyzed a process to design an interagency strategy for managing forest resources (Tuchman et al. 1996). With the issuance of a joint Record of Decision by the Secretaries of Agriculture and Interior, the management direction for federal forest lands and aquatic resources throughout the Pacific Northwest was extensively altered to conserve ecosystems, species, economic viability, and social values (Pipkin 1998). The Forest Plan includes an explicit delineation of a monitoring framework in the Standards and Guidelines accompanying the Record of Decision. Monitoring was given additional strength in the summary judicial decision, where it is stated that "Monitoring is central to the Plan's validity. If it is not funded, or not done for any reason, the plan will have to be reconsidered" (Dwyer 1994).

Ecosystem Function as a Determinant for Ecoregional Monitoring

The influence of hydrological and geochemical drivers in both Everglades and lower Colorado ecosystems is widely recognized. Beginning in the early 1900s, river regulation and channel modification projects brought about a need for understanding the dynamics of water flow in the Colorado River. Because of this, a series of hydrological gaging stations was established and has produced a long-term record of river discharge, the key variable influencing the riverine and riparian environment. Colorado River salinity is also closely monitored due largely to concerns about the quality of water delivered for municipal and agricultural use in the United States and Mexico. Questions about whether the Colorado River's hydrology and salinity influence the riparian forest community were clarified by demonstrating the direct association of the river, the floodplain groundwater system, and plant communities dominated by riparian trees and shrubs (Busch et al. 1992; Busch and Smith 1995). Adding an element of ecological novelty, invasive plants and fire also interact to alter Colorado River ecosystem structure and processes (Busch 1995; Smith et al. 1998).

The monitoring of south Florida ecosystems is similar to that conducted along the lower Colorado River with respect to the utilization of an extensive hydrological record generated during the era of development and more recent restoration. This long-term monitoring record has proved vital to evaluations of the effects of hydroperiod on Everglades vegetation and aquatic biota (Busch et al. 1998; Ewe et al. 1999; Trexler et al. 2000). The effects of fire and exotic vegetation are also substantial in the Everglades (Gunderson and Snyder 1994; White 1994). Accordingly, much of the information on landscape status and trends within these environments comes from monitoring associated with fire and exotic vegetation management programs. In addition, monitoring of species-habitat relationships has furthered the development of landscape models to support Everglades restoration programs (DeAngelis et al. 1998; chapter 6).

Although anthropogenic perturbation of the environment is a common theme in each of the ecosystem initiatives considered, the Northwest Forest Plan differs from the others in terms of the direct effects that human actions have had on terrestrial systems. However, human influence in the form of forest management occurs against a backdrop of biogeochemistry and disturbance similar to that of the other two ecosystems. Whereas reregulation of flows in the Colorado River and the Everglades is envisioned as the principal means of ecosystem restoration, revised forestry practices under the Forest Plan have resulted in a system designed to manage across broad landscapes for ecological complexity and a broad array of goods and services (Kohm and Franklin 1997). This has produced a correspondingly intricate set of ecosystem factors relevant to monitoring. Because of the Forest Plan's roots in the impasse over threatened species, the status and trends of northern spotted owl and marbled murrelet populations were assigned the initial priority in the Plan's system of effectiveness monitoring (Mulder et al. 1999). Given the importance of late-successional and old-growth forests to these and other species, monitoring of forest vegetation was also accorded high priority (Hemstrom et al. 1999; chapter 11). These monitoring elements were the first to be implemented, while monitoring of elements involving arguably greater complexity (i.e., aquatic and riparian ecosystems, forest biodiversity, and social and economic effects) have taken longer to move through the planning and development stage.

Because the overlay of terrestrial land management upon natural ecosystem processes is central to the Forest Plan monitoring strategy, evaluation of the status and trends of forest vegetation is at the heart of the Plan's monitoring program. Predictive monitoring using modeled relationships is an important future step for Forest Plan monitoring, but the development of essential system simulations has not progressed as far as in the Everglades initiative (chapter 6).

An explicit statement of the steps required to design monitoring programs was developed for the Northwest Forest Plan (Mulder et al. 1999):

1. State the goals of the monitoring program.
2. Identify the environmental stressors relating to management goals.
3. Develop a conceptual model linking relevant ecosystem components.
4. Identify candidate indicators most responsive to environmental stressors.
5. Estimate the status and trends of the indicators.
7. Generate expected values for indicator variables.
8. Link monitoring results to decision making.

Although these steps are seemingly straightforward, it has proven necessary to revisit them during monitoring plan development to assure that
the process remains focused. Explicit articulation of the plan development
and implementation process has been critical to progressing toward stated
goals. Nowhere does an explicit statement of intent apply better than to
the development of an ecosystem conceptual framework for monitoring.
Problems with the logical coherence of monitoring programs (National
Research Council 1995) can usually be circumvented by articulating a
clear conceptual framework or model. Moreover, attempts to develop
monitoring programs for ecosystem initiatives have stagnated in the absence of sound conceptual models.

The conceptual model for Forest Plan aquatic and riparian environments serves as one example of how to describe ecosystem attributes pertinent to monitoring (fig. 1-2). The model portrays a physical landscape
template that varies approximately at the scale of the twelve biogeo-
graphic provinces comprising the Forest Plan region. The model also describes upslope, riparian, and stream channel subsystems and the relative
influences these watershed subsystems have on each other. General and
key ecosystem processes are specified, the latter set being the source of indicators for monitoring status and trends. The influence of anthropogenic
and natural stressors on ecosystem processes is taken into account when
selecting indicators, as is the influence of habitat-related processes affecting riparian and aquatic biota. Although seemingly elaborate, this model
is a highly generalized depiction of a complex system. Importantly, frameworks such as this have been shown to serve as a mechanism to keep dialogue focused on ecosystem objectives at all stages of monitoring program development. This importance is highlighted by the presentation of
additional conceptual models throughout this volume.

Conceptual Aspects of Ecoregional Initiative Monitoring

Each of the ecoregion examples given in this book has embraced the paradigm of adaptive ecosystem management wherein ecological monitoring

plays a pivotal role (Holling 1978). Under this concept, the status of the resource monitored is dependent upon stressors that are affected by land management practices (fig. 1-3). Resource management policy and practices are also affected, at least indirectly, by regulatory policy and practice. Feedback affecting both resource management and regulatory policy comes directly from the monitoring and research programs that are integral to ecosystem initiative implementation. Indeed, it has been pointed out that monitoring programs themselves must be adaptive to account for obstacles that limit the ability to design long-term monitoring programs in the face of unpredictable changes certain to occur in the future (Ringold et al. 1996; chapter 3).

Adaptive ecosystem management has been more influential as an idea than as a practical means of gaining insight about ecosystems utilizing a set of questions agreed-upon by initiative stakeholders (Lee 1999). Despite the wide philosophical acceptance of adaptive approaches to ecosystem management, case studies reveal few instances where management uncertainty has been successfully resolved by adaptive management (Walters 1997). While such critiques certainly are valid, most monitoring programs cannot be considered well established given the time lags likely to occur as ecosystem processes (e.g., forest succession, dynamics of rare or endemic species populations) are reestablished. For nearly every dimension of regional ecosystem monitoring, we are still near the beginning of the process to determine system status and trend associated with al-

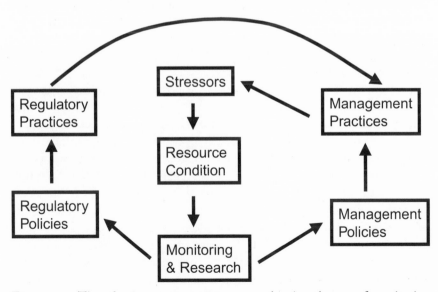

Figure 1-3. The adaptive management process showing the use of monitoring information via both management (direct) and regulatory (indirect) pathways.

tered management policy and practice. Assuming that fledgling monitoring programs can become institutionalized, adaptive management will, over time, have a chance to more fully serve as a unifying theme for regional ecosystem initiatives.

Status and trend detection for the Forest Plan adopted the aforementioned three-phased approach utilizing implementation, effectiveness, and validation monitoring. A similar monitoring framework is integrated into experimental water-release programs for the Grand Canyon. Such program sub-elements have been less well-articulated for Everglades ecoregion monitoring. Concerns have arisen over certain monitoring conventions embodied in this type of monitoring approach. Potential problems have been pointed out with the use of ecological indicators to overcome ecosystem complexity and with scale in stressor-response modeling designs (Morrison and Marcot 1995; Lee and Bradshaw 1998). Despite this, ecoregional monitoring programs generally conform to broadly recognized design recommendations relative to sampling and scaling, trend detection, the identification of stressors, and specification of uncertainty (Dixon et al. 1998), so this approach appears to be valid.

An Outline of Ecoregional Initiative Monitoring

In this chapter, we set the stage for in-depth coverage of various facets of ecosystem monitoring that are more fully developed in later chapters. Our main scale of interest, which is defined as regional, can be clarified by envisioning the context for monitoring programs conducted at this scale that is provided by programs ranging from local- or site-specific to those that are global or continental in scope. Over the last decade, a critical mass of experience on monitoring ecoregional initiatives has provided fertile ground for learning about innovative approaches to monitoring at any scale. Similarities and differences in the ecological, institutional, and adaptive management issues addressed by these initiatives provide opportunities to explore key topics more fully in the chapters that follow.

Following this chapter, the next part of this book is devoted to outlining the conceptual underpinnings for regional-scale ecosystem monitoring. Among the contributors are those who have led efforts aimed at producing monitoring programs that are scientifically credible, comprehensive, and of value to decision makers. In chapter 2, Barry Noon proposes a set of principles that are appropriate for developing ecoregional monitoring programs. A key element of the foundation provided by Noon is a stepwise process intended to guide monitoring program development. This process has been used to develop a number of monitoring programs that are being successfully implemented—a sure sign that Noon's principles are sound. Chapter 3, by Paul Ringold, Barry Mulder, Jim Alegria, Ray Czaplewski, Tim Tolle, and Kelly Burnett, makes a strong case for the idea that just as the idea of adaptive ecosystem management is now widely accepted, ecosystem monitoring can likewise be considered a process that is adaptable. Recognition of the need for a long-term monitoring program for the Colorado River in Glen and Grand Canyons developed from a keen understanding of the need to adapt management of this ecosystem based on findings about the responses of system resources. In chapter 4, Larry Stevens and Barry Gold address monitoring and adaptive management of this system, describing one of our best examples of the application of science to adaptive ecosystem management. Chapter 5, by John Ogden, Steve Davis, and Laura Brandt, provides a glimpse of the extensive infusion of scientific thought into the design of a regional monitoring strategy for the Everglades. The importance of system conceptualization is also highlighted in this chapter.

Analyses of information from monitoring programs is increasingly recognized as a means of predicting trajectories for ecosystem status as well as a means for inducing efficiency in resource management programs. The third part of the book comprises chapters from specialists who are leading efforts to manage data, perform integrative analyses, and build system simulation and prediction into ecoregional monitoring programs. Chapter 6, by Don DeAngelis, Lou Gross, Jane Comiskey, Wolf Mooij, Phil Nott, and Sarah Bellmund, addresses the use of models at multiple scales in the Everglades monitoring program. They describe a sophisticated modeling system designed to link monitored physical and habitat parameters with fish and wildlife population responses. In chapter 7, Keith Reynolds and Gordon Reeves describe the use of knowledge-based systems in the analysis and communication of monitoring data. They offer a solution to the problem confronting monitoring program managers who must manage large volumes of data on disparate sets of indicators from a complex array of ecosystems spanning broad ecoregions. Craig Palmer highlights in chapter 8 lessons learned in the development of information management and quality-assurance systems for initiatives in the arid Southwest, as well as in the wetter ecosystems of the Pacific Northwest. Concluding this part, chapter 9, by John Sauer, William Link, and James Nichols, focuses on trend analysis. Drawing mostly on experience monitoring wildlife populations, these authors focus on a number of quantitative issues that can aid or hinder our ability to detect trends in population parameters.

Consideration of the physical and biotic factors available for monitoring comprises a fourth subset of this book's chapters. This section provides insight on measuring status and detecting trends in the landscape template of upland and aquatic environments. Acute population declines of at-risk species have catalyzed ecosystem restoration initiatives and strongly influenced the direction that these initiatives have taken. Thus, the monitoring of populations and communities of such species makes up an important component of this section. Two chapters cover fundamentals related to developing a characterization, at an ecoregional scale, of the environment monitored. Chapter 10 describes a system for integrating the monitoring of seagrass meadows with water physical and chemical status in the marine environment of Florida Bay and the Florida Keys. Chapter authors Jim Fourqurean and Leanne Rutten further elaborate on the important concept of dividing regional-scale monitoring efforts into spatial pattern and temporal pattern foci. Although a number of species and ecosystems are important to the monitoring program for the Forest Plan in the Pacific Northwest, all program segments rely on the monitoring of late-successional forests. In chapter 11, Miles Hemstrom describes the integration of plot-based and remotely sensed vegetation data with analyses of old-growth forest structure, and he cites results from a pilot test of this regional monitoring strategy that was conducted in the Oregon Coast Range. Charismatic fauna often serve as the catalyst through which public and political will are manifested to formally sanction ecoregional initiatives. However, because of the close scrutiny of such species, expectations for monitoring them can present tremendous challenges. In chapter 12, Peter Frederick and John Ogden discuss monitoring of wading birds, a group of species that has had iconic status since the initiation of the conservation movement in the Everglades. Frederick and Ogden reveal information about the use of wildlife populations as indicators and discuss what monitoring of populations can reveal about the internal and external influences on ecosystems. Joel Trexler, Bill Loftus, and John Chick address the linkage of physical parameters and fish-community monitoring in chapter 13. They point out the importance of long-term databases and the fact that incorrect conclusions about ecoregional initiative effects could be derived from an incomplete monitoring program design. Monitoring all species that ecoregional initiatives are expected to cover generally proves to be overwhelming for program planners. Because of this, monitoring ecoregional biodiversity is alluring, but also frequently misunderstood. In the concluding chapter of this part, Bill Gaines, Richy Harrod, and John Lehmkuhl provide a comprehensive overview of the concept of biodiversity and some critical points of departure that are available to those responsible for designing biodiversity monitoring programs.

It was stated at the beginning of this introductory chapter that we view monitoring and research as key components that must be well integrated to make the scientific basis for ecoregional initiatives functional. Our concluding chapter revisits and expands on those topics we feel will be important to sustaining and improving science-based ecoregional monitoring in the future. In chapter 15, we synthesize information raised in each of the preceding chapters, and we attempt to expand this theme in some key areas. Among the most important messages pervading this volume is the need to pay attention to managing information from monitoring programs as well as the need to develop analytical and communication capabilities. We also put forth a recommendation to better define frameworks linking the monitoring occurring at local, regional, national, and international scales. Consistent with this, our suggestions for taking a "balanced" approach to the monitoring of causes and effects show how efficiency can be heightened and information yield enhanced in implementing ecoregional monitoring programs.

Conclusion

Although this book has its roots in the development of monitoring systems for focal ecoregional initiatives in the United States, the ideas presented transcend this geographic basis. Not only will the lessons presented advance the quality of monitoring programs in the initiatives that are detailed in this volume, but also they should be of great value in similar initiatives both new and well established. However, the concepts, terminology, and protocols presented go well beyond the regional scale. The ideas presented have applicability for determining status and trend at scales ranging from the local or project level to the global or international scale.

Literature Cited

Bailey, R. G. 1989. Ecoregions of the continents. *Environmental Conservation* 16: map supplement.

Bailey, R. G., P. E. Avers, T. King, and W. H. McNab, eds. 1994. *Ecoregions and subregions of the United States*. U.S. Department of Agriculture, Forest Service. Washington D.C.

Busch, D. E. 1995. Effects of fire on southwestern riparian plant community structure. *Southwestern Naturalist* 40:259–267.

Busch, D. E., N. L. Ingraham, and S. D. Smith. 1992. Water uptake in woody riparian phreatophytes of the southwestern United States: A stable isotope study. *Ecological Applications* 2:450–459.

Busch, D. E., and S. D. Smith. 1995. Mechanisms associated with the decline of woody species in riparian ecosystems of the southwestern United States. *Ecological Monographs* 65:347–370.

Busch, D. E., W. F. Loftus, and O. L. Bass Jr. 1998. Long-term hydrologic effects on marsh plant community structure in the southern Everglades. *Wetlands* 18:230–241.

Davis, G. E. 1993. Design elements of monitoring programs: The necessary ingredients for success. *Environmental Monitoring and Assessment* 26:99–105.

DeAngelis, D. L., L. J. Gross, M. A. Huston, W. F. Wolff, D. M. Fleming, E. J. Comiskey, and S. M. Sylvester. 1998. Landscape modeling for Everglades ecosystem restoration. *Ecosystems* 1:64–75.

Dixon, P. M., A. R. Olsen, and B. M. Kahn. 1998. Measuring trends in ecological resources. *Ecological Applications* 8:225–227.

Dwyer, W. L. 1994. *Order on motions for a summary judgment regarding 1994 Forest Plan.* U.S. District Court, Western District of Washington. Seattle.

Ewe, S. M. L., L. S. L. Sternberg, and D. E. Busch. 1999. Water-use patterns of woody species in pineland and hammock communities of south Florida. *Forest Ecology and Management* 118:139–148.

Gunderson, L. H., and J. R. Snyder. 1994. Fire patterns in the southern Everglades. Pp. 291–306 in *Everglades: The ecosystem and its restoration*, edited by S. M. Davis and J. C. Ogden. Delray Beach, Fla.: St. Lucie Press.

Harwell, M. A., J. F. Long, A. M. Bartuska, J. H. Gentile, C. C. Harwell, V. Myers, and J. C. Ogden. 1996. Ecosystem management to achieve ecological sustainability: The case of south Florida. *Environmental Management* 20:497–521.

Hemstrom, M., T. Spies, C. Palmer, R. Kiester, J. Tepley, P. McDonald, and R. Warbington. 1999. *Late-successional and old-growth forest effectiveness monitoring plan for the Northwest Forest Plan.* General Technical Report PNW-GTR-438. U.S. Department of Agriculture Forest Service, Pacific Northwest Research Station. Portland, Ore.

Holling, C. S. 1978. *Adaptive environmental assessment and management.* New York: John Wiley and Sons.

Johnson, K. N., F. J. Swanson, M. Herring, S. Greene. 1999. *Bioregional assessments: Science at the crossroads of management and policy.* Washington, D.C.: Island Press.

Kohm, K. A., and J. F. Franklin. 1997. *Creating a forestry for the twenty-first century.* Washington, D.C.: Island Press.

Lee, D. C., and G. A. Bradshaw. 1998. Making monitoring work for managers. [Online at www.icbemp.gov/spatial/lee_monitor/preface.html]

Lee, K. N. 1999. Appraising adaptive management. *Conservation Ecology* 3:3. [Online at www.consecol.org/vol3/iss2/art3]

Marzolf, G. R., R. A. Valdez, J. C. Schmidt, and R. H. Webb. 1998. Perspectives on river restoration in the Grand Canyon. *Bulletin of the Ecological Society of America* 79:250–254.

McConnell, R. L., and D. C. Abel. 2002. Environmental issues: Measuring, analyzing, and evaluating. Upper Saddle River, New Jersey: Prentice Hall.

Mills, T. J. 2000. Position advocacy by scientists risks science credibility and may be unethical. *Northwest Science* 74:165–168.

Morrison, M. L., and B. G. Marcot. 1995. An evaluation of resource inventory and monitoring program used in national forest planning. *Environmental Management* 19:147–156.

Mulder, B. S., B. R. Noon, T. A. Spies, M. G. Raphael, C. J. Palmer, A. R. Olsen, G. H. Reeves, and H. H. Welsh. 1999. *The strategy and design of the effectiveness monitoring program for the Northwest Forest Plan.* General Technical Report PNW-GTR- 437. U.S. Department of Agriculture, Forest Service, Pacific Northwest Research Station. Portland, Ore.

National Research Council. 1994. *Review of the draft federal long-term monitoring plan for the Colorado River below Glen Canyon Dam.* Washington, D.C.: National Academy of Sciences.

———. 1995. *Review of EPA's environmental monitoring and assessment program: Overall evaluation.* Washington, D.C.: National Academy of Sciences.

Noss, R. F., and A. Y. Cooperrider. 1994. *Saving nature's legacy: Protecting and restoring biodiversity*. Washington, D.C.: Island Press.

Ohmart, R. D., B. W. Anderson, and W. C. Hunter. 1988. *The ecology of the lower Colorado River from Davis Dam to the Mexico-United States international boundary: A community profile*. Biological Report 85(7.19) U.S. Fish and Wildlife Service. Washington, D.C.

Pipkin, J. 1998. *The Northwest Forest Plan revisited*. U.S. Department of the Interior, Office of Policy Analysis. Washington, D.C.

Ricketts, T. H.,E. Dinerstein, D. M. Olson, C. J. Loucks, W. Eichbaum, D. Della Sala, K. Kavanagh, P. Hedao, P. T. Hurley, K. M. Carney, R. Abell, and S. Walters. 1999. *Terrestrial ecoregions of North America: A conservation assessment*. Washington, D.C.: Island Press.

Ringold, P. L., J. Alegria, R. L. Czaplewski, B. S. Mulder, T. Tolle, and K. Burnett. 1996. Adaptive monitoring design for ecosystem management. *Ecological Applications* 6:745–747.

Smith, S. D., D. A. Devitt, A. Sala, J. R. Cleverly, and D. E. Busch. 1998. Water relations of riparian plants from warm desert regions. *Wetlands* 18:687–696.

Smith, T. R., and O. L. Bass Jr. 1994. Landscape, white-tailed deer, and the distribution of Florida panthers in the Everglades. Pp. 693–708 in S. M. Davis and J. C. Ogden, eds. *Everglades, the ecosystem and its restoration*. Delray Beach, Fla.: St. Lucie Press.

Spellerberg, I. F., and M. W. Holdgate. 1991. *Monitoring ecological change*. Cambridge, U.K.: Cambridge University Press.

Suter, G. W., II. 1993. A critique of ecosystem health concepts and indexes. *Environmental Toxicology and Chemistry* 12:1533–1539.

Thomas, J. W., E. D. Forsman, J. B. Lint, E. C. Meslow, B. R. Noon, and J. Verner. 1990. A conservation strategy for the northern spotted owl: Interagency scientific committee to address the conservation of the northern spotted owl. U.S. Department of Agriculture, Forest Service, Pacific Northwest Region. Portland, Ore.

Tolle, T., D. S. Powell, R. Breckenridge, L. Cone, R. Keller, J. Kershner, K. S. Smith, G. J. White, and G. L. Williams. 1999. Managing the monitoring and evaluation process. Pp. 585–602 in *Ecological stewardship: A common reference for ecosystem management*, edited by N. C. Johnson, A. J. Malk, R. C. Szaro, and W. T. Sexton. Kidlington, U.K.: Elsevier.

Tuchman, E. T., K. P. Connaughton, L. E. Freedman, and C. B. Moriwaki. 1996. *The Northwest Forest Plan: A report to the president and congress*. U.S. Department of Agriculture, Forest Service, Pacific Northwest Research Station. Portland, Ore.

Walters, C. 1997. Challenges in adaptive management of riparian and coastal ecosystems. *Conservation Ecology* 1:1–22.

White, P. S. 1994. Synthesis: Vegetation pattern and process in the Everglades ecosystem. Pp. 445–460 in *Everglades, the ecosystem and its restoration*, edited by S. M. Davis and J. C. Ogden. Delray Beach, Fla.: St. Lucie Press.

Wicklum, D., and R. W. Davies. 1995. Ecosystem health and integrity? *Canadian Journal of Botany* 73:997–1000.

Woodward, A., K. J. Jenkins, and E. G. Schreiner. 1999. The role of ecological theory in long-term ecological monitoring: Report on a workshop. *Natural Areas Journal* 19:223–233.

Yaffee, S., A. Phillips, I. Frentz, P. Hardy, S. Maleki, and B. Thorpe. 1996. *Ecosystem management in the United States: An assessment of current experience*. Washington, D.C.: Island Press.

PART II

Principles of Ecosystem Monitoring Design

CHAPTER 2

Conceptual Issues in Monitoring Ecological Resources

Barry R. Noon

Monitoring, also referred to as *environmental surveillance*, is the "measurement of environmental characteristics over an extended period of time to determine status or trends in some aspect of environmental quality" (Suter 1993, 505). The challenge in this definition, and the topic of this chapter, is to clearly understand why monitoring is an important activity, to decide which characteristics of the environment to measure, to determine what information their values convey about environmental quality, and to use this information to make better decisions about the management of natural resources.

Environmental monitoring is purpose oriented and should address specific objectives (Goldsmith 1991). In general, monitoring data are intended to detect long-term environmental change, provide insight to the ecological consequences of these changes, and to help decision makers determine if the observed changes dictate a correction to environmental policy. Monitoring is conducted at regular intervals to assess both the current state and the time trend in various ecological resources. Ecological resources addressed by monitoring programs vary widely in type and extent. They are broadly defined to include any aspects of the environment that have value. Examples include specific ecosystems (e.g., marine estuaries), ecological processes (e.g., photosynthesis), individual species (e.g., sea otters), and habitat elements (e.g., large logs).

Monitoring is a dynamic exercise—it is a continuing activity and its temporal span may be indefinite. The time frame is unspecified because human behavior and continuing human population growth lead to ongoing environmental change with unexpected and "surprise" ecological events as unavoidable consequences. As a result, responsible stewardship requires a continual assessment of the state of the environment and the effects of human behavior.

Before moving on, some clarification of terms is in order. In the following discussion, I consider an environmental attribute to be broadly defined to include any biotic or abiotic feature of the environment that can be measured or estimated. Some of these attributes may be particularly information-rich in the sense that their values are somehow indicative of the quality, health, or integrity of the larger ecological system to which they belong. The convention is to refer to these attributes as *condition indicators* (cf. to definitions in Hunsaker and Carpenter 1990, Olsen 1992, and USEPA 1992). Often an indicator is defined relative to a specific ecological resource. For example, the maximum depth at which a Secchi disk is visible is an indicator of water clarity for freshwater lakes and provides insights into nutrient loadings. In general, indicators include biotic and abiotic measurements that provide quantitative information on the state of an ecological resource.

The most common reason to monitor a specific environmental indicator is to detect differences in its value among locations at a given moment in time (*status*), or changes in value across time at a given location (*trend*). Changes in the values of an indicator are useful and relevant to the extent they provide an assessment of management success or provide an early warning of adverse changes to an ecological system before irreversible loss has occurred. Generally, two specific products arise from a monitoring program. First is an estimate of the state (*value*) of the condition indicator (hereafter simply referred to as an *indicator*) at a given location at a specific moment in time. Second is an assessment of spatial or temporal trend in the indicator. Trend, viewed as the estimated time trajectory of a state variable, is particularly relevant because even if the value of an indicator is currently acceptable, a declining (or increasing) trend may indicate a trajectory toward system degradation, or an undesired state.

The task of detecting and recognizing meaningful change is difficult because natural systems are complex, inherently dynamic, and spatially heterogeneous. Further, many changes that occur in space and time are not a consequence of human-induced impacts, and many are not amenable to management intervention. For example, at least three kinds of changes are intrinsic to natural systems: stochastic variation, successional trends following natural disturbance events, and cyclic variation. Over relevant human time frames ecosystems experience these dynamic variations within stable bounds (Chapin et al. 1996) and management intervention may be inappropriate in response to natural disturbance events. One goal of monitoring is to develop a structural model of an ecological system that incorporates these sources of variation and predicts the distribution of an indicator variable under conditions of natural variation. Observed values of indicator variables could then be viewed in the context of deviations from expectations based on the ecosystem models that assume only natural sources of variation.

Of most interest to monitoring programs are extrinsically driven changes to environmental indicators that arise as a consequence of some human action. Concern arises when extrinsic factors, acting singly or in combination with intrinsic factors, drive ecosystems outside the bounds of sustainable variation. Thus, one key goal of a monitoring program is to discriminate between extrinsic and intrinsic drivers of change; that is, a mechanism to filter out the effects of expected intrinsic variation or cycles (noise) from the effects of additive, human-induced patterns of change (signal).

Intrinsic and extrinsic drivers of change can be collectively referred to as *disturbance events*. Disturbances alter physiological functions or processes and generate changes in the values of the state variables that characterize an organism or an ecosystem. Not all disturbances are undesirable—many ecosystems require disturbance events to maintain essential processes (examples in Pickett and White 1985). Therefore, it is necessary to differentiate disturbances that are accommodated from those that produce adverse effects, and to identify the source of the disturbance. Consistent with the literature on monitoring (e.g., Suter 1993; Thornton et al. 1993, 1994), I define a stressor as any physical, chemical, or biological entity or process that induces adverse effects on individuals, populations, communities, or ecosystems. Further, I focus on stressors that cannot be incorporated within the natural disturbance dynamics of a system, exceed the resilience of the system, and potentially drive an ecosystem to new state.

Stressor effects are evaluated in the context of induced changes in one or more indicators. The magnitude of indicator change that could generate a management response, however, is difficult to determine a priori. This uncertainty arises primarily from an incomplete understanding of the dynamics of ecosystems and the bounds of variation to which they are resilient. Interpretation of the significance of changes in the value of an indicator is also complicated by possible nonlinear, cause-effect relationships between the indicator and its stressor (Holling 1986).

All monitoring programs need to recognize that simple assumptions of linearity fail to acknowledge the possible existence of *thresholds*. Thresholds are regions of change in the value of a stressor that generate precipitous change in the value of the indicator or, more seriously, the larger ecological system. An example relevant to public land management considers the effects of habitat loss and fragmentation on the extinction process. Critical thresholds in the amount and distribution of habitat exist below which species populations rapidly decline (Lande 1987; Lamberson et al. 1992).

Why Monitor?

Monitoring is done to (1) determine if environmental laws have been implemented; (2) determine if specific management actions are having the desired effects; or (3) assess the value and temporal (or spatial) trend of those indicators that characterize the state of an ecological system. All these have a general goal of evaluating and constraining human activities so as to retain some degree of environmental "health." The scientific challenges are greatest for rationales 2 and 3, and I will focus on these.

On multiple-use public lands, for example, management actions are subject to many environmental standards. The public demands information on whether these standards are being realized and resources sustained. Monitoring is mandated on national forest lands to ascertain the degree of compliance with the population viability requirement of the National Forest Management Act (1976) and the minimum water-quality standards of the Clean Water Act (1977). Even for lands reserved from resource extraction and multiple use, such as national parks, compliance with the broad mandate to sustain "wild" resources for the enjoyment of future human generations must be assessed.

To ascertain compliance to a monitoring goal requires an appropriate set of indicators and a predetermined standard, or norm, of comparison for each indicator. The degree of deviation of the indicator from its desired value serves as a signal of noncompliance or a measure of environmental degradation. Normative or benchmark values and distributions are particularly important when monitoring is part of an overall restoration project. In highly degraded ecosystems, for example coastal watersheds in the Pacific Northwest (Reeves et al. 1995; Bisson et al. 1997), some time may elapse before indicator values begin to approach the standard. However, evidence that the indicator is changing in the direction of the normative value is evidence that the restoration effort is working.

One way to estimate the benchmark value for an indicator is to make reference to documented historical values or to conduct preliminary, baseline monitoring of a nonimpacted ("pristine") system. Given the scarcity of truly pristine systems, however, benchmarks may have to be based on some concept of a "desired condition" (see discussions in Bisson et al. 1997; Landres et al. 1999).

Environmental monitoring programs also have great value as early warning systems. By providing measures of those attributes indicative of ecological change in the early stages of decline, monitoring can result in prompt intervention before environmental losses are irreversible. This type of monitoring program is predictive and requires indicators whose values are somehow anticipatory of future trends.

Thus monitoring, whether for compliance, assessment, or early warning, is undertaken to ascertain whether the current state of the system matches some expected value or lies within some acceptable confidence region about the expected value. If monitoring results indicate that conditions lie outside the acceptance region, then some specific change in a management practice or resource policy is triggered. Alternatively, the information from monitoring can be used to investigate the response of the system to specific management actions. In this case management is viewed as a specific manipulation of a system with a desired outcome. The system is monitored to determine if management has produced the desired outcome. To the extent that the outcome has not been realized, management practice is adjusted in an adaptive fashion or some different management practice adopted (Walters 1986).

What Can a Monitoring Program Tell Us?

In general, an ecological monitoring program can help to (1) determine if a stressor is affecting one or more resources, (2) evaluate and contrast the effectiveness of current management practices, (3) develop a predictive understanding (in the form of one or more testable hypotheses) of why an environmental indicator is changing, and (4) decide when more active management and intervention are required. In addition, if the purpose is early warning of system change and an appropriate indicator has been selected, then a monitoring program can alert us to impending environmental decline.

By itself, however, a monitoring program cannot unambiguously ascertain the cause of an observed change, help decide on how much change is acceptable (i.e., is the observed change still within the range of acceptable variation?), decide on threshold values of the indicator that trigger specific management actions, or avoid Type I errors—that is, concluding the state of the system has changed significantly when no meaningful change has occurred.

Interpreting changes in the state of an indicator is difficult without prior knowledge of cause-effect relationships between a stressor or management practice and the indicator's value. However, causation can seldom be inferred from the results of monitoring studies alone. Usually, controlled experiments are required to establish the specific relationship between the stressor (or management) action and system response. These are research issues and demonstrate the complementary nature of research and monitoring programs. Unfortunately, the possibility of incorrectly interpreting the monitoring signal can never be totally eliminated. For a fixed sampling effort, limiting false positives (type I errors) occurs at the cost of increasing the likelihood of type II errors (i.e., failing to detect a significant biological effect when it has occurred). The trade-off between these risks is determined by which error is considered most important to avoid.

The Legacy of Environmental Monitoring Programs

Despite the obvious value of environmental monitoring, few examples exist of successful monitoring programs at the ecosystem scale. Unfortunately, little evidence supports the idea that such programs have contributed to informed management decisions or proven valuable in averting biological crises (GAO 1988; NRC 1990). In fact, the most ambitious (and expensive) monitoring program to date, the Environmental Monitoring and Assessment Program (EMAP) of the Environmental Protection Agency, has little tangible evidence of success and has been heavily criticized on both scientific and technical grounds (NRC 1990, 1994a, 1994b, 1995). Given the obvious importance of knowledge of the status and trends of our nation's natural resources and the integrity of the ecosystems that provide these resources, why have monitoring programs contributed so little to environmental decision making or policy formulation?

One fundamental reason for consistent failure is that monitoring costs are perceived to be prohibitively high. As a result, decision makers and the public are reluctant to commit to their implementation. In addition, environmental monitoring programs are often discussed in abstract terms, have little theoretical foundation, try to measure too many attributes, have vague objectives, and have no institutionalized connections to the decision-making process. The result has been a shallow comprehension of the need for, and components of, effective monitoring programs. Further, almost all previous programs have been given low priority (particularly compared to resource exploitation), seldom fully implemented, and insufficiently funded. In times of budget reductions, monitoring programs are often the first to be eliminated. The primary reasons for the failure of monitoring programs are summarized in box 2-1.

Box 2-1. Deficiencies of past environmental monitoring programs.

- A minimal foundation in ecological theory or empiricism
- Little or no logic justified the selection of the condition indicators
- No obvious linkage to a cause-effect interpretation of the monitoring signal
- Critical indicator values that would trigger a policy response were not identified
- No connection between the results of monitoring and decision making
- Inadequate or highly variable funding

To gain institutional support, the concept of environmental monitoring must become less abstract, its purposes more relevant, and its contributions more apparent. At a minimum, a defensible monitoring program should address the following: First, provide a clear statement of why the monitoring program has value, what information it will provide, and how the interpretation of that information will lead to a more responsible management. Second, provide a clear exposition of the logic and rationale that underlies the selection of the environmental attributes (indicators) to be measured. Recognizing that every species or physical or biological process of interest cannot (and need not) be measured, it is important to clearly articulate the logic used to select the attributes to be monitored. Inherent in this step is the need to select indicators that can be measured in a simple and cost-effective manner. Third, outline the sampling design and methods of measurement to estimate the value of the indicator variable. This element includes, but should not be limited to, sampling and measurement protocols. For example, the sampling design must address the necessary precision of indicator estimation to detect a given magnitude of change and the likelihood of detecting this change should it occur (e.g., see Zielinski and Stauffer 1996). The fourth component includes those procedures that connect the monitoring results to the decision-making process. For example, what magnitude of change in a given indicator should trigger a management response, and what should that response be?

Most existing monitoring programs fail to address one or more of these elements. Most attention has been given to an exhaustive discussion of the sampling and measurement protocols. It is not unusual to discover that great thought and deliberation have gone into how, when, and where to measure a given indicator, but little discussion of why that particular attribute is being measured or what magnitude of change needs to be detected (i.e., issues of survey design and statistical power).

Prospective (Predictive) or Retrospective Monitoring?

To be most meaningful, a monitoring program should provide insights into cause-and-effect relations between environmental stressors or between specific management practices and anticipated ecosystem responses. Prior knowledge of the factors likely to stress an ecological system or the expected outcomes from management should be incorporated into the selection of variables to measure and the sampling design. Indicators should be chosen based on a conceptual model that clearly links stressors (e.g., pollutants, management practices) and indicators with pathways that lead to effects on the structure and function of ecological systems (NRC 1995, 2000). This process enables the monitoring program to investigate the relations between anticipated stressors, or between management practices and environmental consequences, and provides the opportunity to develop predictive models.

In epidemiology, a prospective study begins by selecting cases with and without a suspected antecedent cause (stressor exposure) and follows cases to determine if the anticipated effect is associated with the antecedent cause. Conversely, a retrospective study begins by selecting cases with and without an effect and tracing back the cases to determine whether the effect is associated with the suspected antecedent cause. Both approaches have their foundation in identifying putative causal relationships between an antecedent cause and its expected effect. The two perspectives differ only on whether the study begins with a set of cases with or without a suspected antecedent (e.g., stressor) or with a set of cases with or without an anticipated effect. Prospective and retrospective studies focus on determining if a cause-and-effect relationship exists as postulated.

The National Research Council (NRC 1995) states that retrospective or effects-oriented monitoring seeks to find effects after they have occurred—for example, "by detecting changes in status or condition of some organism, population, or community." In contrast, "predictive or stress-oriented monitoring seeks to detect the known or suspected cause of an undesirable effect before the effect has had a chance to occur or to become serious." Thus, stress-oriented monitoring, unlike effects monitoring, assumes prior knowledge of a cause-effect relationship (Thornton et al. 1994). However, cause-effect relationships are seldom known with certainty and are usually only suspected. In this case, a hybrid approach is in order. The hybrid approach emphasizes simultaneous indicator and stressor measurement, and modeling the relationship between stressor action, change in state of the indicator, and subsequent ecological effects.

The design chosen for effectiveness monitoring of the Northwest Forest Plan (USDA et al. 1993) is an example of a hybrid approach combining effects-oriented and stressor-oriented monitoring (Mulder et al. 1999). It incorporates hypothesized causal relationships between effects and stressors through the judicious selection of indicators. The design process begins with a listing of threats (stressors) or specific management practices to which the land unit will be exposed. A conceptual model then outlines the pathways from the stressor or management action to the putative ecological effects. Attributes that are indicative *and* predictive of the anticipated changes in specific ecological conditions are then selected for measurement. These are the indicators. The success of this approach depends on the validity of the assumed causal relationship between the stressor or management action, their ecological effects, and the selected indicators. Given the correct set of indicators, this approach to monitoring does not require an ongoing assessment of ecological condition. It detects initial ecological effects as they are occurring by measuring indicators that are anticipatory rather than describing effects after they have occurred (fig. 2-1).

An advantage of this approach is that the emphasis on anticipated cause-effect relationships, expressed as changes in the value of the

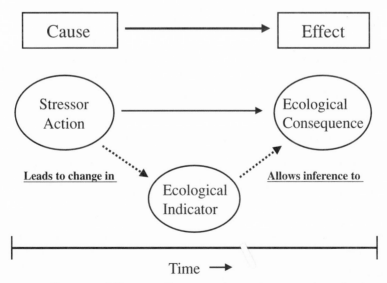

Figure 2-1. Conceptual diagram of a prospective environmental monitoring program. Indicators are selected in the context of known or suspected stressors to the ecological system.

indicator(s), allows an earlier and more focused management response to environmental change. However, given that all potential stressors or management outcomes cannot be identified, complete adoption of this approach is not without risk. Without a thorough assessment of ecological condition, the possibility exists of failing to detect the ecological impacts of significant but unanticipated stressors or management consequences if a wrong or incomplete set of indicators has been selected.

Focus of This Chapter

A key premise of this chapter is that monitoring to assess the integrity of ecological systems is an essential responsibility of land managers. My belief is that by clarifying the components of ecological monitoring, specifically its foundation in ecological theory and empiricism, it is more likely to be implemented. The value of monitoring should become more apparent because the process will become less abstract and better focused; the information content of the indicators relative to system integrity or management success will be more obvious; and a stronger theoretical framework should increase the relevance of monitoring data to policy decisions.

Many of the principles and concepts discussed below are general— applicable to environmental monitoring programs regardless of spatial scale—local, regional, or national. The process and recommendations expressed here, however, may be most applicable to regional-scale monitoring programs. I have developed the ideas presented below based on my experiences with the management of large landscape units, such as national forests and national parks. Monitoring programs to evaluate the effects of specific projects at the local scale are more amenable to the before-after experimental designs of environmental impact studies (see Schmitt and Osenberg 1996).

Key Components of a Monitoring Program

This section presents, in a step-down fashion, the key components of a hybrid monitoring program. Because monitoring is an ongoing, active process, however, the implementation of these components is never completely finished. They must constantly be revisited and revised as scientific knowledge is acquired and as the threats to the focal ecosystem change.

State the Goals of the Monitoring Program

In general, environmental monitoring programs have focused at one of two scales in the ecological hierarchy—either at the level of the ecosystem or at the level of an individual species or small set of species. In the discussion that follows, I retain this dichotomy but attempt to link them by expanding on the concept of focal species (Davis 1996; Lambeck 1997; Noss 1999). Focal species are those that play significant functional roles in ecological systems by their disproportionate contribution to the transfer of matter and energy, by structuring the environment and creating opportunities for additional species, or by exercising control over competitive dominants and thereby promoting increased biological diversity.

Rapport and others (1998, 1999) suggest devising "fitness tests" for ecosystems similar to those used by doctors. These are based on the hypothesis that "healthy" ecosystems are resistant to stress and recover more quickly following disturbance. They propose the use of measures based upon speed of recovery after perturbation, adjusted for relative stress load, as a preliminary assay of ecosystem health. If the measures indicate that the ecosystem could be at risk, the functions thought to be at risk can be further evaluated.

No universal standard exists that defines a "healthy" ecosystem. An important step toward making the concept of "environmental quality" more general follows from a definition of *ecological integrity*—the capacity to support and maintain a balanced, integrated, adaptive community of organisms having a species composition, diversity, and functional organization comparable to that of natural habitats of the region (Karr 1987, 1991). Ecological integrity has been proposed by the scientific community as an appropriate metric to assess the state of ecological systems, and measurable definitions have been provided (e.g., Karr 1987, 1991, 1996; DeLeo and Levin 1997). Unfortunately, this definition is still equivocal because several key terms, such as "natural" remain vague. Given our state of knowledge about the functioning of complex ecosystems, however, such equivocations are unavoidable at this time.

According to the dictionary, "integrity" is "the state of being unimpaired, sound," "the quality or condition of being whole or complete." A variety of definitions of ecological integrity exist, most differing in the scale of assessment. A fine-scale approach stresses the structural and compositional aspects of ecological systems focusing on individual species and their dynamics within specific ecosystems. A more coarse-scale approach focuses on macro-scale processes (i.e., primary productivity, nutrient cycling, hydrological regimes), and pays considerably less attention to the structure and composition of the systems from which these processes emerge (COS 1999).

Implicit in Karr's definition as a necessary precondition is the physical and chemical integrity of the environment. If this integrity is compromised, then measurable ecological effects should arise. Some attribute measures will be more comprehensive and anticipatory than others. Obviously, the concept of ecological integrity is complex—integrity cannot be assessed through a single indicator but will require a set of indicators measured at different spatial and temporal scales and at different levels in the hierarchy of ecological systems.

Invoking the concept of ecological integrity puts the problem in the context of an ecological system composed of integrated biological components—individual organisms, populations, species, and communities—connected by exchanges of matter and energy. This model represents the traditional notion of an ecological hierarchy and is a familiar starting point for most ecologists. To have societal support, however, a connection must be made between measured biological and physical attributes and what society values. This linkage requires a conceptual framework that identifies the relations between societal values (the ultimate assessment endpoints for an environmental monitoring program) and ecological integrity.

Assessment endpoints relate directly to societal values but may not be easily measured (Suter 1990). Measurement endpoints—the indicator variables that are actually estimated in the field—act as surrogates for those values that are difficult or impossible to measure directly (Thornton et al. 1993, 1994). As an example, consider the following societal value: the probability of catching a fish in a freshwater lake given a certain effort. This value could be measured directly, but it may be difficult or costly to do so. A surrogate variable, related to fish survival and population size, is pH of the lake at fall turnover. The distribution of this indicator variable could serve as the measurement endpoint (Messer 1992). It has the advantage of cost effectiveness, but in addition is applicable to a much broader range of ecological resources than just game fish.

The monitoring goals established for public lands are often based on legal mandates to maintain environmental quality or retain biological diversity. For example, the legal mandate to the National Park Service is to maintain each park, its resources and ecosystems, for the enjoyment of future human generations. Thus, the monitoring goals would be to assess how well the ecosystems of the park were being sustained by current management practices (biological monitoring) and if the public was experiencing full enjoyment (social monitoring). On national forest lands, monitoring is required to demonstrate compliance to the many requirements of the National Forest Management Act (NFMA), including that of maintaining biological diversity (plants and animals) in perpetuity. Considering just the biological objectives of the public land management agencies, these goal statements are far too vague to develop an effective monitoring program. To be useful, they need to be further refined and made more specific.

The Northwest Forest Plan (USDA et al. 1993) provides an example of refining the broad management objectives on U.S. Forest Service and U.S. Bureau of Land Management (BLM) lands into a set of monitoring goals. A key ecological resource identified in the plan, and mandated for monitoring in subsequent legal decisions (ROD 1994), is the status and trend in late-seral old-growth (LSOG) forests. Monitoring goals, stated in the form of questions that can be addressed with monitoring data from late successional forests, include:

- What is the amount and distribution of forest age classes [including LSOG] at the landscape scale?
- What are the patch size distribution, patch interior area distribution, and interpatch distance distribution for LSOG forests at the landscape scale?
- Based on stand sample data, what changes have been produced by stressors in the amount and distribution of forest age classes, beginning with data collected for the 1993 Forest Service and BLM assessments?
- What are the effects of silvicultural treatment and salvage logging on LSOG structure and composition at the stand scale?

These questions refine the monitoring goals and suggest attributes (indicators) to measure based on those components of late-successional forests assumed to reflect successful implementation of standards and guidelines of the Northwest Forest Plan.

Identify Barriers to Attaining the Management Goals

This step will usually take the form of identifying the anticipated extrinsic environmental stressors that may compromise the integrity of the management unit and its component species and ecosystems. From historic studies of disturbed ecological systems (e.g., Delcourt et al. 1983), we know if the effects of an extrinsic stressor exceed the resilience or adaptational limits of the constituting processes or species, change occurs, the ecosystem moves to a new state, and ecological integrity may be compromised.

Stressors, as envisioned here, can be both human-induced and "natural" disturbances. Examples (Barber 1994) include:

- Loss of late-seral habitat by fire
- Alterations of hydrologic cycles due to dams or water diversions
- Reduction, loss, or fragmentation of critical habitat
- Increased sediment loads to streams after storm events
- Overharvest of game species

Based on retrospective studies of historic patterns of change in ecological systems (e.g., Foster et al. 1996; Gauthier et al. 1996; Swetnam et al.

1999), it quickly becomes apparent that the term "disturbance" is far too generic. Events that lead to changes in the rates of ecological processes or in the composition of ecological systems vary extensively in extent and consequence (see White et al. 1999). Therefore, it is important to characterize disturbance events in terms of several defining attributes that allow them to be discriminated and categorized (box 2-2). In addition, the consistent use of these defining attributes is important when human-induced disturbances are assumed to mimic natural disturbance events.

To retain the possibility of establishing cause-effect relations from the monitoring program, the status of the stressor must also be periodically estimated. That is, to infer causation from an observed change in the value of an indicator requires concurrent estimates of the status of the indicator and the magnitude of the putative stressor.

To aid the process of indicator selection, it is important to identify the ecological resource(s) likely to be affected by a given stressor. A resource is broadly defined as an ecological entity subject to stressor effects. In practice, a resource is considered to be a key component of a larger ecological system or management unit with social or ecological value. Examples include freshwater lakes and montane meadows within a national forest, rates of primary productivity within a grassland ecosystem, and the community of leaf litter arthropods in a deciduous forest. A resource can be either discrete or extensive (USEPA 1993). Examples from the Northwest Forest Plan of extensive resources include late-seral forests and aquatic/riparian ecosystems; discrete resources include spotted owls, marbled murrelets, and anadromous fish. Establishing the functional relations between stressors (natural or human-induced) and resources is an essential first step toward developing the conceptual model (see below).

Box 2-2. Characteristics of natural disturbances and human-induced stressors acting on ecological systems.

- Frequency (number of occurrences per unit time)
- Extent (area over which the event occurs)
- Magnitude:
    - Intensity (degree of effect on the biota)
    - Duration (length of stressor event)
- Selectivity (portion of the biota affected)
- Variability (probability distribution for each of the above)

Following from the previous example, the management goal may be to have fishermen experience a high likelihood of catching a fish in a national park. Various types of water pollution may significantly compromise attainment of this goal, however. Possible stressors could be atmospheric deposition or mining activities outside the boundaries of the park that adversely affect the fish resource. The conceptual model would attempt to show the relations between increasing atmospheric deposition, for example, and attributes that affect the fish resource.

Develop a Conceptual Model Linking Relevant Ecosystem Components

To select indicators that reflect underlying ecological structure and function requires well-developed conceptual models of the ecological system(s) being managed (Barber 1994; NRC 1995, 2000; Manley et al. 2000). The conceptual model outlines the interconnections among ecosystem components, the strength and direction of those linkages, and the attributes that characterize the state of the system. The model should demonstrate how the system "works," with particular emphasis on anticipated system responses to stressor action. The model should also indicate the pathways by which the system accommodates natural disturbances and what system attributes provide resilience to disturbance. These processes could be portrayed by illustrating the acceptable bounds of variation of system components, and normal patterns of variation in input and output among the model elements.

As a general goal, management will strive to sustain fundamental ecological processes. These processes, however, are often difficult or impossible to measure directly. Conceptual models should therefore identify tangible structural and compositional elements of the system that reflect the state of underlying processes. A heuristic to guide the model development links process and function to measurable aspects of structure and composition. These elements, in turn, can be used to make predictions of an expected biological response (fig. 2-2). The model shown in figure 2-2 presumes that structure and composition both influence and are influenced by process and function (Spies and Franklin 1988). For example, disturbance events change the stand and landscape structure of the vegetation that in turn affects the likelihood and pattern of subsequent disturbance across the landscape (Turner 1989). In addition, this simple model assumes a strong relationship between structure and composition and biological diversity. Species diversity is the biological foundation of structure and composition, and structure and composition, in turn, provide habitat conditions to support species diversity (Hemstrom et al. 1998).

Figure 2-2. Conceptual model proposed as a basis for identifying indicators based on structural and compositional attributes of ecological systems. Stressors (fig. 2-1) are integrated into this model based on where their effects are most readily observed.

Measurement of, and inference from, ecological systems is affected by the scale of observation. Therefore, to determine the appropriate scale for measuring an indicator, the temporal and spatial scales at which processes operate and resources respond must be estimated (at least to a first approximation) and clearly identified in the conceptual model. As a result, the most useful conceptual models will have a hierarchical structure (Allen and Starr 1982). That is, a given structural or compositional resource in the model will reflect processes operative at smaller temporal and spatial scales and implicate the constraints operating at larger scales (Allen and Hoekstra 1992).

Indicators are measurable attributes that characterize the state of unmeasured structural and compositional resources and system processes—that is, they are indicative of the state of some component (e.g., habitat, species, community) of a complex ecological system. In a prospective monitoring program, indicators selected for measurement should reflect known or suspected cause-effect relations among system components as identified in the model. Ecological resources occupying central positions in the model should receive increased weight when the indicators are selected. As a result, the model justifies, in terms of contemporary ecological principles and theory, the indicator or indicators selected for monitoring, and how knowledge of the status and trend of the indicator reflects the state of the underlying processes and functions of the ecological system. In most cases, it will be sufficient to model a restricted, but relevant, component of the system. Thus, a complete model of an ecological system is seldom necessary to proceed with a reliable monitoring program.

Identify Candidate Indicators

On the basis of the conceptual model and characterization of its central components, candidate indicators are proposed for monitoring and subsequent field testing. At this point, the primary criteria for selecting indicators are that they

- Reflect underlying ecological processes *and* are sensitive to changes in stressor levels
- Provide information on the state of the unmeasured resources and processes of the focal ecological system
- Are measurable in a cost-effective manner

Before field or simulation testing, the list of candidate indicators can be narrowed by focusing on those with the following properties:

- Their dynamics parallel those of the larger environmental component or system of ultimate interest
- They each show a short-term but persistent response to change in the status of the environment
- They can be accurately and precisely estimated (i.e., a high signal-to-noise ratio)
- The likelihood of detecting a change in their magnitude is high given a change in the status of the system being monitored
- Each demonstrates low, or well-understood, natural variability, and changes in their values can readily be distinguished from background variation
- They are clearly relevant to some important societal value

Additional evaluative criteria for screening candidate indicators are given in Barber (1994).

Even if a monitoring program is fully funded and implemented for many years, it will fail if the wrong indicators were selected. Thus, *the ultimate success or failure of the program may be determined by this one step*. The likelihood of choosing appropriate indicators is greatly improved if the conceptual model thoroughly characterizes the system's dynamics, and accurately reflects stressor inputs.

Estimate the Status and Trend of the Indicator

In general, determining the value of an indicator is a problem in estimating an unknown parameter within some specified bounds of precision. Estimates of trend address the pattern of change over time in the value of the indicator. The topic of parameter estimation and trend analysis is broad, and proper design requires substantial statistical expertise. Fortunately, there exists a large body of statistical literature on parameter estimation, hypothesis testing, and trend estimation that is relevant to this problem (e.g., Sauer and Droege 1990; Thompson et al. 1998; Olsen et al. 1999; Manly 2001).

Some debate exists over what is the correct statistical framework for monitoring—parameter estimation or hypothesis testing (e.g., Stewart-Oaten 1996). For the moment, I will frame the monitoring question in terms of a statistical null hypothesis of no difference between the estimated value of the indicator and its hypothesized normative value. The choice of significance level (α) for tests of the null hypothesis of no difference in the status of an indicator must be balanced against the likelihood of failing to detect a significant biological difference. Determining the α risk level is a burden borne by decision makers. The ß risk level, in contrast, is a burden borne by those charged with maintaining ecological integrity. The managers' responsibility, therefore, is to implement an environmental monitoring program with sufficient statistical power (i.e., an acceptable value of 1-ß) to detect meaningful changes in the values of the indicators. For the monitoring design and analyses to be meaningful, statistical power must be considered a priori when determining sample sizes, and post hoc to interpret the result of statistical tests that failed to reject the null hypothesis (Skalski 1995; Zielinski and Stauffer 1996). A comprehensive discussion of statistical power, and its relevance to decision making in the context of responsible management of natural resources, is found in Peterman (1990) and Osenberg et al. (1994).

Determine Values That Trigger a Response

One of the most difficult challenges is to determine the value of an indicator or the magnitude of change in its value over some time interval that indicates a significant ecological effect. In statistical terms, this amount is referred to as the effect size (Δ), or magnitude of change in the value of the indicator that the monitoring program should be able to detect. Initial estimates of an appropriate effect size can be based on the spatial or temporal variation in the indicator (σ^2) under baseline or reference conditions (Skalski 1995). In sum, specification of acceptable levels for type I and II errors (α and ß), natural variability of the indicator (σ), and the sensitivity of the test (Δ), determines the sampling effort for a given effect size.

Once the environmental indicators for monitoring have been selected, threshold values must be defined for those indicators that will lead to ("trigger") a response by managers or policy makers. Example responses include moratoria on harvest, direct ecosystem intervention, changes in land management practices, or proposals for endangered species consideration. For a given monitored location at a specific time, however, it may be inappropriate to assess the indicator value relative to a specific trigger point. Instead, the site-specific indicator value must be viewed in a larger spatial context, including estimates of the value of the indicator from many locations throughout the assessment region. In this case, we are interested in the *distribution* of indicator values (fig. 2-3). A sufficient change in the distribution of indicator values across time will provide the collective signal needed to interpret ecosystem integrity at broad spatial scales.

Defining the specific value of environmental indicators that dictate a management response is difficult and complex. It is important to clearly state the limits of our current understanding of ecological processes, population dynamics, and the unavoidable uncertainty associated with defining precise threshold values for environmental indicators. Acknowledging uncertainty up front is critical because such incomplete knowledge is an explicit risk that can be addressed in the decision process. Despite this uncertainty, monitoring programs are of limited value if trigger points or critical distributions have not been identified. As a general rule, the more irreversible the potential environmental loss, the more sensitive should be the trigger point. That is, the burden should be on minimizing the risks of type II errors (concluding no adverse effect when one exists), particularly when resource declines are irreversible (Shrader-Frechette and McCoy 1993).

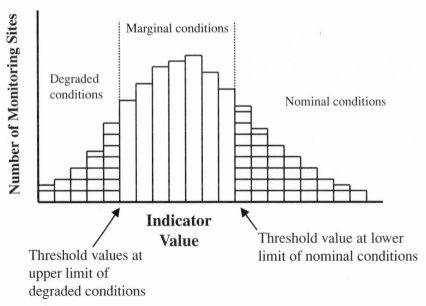

Figure 2-3. Distribution of indicator values based on sampling numerous sites at various locations on the landscape. Individual sites can be interpreted as representing degraded, marginal, or nominal ecological conditions, but the overall inference is to the ecological state at the landscape scale.

Linking Monitoring Results to Decision Making

The results of monitoring programs are of value to the extent that they inform the management decision-making process, provide early warnings of ecosystem change, and result in a meaningful response. Given the usual uncertainty between the value of an indicator and its ecological interpretation, the linkage between monitoring results and decisions should be framed in the context of statistical decision theory (Lindley 1985; Clemen 1996). This process entails a listing of possible understandings of the monitoring data (i.e., possible ecological interpretations of indicator values), the probability of the various interpretations being "true," the range of management decisions being considered, and the utilities (e.g., societal values) associated with each combination of decision and interpretation. Example applications of decision theory to species conservation are found in Conroy and Noon (1996), Guikema and Milke (1999), and Possingham et al. (2001). This is an important topic that goes beyond the scope of this chapter.

Developing the Conceptual Model

Perhaps the most challenging task is to develop a conceptual model(s) of the ecological system to be monitored. Conceptual models demonstrate the linkages between anticipated environmental stressors, system components and resources, and expected ecologic consequences (Thornton et al. 1994). They are necessary to guide the selection of environmental indicators (NRC 1995, 2000).

For the purposes of model development, I will focus further discussion on stressors arising from human-induced disturbances that alter ecological resources or processes and lead to undesired ecological conditions. Stressor action may result in structural and compositional changes in the biota or changes in the rates or patterns of flow of matter and energy. An exhaustive listing of the anticipated stressors, the ecological consequences of stressor action, and how they would be measured, are the first steps toward indicator identification and eventual selection. The linkages between stressors and consequences are illustrated by use of a "model" that summarizes how a given component of an ecological system functions, at what level in the system stressor action is expressed, the anticipated outcomes of stressor action, and the expected change in one or more environmental attributes.

Developing models is always contingent on current understanding of ecological system processes and cause-and-effect relations. Because this understanding is incomplete, models are always a "work in progress" subject to revision as our understanding of nature progresses. To further complicate matters, the process of model development is inherently vague, and few clear examples exist to show how to do it. Nevertheless, such models are recognized as an essential step toward identifying and selecting environmental indicators (NRC 1995, 2000).

To make the process of model development useful and repeatable requires developing a step-down procedure grounded in current ecological theory. Toward this end, I propose a process that explicitly brings in aspects of spatial and temporal scale, and several levels of the ecological hierarchy. This process is followed with full knowledge that the boundaries among scales and positions in any spatial or temporal hierarchy are indistinct (O'Neill et al. 1986). To the extent possible, however, we must attempt to make them concrete, at least in an operational sense. Only by doing so can we justify making inferences from patterns observed at one scale to biotic consequences at higher or lower scales.

Noss (1990) proposed a useful hierarchical approach to monitoring based on four scales of ecological organization—landscape, community-ecosystem, population-species, and genetic—and three primary attributes of biological diversity (after Franklin et al. 1981)—function, structure, and composition. The organizational scales are not discrete but exist along a continuum. The need to categorize otherwise continuous phenomena into discrete categories arises because of the nature of environmental impacts. Land-use changes that generate impacts or stresses are by definition explicit in time and space. When viewed in the context of a specific environmental stressor, a useful conceptual model requires a framework that makes explicit the space-by-time position of each structural, compositional, or functional element subject to stressor effects.

The Initial Conceptual Model

Conceptual models of ecological systems are difficult to develop because of limits to our understandings of complex ecological systems and the dangers of omitting key system components. Limits to our knowledge first become apparent when we attempt to specify the "extent of the system," essential system components, the linkages among components (direction and magnitude of the exchange of matter and energy), and the appropriate scale (temporal and spatial) to observe and measure the components (Manley et al. 2000). Because this process is characterized by uncertainty and incomplete information, in some sense it is always a "work in progress." The conceptual model should be continually refined and updated as new ecological knowledge is acquired.

A first step in conceptual model development could follow the heuristic model shown in figure 2-2 that suggests logical links between process, structure and composition, and biological expression. Both natural and human-induced processes and functions are elaborated on the left-hand side of the diagram. It is useful to think of these processes in terms of phenomena that ultimately generate observable change in the biota. For example, key processes and functions might include local dynamics such as vegetation growth and regeneration, ecosystem scale dynamics such as nutrient cycling, hydrology, and trophic interactions, and regional effects including climate and topography. These processes are distinguished by characteristic rates and magnitudes whose values are affected by external disturbance events. Some disturbances are part of the inherent variability of the environment and others are artificial, in the sense that they are a consequence of human behavior. Our understanding is that species, communities, and ecosystems have evolved in ways that accommodate most "natural" disturbance. Our concern is whether human-induced disturbances are incorporated as natural events, or whether they exceed the adaptation limits of some components of the biota and move the ecological system to a new state.

A key assumption of the preliminary model (fig. 2-2) is that ecological processes have tangible expressions in the form of structural and compositional elements that can be mapped and measured at local and regional scales. These elements, in turn, provide feedback to the generating processes in a set of reciprocal interactions (i.e., a two-headed arrow). Note that structural and compositional elements exist at many spatial and temporal scales. For example, at a landscape scale we can measure structure in terms of the number, size, and distribution of patches of forest vegetation differing in composition or configuration. At a local scale we can measure structure in terms of the size distribution of trees or the amount of coarse-woody debris on the forest floor. In addition, species respond directly to structural and compositional elements of a landscape and indirectly to fundamental processes and functions that give rise to these elements. The premise is that the dynamics of some small subset of species may closely reflect important changes in underlying ecological processes.

For a specific ecological system, the key physical and biological processes likely to be affected by stressor action should be identified. Second, those measurable structural and compositional attributes most affected by the changed processes should be determined. Measurable attributes should be explored at several spatial scales ranging from microsite to landscape scales. Finally, a small set of plant or animal species most likely to respond to changed habitat conditions or directly to changed processes should be proposed. Those measurable structural/compositional attributes and individual species that provide information about the state of the system represent possible indicators of stressor action.

An example of the approach outlined in figure 2-2 was adopted by Northwest Forest Plan assessment team (USDA et al. 1993) to aid selection of indicator variables for late-seral old-growth (LSOG) forests in the Pacific Northwest (Hemstrom et al. 1998, chapter 11). This approach began by equating stressors to the key dynamic processes of change in Pacific Northwest forests. These included wind, fire, exotic species, pathogens and disease, growth and succession, and human land uses and their possible synergisms with natural disturbance events. With the aid of a conceptual model, this combination of stressors/processes was linked to specific structural and compositional elements assessed at multiple spatial scales. The relationship between process-induced changes to structure and biological diversity considered three components of structure: scale, heterogeneity, and complexity (Bell et al. 1991). Structure depends on spatial scale and was assumed to have different associations with biological diversity at each scale. Hemstrom et al. used the logic summarized in their conceptual model of LSOG forests (see chap. 11) to propose a series of indicator variables for long-term monitoring (box 2-3).

Conceptual models to guide indicator selection were developed as part of the EMAP process (Thornton et al. 1994) and recently for Forest Service lands in the Sierra Nevada (Manley et al. 2000). These examples

---

Box 2-3. Proposed indicators for monitoring the state of late-seral old-growth (LSOG) forests in the Pacific Northwest.

- Large landscape-scale indicators
  - –Amount of LSOG forest
  - –Distribution of LSOG forest
  - –Average stand size
  - –Average inter-stand distance
- Stand-scale indicators
  - –Structure and composition
  - –Tree diameter class distribution by species
  - –Canopy structure and height class by species
  - –Snag height and diameter class by species
  - –Down wood per acre
- Change-agent indicators
  - –Large landscape change indicators
  - –Stand change indicators

demonstrate that many methods can be used to develop conceptual models to guide the selection of environmental indicators—there is no right or wrong method. Some common methods are found across useful models, however. One of the most important is a clear representation of the appropriate temporal and spatial scale for observing and measuring an environmental component or indicator. Problems arise if the scale of monitoring and indicator selection is too coarse (or fine) relative to the scale of management action required to ameliorate the effects of a stressor. If too coarse, significant effects may go undetected. If too fine, too many false positives may appear.

Linking Stressors to Indicators

Given an initial conceptual model, the next step is to identify where a given stressor is most likely to affect the state of the system and what measurable changes may be caused by stressor action. For example, does the stressor directly affect the rate of some key process such as primary productivity, or alter the structure and composition of the dominant vegetation, or does it lead to a direct reduction in the size and distribution of some biological population?

To infer causal relations the putative stressors must be closely linked to the system components they are most likely to affect. This linking requires both prior knowledge of the stressors and a preliminary understanding or hypothesis of the cause-effect relations in the ecological system being monitored. For most ecological systems, the cause-and-effect understanding will be incomplete, and assumed relationships should be treated as hypotheses to be tested. To reduce the uncertainty, a first step should be a thorough synthesis of published studies that provide insights to the functioning of the system of concern.

The goals at this stage are to establish how a given stressor is expected to affect an ecological system in terms of specific processes, structural or compositional elements, and species populations. The relations between stressor action, likely system response, and expected stressor manifestation will hopefully guide the indicator selection process.

Developing a List of Candidate Indicators

Viewing ecological systems as hierarchically structured is a useful first step toward developing conceptual models and identifying appropriate scales for measurement (Allen and Starr 1982; O'Neill et al. 1986). For this reason, a hierarchical framework was specifically invoked as part of developing the example conceptual model for the Pacific Northwest (chap. 11).

Importantly, the position of an indicator in the hierarchy must be correctly identified to apply the appropriate measurement scale. Once position in the hierarchy is identified, "it is necessary to look both to larger scales to understand the context and to smaller scales to understand the mechanism; anything else would be incomplete." (Allen and Hoekstra 1992, 8). To have a predictive understanding of causation in a stressor-indicator association, for example, requires understanding the scales at which the stressor is expressed; assigning the correct level in the hierarchy to the indicator; understanding the scale below to provide a mechanistic understanding of the indicator's behavior; and observing the scale above to understand the context in which indicator change is expressed.

Once candidate indicators have been listed, the next step is to select a subset to measure. This step is usually necessary for pragmatic reasons—the list of candidate indicators often exceeds the funds available for monitoring. The task is to establish the criteria that will act as a "filter," retaining those indicators that are somehow "best." The scientific challenge is to design the logical fabric of the filter so as to retain those indicators that allow the most reliable induction to the state of the ecological system. Critical to the design of the filter is the concept of the indicator value (or distribution) as a "signal" that allows inductive inference to the "true" state of the ecological system with minimal uncertainty.

Developing the criteria that will constitute the fabric of the filter is a major challenge. Unfortunately, the scientific knowledge needed to guide this process is either unknown or has not been synthesized with monitoring questions in mind. As a result, it is not possible to simply go to the "shelf of scientific understanding" and pull off what is needed. It is clear, however, that the nature of the criteria will vary according the stressors acting on the ecological system and their defining attributes (box 2-2).

Indicators of Ecological Sustainability

Ecological systems consist of a varied array of processes and elements that exist at multiple spatial scales and function at diverse temporal scales. Given this complexity, the goal is to develop a small subset of measurable attributes of these systems that allow inference to their sustainability. That is, we seek simple, reliable measures that indicate the degree to which management practices are sustaining ecological systems. However, the concept of environmental indicator, or indicator species, has been much criticized by scientists (Landres et al. 1988; Morrison and Marcot 1995). The hard reality, however, is that all possible attributes of an ecological system cannot feasibly be measured. Therefore, a small subset of particularly information-rich attributes must be selected. The challenge is in the selection process—that is, the logic and rationale that is brought to bear to justify the environmental attributes chosen for measurement.

The initial decision that has to be made regarding indicator selection is choosing between function-based, structure-based and species-based indicators (see Lindenmayer et al. 2000; fig. 2-4). Function-based indicators include direct measures of processes and their rates. Examples include primary productivity, rates of nutrient cycling, and water flows. Structure-based indicators, measured at local and landscape scales, include elements such as vegetation structural complexity, among-patch vegetation heterogeneity, landscape connectivity, and landscape pattern (i.e., the distribution and abundance of different patch types). These metrics are often assumed to constitute a "coarse filter" because of their ability to predict broadscale patterns of biological diversity (Hunter et al. 1989; Haufler et al. 1996). Both function- and structure-based indicators can be measured at multiple spatial scales ranging from local, to landscape, to regional. In addition, there are composition-based indicators that include the direct measurement of some aspects of a species' life history, demography, or behavior. These are often referred to as "fine filter" assessments because they evaluate the effects of management practices on individual species (Haufler et al. 1996).

As previously discussed, the process of indicator identification begins with the development of conceptual models of the ecological systems

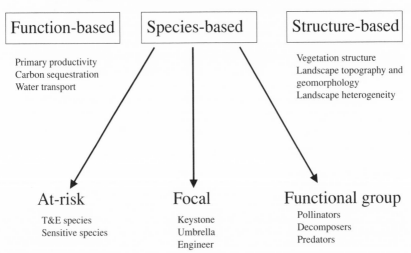

Figure 2-4. Possible categories of condition indicators for monitoring ecological systems. Emphasis in this diagram is placed on species-based indicators.

being managed (NRC 1995). Even with a stressor orientation, this process will undoubtedly identify a large number of function-, structure-, and species-based indicators. Once the list of possible candidate indicators is identified, however, it is often so large that some sort of filtering process is required. Given the complexity of ecological systems, however, a useful strategy is to retain two or more indicators from each indicator category. This strategy recognizes that inference to the state of any ecological system will usually require a collective evaluation of all indicators.

During the indicator selection process, change in some attribute of a species' population may be selected as an appropriate early warning of a stressor affecting an ecosystem. Even if species populations do not "fall out" as appropriate measures during the indicator selection process, there will often be legal (listed species under the ESA) or social (economically valuable species) mandates to directly monitor the population status of some species. There has been substantial debate over the usefulness of "indicator species." Most of this criticism has been directed at the reliability of drawing inferences to the status of unmonitored species from the status of a monitored species (Landres et al. 1988; Landres 1992). For the most part, criticism has not been directed at the use of species as sentinels of system change in response to the action of one or more stressors.

Current research has revived the value of measuring aspects of species populations as useful indicators of ecological integrity. For example, multispecies conservation issues in Africa and Australia have been addressed by concentrating on "focal species" (Davis 1996; Lambeck 1997). Often these species are simply those with large area requirements in habitats experiencing loss and fragmentation. By sustaining the populations of species with large area requirements, presumably those species with similar habitat requirements, but requiring less area, will also be conserved. Recently, a more theoretical approach has been used to compare and rank species with similar life histories in terms of their sensitivities to habitat loss and fragmentation (Noon et al. 1996). This approach represents an important step toward moving from single- to multispecies conservation planning. Of most interest to ecological monitoring, however, may be the recent quantitative formulation of the focal species concept into an expression that formally links species dynamics to ecological processes (COS 1999).

The Focal Species Concept

The key characteristic of a focal species that make it particularly relevant to monitoring is that its status and trend are hypothesized to provide insights to the integrity of the larger ecological system to which it belongs.

"Focal species serve an umbrella function in terms of encompassing habitats needed for many other species, play a key role in maintaining community structure or processes, are sensitive to changes likely to occur in the area, or otherwise serve as an indicator of ecological sustainability" (COS 1999). Attempts to streamline the assessment of ecological systems by monitoring focal species can be seen as a pragmatic response to ecosystem complexity.

The focal species concept differs subtly, but significantly, from the previous "management indicator species" concept. Rather than acting as an indicator of a specific management prescription or an indicator to the abundance of less easily surveyed species and their associated habitats, focal species allow induction to the state of the ecological system to which they belong. In this sense, the focal species concept is more closely aligned with the concept of a condition indicator (see discussion in Zacharias and Roff 2001). The concept is inclusive of several categories of species being discussed by ecologists, including keystone (Power et al. 1996), umbrella (Wilcox 1984; Berger 1997; Fleishman et al. 2000), traditional indicators (Landres 1992), and engineering species (Jones and Lawton 1994; Jones et al. 1994, 1997; Brown 1995).

Even if species are not necessarily the most comprehensive indicators of stressed ecological systems, there will often be considerable social pressure to include species as part of a comprehensive monitoring program. In addition, arguments in favor of assessing species as part of a monitoring program are based on the fact that biological populations often integrate conditions over longer time periods and are therefore better measures of chronic stresses than more rapid response indicators. For example, a freshwater river system exposed to an acute toxicant may show no chemical evidence of the stress after a short period of time but may retain a long-term record of the stress in terms of changes in its biological populations.

Conceptual models may also prove useful for the selection of indicator species. The goal of such models is to identify how changes in the status of a species should be interpreted in terms of the processes, structures, and compositional components of ecological systems that support them. As part of its monitoring requirements under the Northwest Forest Plan (USDA et al. 1993), for example, the U.S. Forest Service and U.S. Bureau of Land Management adopted a focal-species approach based on northern spotted owls (Lint et al. 1999) and marbled murrelets (Madsen et al. 1999). As a framework for conceptual model development for the spotted owl, Lint et al. (1999) characterized the owl's ecological web as an *envirogram* (Andrewartha and Birch 1954, 1984) to distinguish sources of stress, identify components of the species life history most sensitive to stress, and link aspects of the animal's ecology to structural and compositional elements of the environment affected by current management practices. Another example of focal species monitoring to aid assessment at the ecosystem scale is the case of the red-cockaded woodpecker. James et al. (1997) developed a detailed envirogram that clearly links the dynamics of the woodpecker population to changes in structural and compositional elements at both the stand and landscape scales.

An example monitoring program that links the focal species to its ecological web is the indirect assessment of the status of an indicator species based on habitat measurements. The set of habitat attributes selected for measurement are assumed to be the indicators of habitat amount and quality for the focal species. Based on these habitat characteristics, a model is developed to predict the probable distribution or population status of the focal species on existing or future landscapes. The model may assume constant habitat conditions, hence predicting presence under steady-state conditions, or it may be dynamic simulating growth, disturbance, or management outcomes to enable a prediction of population trend. Given a perfect model (an impossibility), no direct measures of the focal species populations are required. In practice, however, direct assessment of populations should be conducted periodically for model validation and refinement. For an example of this approach, see Lint et al. (1999).

Developing a logical foundation for focal species selection is critical but poorly developed at this time. Past efforts to identify species that act as surrogates of the states of ecological systems have met with mixed success or were logically flawed (Landres 1992; Landres et al. 1988). Some of this failure is due to the lack of clearly defined and defensible criteria associated with their selection and measurement. Currently, most land management agencies have a legal mandate to assess biological diversity but they do not have a coherent, defensible, and pragmatic approach to address this mandate. A monitoring program based on a small set of indicators that provides direct or indirect inference to the state of overall biological diversity remains a significant scientific challenge.

Expected Values and Trends

An essential component of a monitoring program is a theoretical or empirical foundation for the expected values or expected time trends of the indicator variables. That is, when the state of an ecological system is measured at time t (following stressor exposure or management action), the value of the indicator is compared to its expected value or trend under an assumption of continuing system integrity. Such a comparison is needed to assess the effectiveness of management practices. Significant deviation of an indicator from its expected value should be the threshold point that triggers a change in management practices. Estimating expected values (i.e., benchmark conditions), however, is difficult, imprecise, and controversial for several reasons, including: (1) the limited availability of pristine, undisturbed ecosystems to provide insights to benchmark conditions; (2) an incomplete understanding of the relationship between the value of an indicator and the desired state of the ecological system; (3) inadequate knowledge of the expected variability, over time and space, of the indicator of ecological state (or species status); (4) the nonlinear relationships between indicator values and ecological processes (including the existence of sharp threshold regions); and (5) the fact that indicator benchmarks may be best represented by probability distributions rather than single target values.

Expected values and thresholds implicitly assume that an ecological system will evolve to (or was historically at) an equilibrium with regard to system state. This assumption is clearly false. The dynamic nature of ecological systems argues for specifying a distribution of expected values (e.g., fig. 2-3) rather than an expected value at a single point in time or space. Even though the range of possible states for an ecological system is very broad, these states differ widely in their probabilities of occurrence under natural disturbance regimes (Holling 1986; Chapin et al. 1996, 2000; Turner et al. 1993; Rapport and Whitford 1999). Knowledge of this distribution allows one to make a probability statement about the likelihood of a given system state—an unlikely state may be justification for a change in management practices.

A second aspect of the concept of a distribution of indicator values concerns the predicted trajectory for an indicator given an initial degraded ecosystem. Given the long time delay between management actions and the response of the ecosystem to those actions, a monitoring program should provide insights into the time trend or future expected value of the indicator. This will require a model that simulates the system response to management and projects the future outcome of current management practices.

Determining threshold values requires the selection of a spatial scale to observe the ecological system. If the spatial scale is a point in space, an indicator threshold may be specified as a single value. An example would be a maximum water temperature beyond which conditions become lethal for cutthroat trout (temperature greater than 22 degrees C). However, if the spatial scale includes a complete watershed, or the range of the species, then it may be unreasonable to expect the water temperature of all stream reaches within this area to be 22 degrees C or less. Specifying an expected distribution of temperature values over the entire area would be more appropriate. Thus, there are two different categories of indicators—those that lend themselves to threshold values (e.g., water temperature for some fish and amphibians) and those best categorized by a target distribution (e.g., average number of snags and logs per hectare). In practice, few indicators will be characterized by a single target value.

In addition, because the physical and biological processes and structural/compositional elements that characterize ecosystems vary in space and time, most indicator variables are best characterized by frequency distributions. That is, when integrated across space (time), at a given point in time (space), a specific process or landscape element is characterized by a dynamic distribution. To illustrate, assume that we have selected "forest stand age" as our measured indicator. We know that under a natural disturbance regime, there would be a dynamic distribution of stand ages that would vary according to time and the spatial scale of aggregation. If the goal of management is to mimic natural disturbance processes, then the scientific challenge is to estimate the benchmark distribution of stand ages that management should aspire to achieve. This distribution, however, is dependent upon spatial scale. The age distribution would change as it is estimated for different sized areas. In general, a threshold cannot be specified without having some idea of the "correct" spatial and temporal scale for measuring the indicator and the "correct" spatial scale for aggregating the measurements.

Once the scale of observation has been determined, it is possible to aggregate indicator values into a frequency distribution. The observed distribution of indicator values could be compared directly to an expected distribution, perhaps most easily evaluated by computing the cumulative distribution of indicator scores. This distribution allows a direct determination of the proportion of sites below (or above) a given indicator value (i.e., the lower or upper acceptable value of the indicator). It is worth reiterating that the concept of a spatial distribution of indicator values as the appropriate evaluative statistic is critical to the monitoring of most ecological systems.

Given the inherent dynamical nature of ecological systems, the value of a given ecosystem indicator (e.g., process or landscape element) will follow a probability distribution. Based on this understanding, a monitoring program must address two distinct questions: (1) is the observed value of the indicator at a specific location, or at some point in a time series, within acceptable bounds of its expected distribution, and (2) when the observed value of the indicator, at a given point in time and space, is considered in the context of neighboring locations on the landscape or in the context of a longer time series, does the expected distribution of indicator values result? For a given resource (e.g., a segment of stream, a forest stand, a riparian corridor) at a particular point in time, it may be appropriate to establish a target value for an indicator. However, when evaluating deviation from the desired ecosystem state at the landscape scale, inferences from an indicator's value at a single site are of limited use without considering that signal in the broader context of values from neighboring landscape locations.

The concept of the distribution of indicator values as a collective index of ecological state at the landscape scale is illustrated in figure 2-5. This figure shows a reconstructed historic distribution used as a benchmark, the current distribution of indicator values, and the future, desired distribution. The desired distribution is not equivalent to the historic distribution (fig. 2-5) in recognition of the frequent impossibility of returning to preindustrial, pristine conditions. Despite the need to establish benchmark distributions, the process of establishing such distributions is subject to considerable debate (e.g., Landres et al. 1999). For example, because the width of a distribution of states is positively related to the length of the historic interval, there is no clear guidance on how far back in time one should go to find an appropriate point of reference. And it is unclear how one can reconcile the concept of benchmarks with the dynamical nature of landscapes, especially when viewed over long time scales. For the time being, estimating benchmark conditions will remain controversial—however, useful examples exist of its application to managed ecological systems (Cissel et al. 1999; Stephenson 1999; Swetnam et al. 1999).

In the interim, benchmark distributions, and the critical values that separate degraded from nominal conditions, will be based on best available information. Evaluating local conditions relative to these threshold points will be the basis of management decisions even though the locations of threshold points and expected distributions are subject to change as ecological understanding increases. In the absence of decision thresholds or explicit objectives that management seeks to achieve, monitoring becomes completely disconnected from management and policy formulation.

Most natural systems and resources recover slowly and will be slow to respond to changes in management practices. In the interim, while ecological systems are moving in the direction of a more desired distribution of states, it is useful to identify appropriate trajectories of change in indicator values that, if continued, would lead to the target distribution (fig. 2-5). Thus, periodic estimates of the direction and magnitude of indicator change provides an ongoing evaluation of the appropriateness of the management strategy.

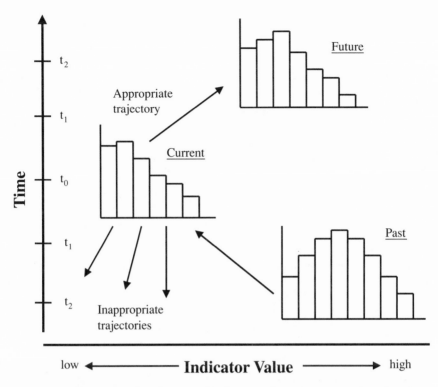

Figure 2-5. Frequency distribution of indicator values shown at various locations on the landscape. The past benchmark distribution, current distribution, and future desired distribution are shown as they change through time. Current-to-future changes are a consequence of management intervention.

Some Practical Statistical Issues

The sampling plan for a monitoring program must thoroughly incorporate several statistical concepts of experimental design—accuracy and precision of indicator estimation, and the power to detect change. That is, the indicator must be estimated with sufficient precision and accuracy so that when an ecologically significant change has occurred, the likelihood it will be detected is high. Concluding no change in ecosystem state when change has occurred is referred to as a type II error, which occurs with probability equal to ß. Determining when a change is ecologically significant is difficult, and the ability to recognize such a change will often depend on results arising from parallel research programs. Further complexity arises when the value of the indicator changes in a nonlinear fashion with incremental changes in the value of the stressor. Anticipating nonlinear changes in ecological processes with increasing stress represents one of the most significant challenges to an environmental monitoring program (Holling 1986; Rapport and Whitford 1999).

The time span of the monitoring program should be discussed, and its duration should bear an obvious relation to the temporal dynamics of the indicator being measured. For example, assessing the trend of any biological population will often require sampling for several generations. Thus, if generation length of the monitored species is measured in years, the monitoring program will need to extend over many tens of years. When species are used as indicators, body size may provide a general index for monitoring duration by invoking the argument that body size is proportional to generation length, which in turn scales the length of the monitoring program.

Several practical statistical issues must be addressed to have confidence that the sampling design and indicator measurement is providing reliable information. Some of the more important include the following:

- What are the defining boundaries of the ecological system? This step is essential to establish the limits, in both time and space, for the allocation of sample locations and to the area of inference.
- What is the appropriate temporal frame for sampling?
- What is the appropriate time interval between samples? That is, how often do indicator variables need to be reestimated to have sufficient power to test for change or time trend?
- What sample size is necessary to estimate the value of the indicator? To thoroughly answer this question requires a discussion of the necessary precision with which the indicator needs to be estimated (α), the costs and acceptable risk of failing to reject the null hypothesis of no change when it is false (the power of the test, 1-ß), and the biologically significant effect size (Δ), specified by the alternative hypothesis. To resolve these issues will require an estimate of the cost of being wrong and the likelihood of making a wrong decision.
- What survey design is most efficient (random, systematic, stratified random, etc.)? If there are obvious sources of spatial heterogeneity in the system, then some sort of prior stratification is probably justified.
- What is the appropriate unit of measure for the indicator variable? This choice ultimately affects the limits to precision associated with each individual measurement.
- Is there an optimal sample unit size and shape for estimating the value of the indicator? This question must be carefully considered because the value of an indicator may be affected by the scale at which it is measured.

- What are the trade-offs between gains in precision and power and the additional costs per sample unit? That is, when does the increase in cost arising from incremental increases in sampling effort begin to exceed the marginal gains in statistical power?
- How can the monitoring program best be designed so that sources of uncertainty about the true state of the ecological system are minimized?

Monitoring as a Catalyst for Change

Monitoring programs do not end with the collection of data, its analysis and synthesis, or even with summary reports. The results of monitoring programs are of value to the extent that they inform the management decision-making process and provide early warnings of ecosystem degradation. The link between monitoring and decision making begins with the formulation of, and agreement on, the monitoring questions. The "correct" questions enable monitoring to be directed at areas where management requires information to adjust activities to mitigate unplanned and undesirable outcomes. Because the behaviors of complex systems are frequently unpredictable, the linkage between decision making and monitoring is essential.

Decision makers begin by asking questions such as "Is the Northwest Forest Plan achieving its objectives for late-successional old-growth forests?" Recognizing a simple yes-or-no answer is not necessarily useful; instead, a process must be instituted to connect the decision makers' questions to the analysis and summary of the monitoring data. One formalization uses principles of statistical decision theory (Lindley 1985; Maguire and Boiney 1994; Clemen 1996).

By posing the decision makers' question in the simple yes/no form, one may formulate the problem in the form of a statistical hypothesis test. This requires setting up a null hypothesis and one or more alternative hypotheses. Typically, the null hypothesis is stated such that the management plan is achieving its desired goal(s), while the alternative is not. As discussed earlier, formulating the decision-making problem in this manner raises questions about the costs associated with the two types of incorrect decisions (i.e., α-level and ß-level risks, see Peterman 1990). Mapstone (1995) proposes a method that attempts to better balance the two risk types and focuses on the magnitude of the effects considered important. The hypothesis testing approach not only is problematic in balancing the risks of incorrect decisions but also is problematic concerning a debate over whether hypothesis testing is appropriate for survey-design monitoring programs (e.g., Stewart-Oaten 1996). More importantly, the outcome of the hypothesis test often provides minimal information to the decision maker. Hence it is more informative to link monitoring results to decision making by focusing on the reliability of the estimates of ecological indicators (measures of precision and bias).

Two general statistical approaches for providing estimates of indicator values and their uncertainty are *frequentist parameter estimation* and *Bayesian parameter estimation* (Ellison 1996). Most ecologists and decision makers are familiar with the frequentist approach. The frequentist approach provides an estimate of the true value of the indicator for the population, for example, the number of hectares of old-growth forest or proportion of federal land suitable as owl habitat, and a confidence interval for the estimate. Often, we have other information available that may increase the likelihood of making a correct decision. Bayesian parameter estimation leads to the calculation of an a posteriori distribution for the parameter based on combining the monitoring data with auxiliary information, expressed as an assumed prior distribution for the parameter. Typically, the parameter estimate is taken as the most probable value of the posterior distribution and 95 percent credibility (confidence) region determined from the posterior distribution as well. The Bayesian approach requires the explicit statement of assumptions and use of preexisting quantitative information in order to define the prior distribution. With either approach, decision makers are given explicit estimates and associated statistics reflecting confidence in the estimates. How the decision maker uses the information is not explicitly considered in the parameter estimation process.

Predictive models can be evaluated in the context of Bayesian parameter estimation. The Northwest Forest Plan process (USDA et al. 1993) provided general expected trajectories for future amount of late-seral and old-growth forest in future decades. Using monitoring information from the Plan, forest growth models, and future land-use information, a predictive model can be constructed to estimate the likely trajectory in the area of late-seral forest. The model requires specification of uncertainty in terms of input parameters as well as monitoring information. Using Bayesian analyses, an a posteriori distribution can be estimated from monitoring data, a process that explicitly incorporates a priori information and assumed relationships among variables. An example of this approach, applied to a deterministic population dynamic model for bowhead whales and used by the International Whaling Commission, is given by Raftery et al. (1995). Others have explicitly used quantitative approaches in the selection of a decision from among alternatives. Statistical decision theory involves determining the potential alternative ecological outcomes, assessing the probability each of these outcomes is "true," describing the management decisions under consideration, and associating a "utility" with each combination of decision and outcome (box 2-4). Several example applications of decision theory to the management of ecological systems have been published. Conroy and Noon (1996) applied decision theory to the question of reserve selection and species conservation. Maguire and Boiney (1994) used decision analysis in conjunction with dispute resolution techniques to resolve a public policy dispute in Zaire over the best policy for management of an endangered species. And, Possingham (1997) and Possingham et al. (2001) demonstrated how decision theory can lead to better conservation decisions.

Application of decision analysis under uncertainty involves (1) specifying management objectives and criteria for measuring success in achieving them; (2) identifying alternatives to achieving the objectives; (3) describing the uncertain events in ecological and sociopolitical environment that influence the outcome of actions taken; (4) assessing the outcome of each combination of management alternative and uncertain events in terms of the decision criteria; (5) estimating the likelihood, or probability, of each uncertain event; (6) calculating the expected values of the decision criteria for each alternative; (7) resolving any tradeoffs among conflicting criteria; and (8) reexamining the "optimal" decision by analyzing its sensitivity to changes in input parameters (Maguire et al. 1988).

Additional discussion of decision making given uncertainty in the monitoring signal is warranted but goes beyond the scope of this chapter.

Box 2-4. Necessary steps to link the results of environmental monitoring to the decision-making process.

- Determine the bounds of the management decision space
- Provide a range of possible management responses to the monitoring data
- Estimate the probabilities associated with each possible interpretation of the monitoring data
- Estimate the utilities associated with each possible combination of decision and monitoring data interpretation (i.e., the costs of wrong decisions and misinterpretation of the monitoring signal)
- Determine the decision that maximizes the overall utility

The references mention above, however, should provide entry into to this important area of research.

Conclusion

For many years, environmental scientists considered the monitoring of ecological systems to be the purview of management. There was a general appreciation of the importance of monitoring but little appreciation of its complexity. Though monitoring was, and continues to be, characterized by significant areas of scientific ignorance, these knowledge gaps were of limited interest to the majority of researchers. Some of this lack of interest may be attributable to the poor reputation that purely descriptive studies received during the emergence of experimental approaches in ecology in the 1960s. Also, many monitoring studies were so poorly designed that many researchers did not consider them to be science.

Fortunately, within the past decade or so that perspective has changed. Many environmental scientists now recognize both the relevance and intellectual challenges of monitoring complex ecological systems. A partial list of these important, but largely unresolved, challenges includes:

- *Indicator selection.* What to measure and how to measure it have been dominant themes of this chapter. This remains the most significant challenge to the effective monitoring of ecological systems.
- *Threshold values or distributions for environmental indicators.* What magnitude of change in an indicator signals an unacceptable state for an ecological system?
- *Estimating normal rates of change in composition, structure, and processes of ecological systems.* Ecological systems are inherently dynamic and we expect their attributes to change through space and time. However, what rates or patterns of change are harbingers of ecological decline?
- *Defining a "desired future condition" for dynamic ecological systems.* Given that ecological systems are inherently dynamic, how do we define the appropriate distribution of systems states that we accept as "normal"?
- *Detecting causation given time lags and synergistic effects.* Monitoring systems are most easily designed to detect the action of a single stressor leading to a change in the state of a single indicator. What of the problem where the state of the indicator (and the ecological system) reflects the joint effect of multiple stressors, each considered innocuous when evaluated separately?
- *Addressing uncertainty in the monitoring signal.* The monitoring signal is subject to inherent variation due to measurement and sampling, true

process variation, and interpretational error. How does the environmental scientist make recommendations in the context of this uncertainty?

• *Predicting across spatial and temporal scales on the basis of the monitoring signal.* Monitoring programs are limited in space and time. However, our interests extend beyond the scales of measurement. How do we reliably make predictions from systems measured at local scales to broader spatial and temporal scales?

In addition to these scientific challenges, there are practical issues that must be addressed for monitoring to be successful. An additional theme of this chapter is that the design of a monitoring program requires a religious adherence to a logical, step-down process (box 2-5). In order to gain and sustain public support for expensive monitoring programs, the logical foundation must be strong. The absence of any critical component will make it impossible for monitoring to serve its role—that is, as a comprehensive and sensitive assay of our ability to sustain ecological systems for future human generations.

Box 2-5. Steps in the design of a prospective environmental monitoring program.

• Characterize the anticipated stressors and disturbances according to the attributes listed in box 2-2
• List the ecological processes and resources affected by the stressors or disturbances
• Ordinally rank the stressors according to their degree of impact and/or degree of irreversible consequences
• Develop conceptual models of the ecological system. Outline the pathways from stressors to the ecological effects on one or more resources
• Select an "optimal" set of condition indicators (detects stressors acting on resources)
• Determine detection limits for the condition indicators
• Establish critical decision values for the indicators (trigger points)
• Establish clear connections to the management decision making process

LITERATURE CITED

Allen, T. F. H., and T. W. Hoekstra. 1992. *Toward a unified ecology.* New York: Columbia University Press.

Allen, T. F. H., and T. B. Starr. 1982. *Hierarchy: Perspectives for ecological complexity.* Chicago: University of Chicago Press.

Andrewartha, H. G., and L. C. Birch. 1954. *The distribution and abundance of animals.* Chicago: University of Chicago Press.

———. 1984. *The ecological web: More on the distribution and abundance of animals.* Chicago: University of Chicago Press.

Barber, M. C., edited by. 1994. *Environmental monitoring and assessment program indicator development strategy.* EPA/620/R-94/XXX. Athens, Ga.: U.S. Environmental Protection Agency, Office of Research and Development, Environmental Research Laboratory.

Bell, S. S., E. D. McCoy, and H. R. Mushinsky, eds. 1991. *Habitat structure: The physical arrangement of objects in space.* New York: Chapman and Hall.

Berger, J. 1997. Population constraints associated with the use of black rhinos as an umbrella species for desert herbivores. *Conservation Biology* 11:69–78.

Bisson, P. A., G. H. Reeves, R. E. Bilby, and R. J. Naiman. 1997. Watershed management and desired future conditions. Pp. 447–474 in *Pacific salmon and their ecosystems: Status and future options,* edited by D. J. Stouder, P. A. Bisson, and R. J. Naiman. New York: Chapman and Hall.

Brown, J. H. 1995. Organisms as engineers: A useful framework for studying effects on ecosystems? *Trends in Ecology and Evolution* 10:51–52.

Chapin, F. S. III, M. S. Torn, and M. Tateno. 1996. Principles of ecosystem sustainability. *American Naturalist* 148:1016–1037.

Chapin, F. S. III., E. S. Zaveleta, V. T. Eviner, R. L. Naylor, P. M. Vitousek, H. L. Reynolds, D. U. Hooper, S. Lavorel, O. E. Sala, S. E. Hobbie, M. C. Mack, and S. Diaz. 2000. Consequences of changing biodiversity. *Nature* 405:234–242.

Cissel, J. H., F. H. Swanson, W. A. McKee, and A. L. Burditt. 1999. Using the past to plan the future in the Pacific Northwest. *Ecological Applications* 9:1217–1231.

Clemen, R. T. 1996. *Making hard decisions: An introduction to decision analysis.* Duxbury Press: Pacific Grove, Calif.

COS (Committee of Scientists). 1999. *Saving the people's land: Stewardship into the next century.* U.S. Department of Agriculture, Forest Service. Washington, D.C.: Government Printing Office.

Conroy, M. J., and B. R. Noon. 1996. Mapping of species richness as a tool for the conservation of biological diversity: Conceptual and methodological issues. *Ecological Applications* 6:763–773.

Davis, W. J. 1996. Focal species offer management tool. *Science* 271:1367–1368.

DelCourt, H. R., P. A. DelCourt, and T. Webb. 1983. Dynamic plant ecology: The spectrum of vegetational change in space and time. *Quaternary Science Review* 1:153–175.

DeLeo, G. A., and S. A. Levin. 1997. The multifaceted aspects of ecosystem integrity. *Conservation Ecology* 1:1–20. [Online at www.consecol.org/vol1/iss1/art3]

Ellison, A. M. 1996. An introduction to Bayesian inference for ecological research and environmental decision-making. *Ecological Applications* 6:1036–1046.

Fleishman, E., D. D. Murphy, and P. F. Brussard. 2000. A new method for selection of umbrella species for conservation planning. *Ecological Applications* 10:569–579.

Foster, D. R., D. A. Orwig, and J. S. McLachlan. 1996. Ecological and conservation insights from reconstruction studies of temperate old-growth forests. *Trends in Ecology and Evolution* 11:419–425.

Franklin, J. F., K. Cromack, W. Denison et al. 1981. *Ecological characteristics of old-growth Douglas-fir forests.* General Technical Report PNW-GTR-118. U.S. Department of Agriculture, Forest Service, Portland, Ore.

Fule, P. Z., W. W. Covington, and M. M. Moore. 1997. Determining reference conditions for ecosystem management and southwestern ponderosa pine forests. *Ecological Applications* 7:895–908.

GAO (United States General Accounting Office). 1988. *Environmental Protection Agency. Protecting human health and the environment.* GAO/RCED-880-101. Washington, D.C; U.S. General Accounting Office.

Goldsmith, F. B., edited by. 1991. *Monitoring for conservation and ecology.* New York: Chapman and Hall.

Guikema, S., and M. Milke. 1999. Quantitative decision tools for conservation programme planning. *Environmental Conservation* 26:179–189.

Haufler, J. B., C. A. Mehl, and G. J. Roloff. 1996. Using a coarse-filter approach with species assessments for ecosystem management. *Wildlife Society Bulletin* 24:200–208.

Hemstrom, M., T. Spies, C. Palmer, R. Keister, J. Teply, P. McDonald, and R. Warbington. 1998. *Late-successional and old-growth forest effectiveness monitoring plan for the Northwest Forest Plan.* General Technical Report PNW-GTR-438. Portland, OR: U.S. Department of Agriculture, Forest Service.

Holling, C. S. 1986. The resilience of terrestrial ecosystems: Local surprise and global change. Pp. 292–313 in *Sustainable development of the biosphere,* edited by W. C. Clark and R. E. Munn. Cambridge: Cambridge University Press.

Hunsaker, C. T., and D. E. Carpenter, eds. 1990. *Environmental monitoring and assessment program: Ecological indicators.* EPA/600/3-90/060, BP91-14196. Research Triangle Park, N.C. Environmental Protection Agency.

Hunter, M. L., G. L. Jacobson, and T. Webb. 1989. Paleoecology and the coarse-filter approach to maintaining biological diversity. *Conservation Biology* 2:375–385.

James, F. C., C. A. Hess, and D. Kufrin. 1997. Species-centered environmental analysis: Indirect effects of fire history on red-cockaded woodpeckers. *Ecological Applications* 7:118–129.

Jones, C. G., and J. H. Lawton, eds. 1994. *Linking species and ecosystems.* New York: Chapman and Hall.

Jones, C. G., J. H. Lawton, and M. Shachak. 1994. Organisms as ecosystem engineers. *Oikos* 69:373–386.

———. 1997. Positive and negative effects of organisms as physical ecosystem engineers. *Oikos* 86:453–462.

Karr, J. R. 1987. Biological monitoring and environmental assessment: A conceptual framework. *Environmental Management* 11:249–256.

———. 1991. Biological integrity: A long neglected aspect of water resources management. *Ecological Applications* 1:66–84.

———. 1996. Ecological integrity and ecological health are not the same. Pp. 97–109 in *Engineering within ecological constraints,* edited by P. C. Schultz. Washington, D.C.: National Academy Press.

Lambeck, R. J. 1997. Focal species: A multi-species umbrella for conservation. *Conservation Biology* 11:849–856.

Lamberson, R. H., R. McKelvey, B. R. Noon, and C. Voss. 1992. A dynamic analysis of northern spotted owl viability in a fragmented forest landscape. *Conservation Biology* 6:505–512.

Lande, R. 1987. Extinction thresholds in demographic models of territorial populations. *American Naturalist* 130:624–635.

Landres, P. B. 1992. Ecological indicators: Panacea or liability? Pp. 1295–1318 in *Ecological indicators*. Vol. 2. Edited by D. H. McKenzie, D. E. Hyatt, and V. J. McDonald. New York: Elsevier Applied Science.

Landres, P. B., J. Verner, and J. W. Thomas. 1988. Ecological use of vertebrate indicator species: A critique. *Conservation Biology* 2:316–328.

Landres, P. B., P. Morgan, and F. J. Swanson. 1999. Overview of the use of natural variability concepts in managing ecological systems. *Ecological Applications* 9:1179–1188.

Lindenmayer, D. B., C. R. Margules, and D. B. Botkin. 2000. Indicators of biodiversity for ecological sustainable management. *Conservation Biology* 14:941–950.

Lindley, D. V. 1985. *Making decisions*. New York: John Wiley and Sons.

Lint, J., B. Noon, R. Anthony, E. Forsman, M. Raphael, M. Collopy, and E. Starkey. 1999. *Northern spotted owl effectiveness monitoring plan for the Northwest Forest Plan*. General Technical Report PNW-GTR-440. U.S. Department of Agriculture, Forest Service, Portland, Ore.

Madsen, S., D. Evans, T. Hamer, P. Henson, S. Miller, S. K. Nelson, D. Roby, and M. Stapanian. 1999. *Marbled murrelet effectiveness monitoring plan for the Northwest Forest Plan*. General Technical Report PNW-GTR-439. U.S. Department of Agriculture, Forest Service, Portland, Ore.

Mapstone, B. D. 1995. Scalable decision rules for environmental impact studies: Effect size, type I, and type II errors. *Ecological Applications* 5(3): 401–410.

Maguire, L. A., and L G. Boiney. 1994. Resolving environmental disputes: A framework incorporating decision analysis and dispute resolution techniques. *Journal of Environmental Management* 42:31–48.

Maguire, L. A., T. W. Clark, R. Crete, J. Cada, C. Groves, and M. L. Shaffer. 1988. Black-footed ferret recovery in Montana: A decision analysis. *Wildlife Society Bulletin* 16:111–120.

Manly, B. F. J. 2001. *Statistics for environmental science and management*. New York: Chapman and Hall/CRC.

Manley, P. N., W. J. Zielinski, C. M. Stuart, J. J. Keane, A. J. Lind, C. Brown, B. L. Plymale, and C. O. Napper. 2000. Monitoring ecosystems in the Sierra Nevada: The conceptual model foundations. *Environmental Monitoring and Assessment* 64:139–152.

Messer, J. J. 1992. Indicators in regional ecological monitoring and risk assessment. Pp. 135–146, in *Ecological Indicators*. Vol. 1. Edited by D. H. McKenzie, D. E. Hyatt, and V. J. McDonald. New York: Elsevier Applied Science.

Morrison, M. L., and B. G. Marcot. 1995. An evaluation of resource inventory and monitoring program used in national forest planning. *Environmental Management* 19:147–156.

Mulder, B. S., B. R. Noon, C. J. Palmer, T. A. Spies, M. G. Raphael, C. J. Palmer, A. R. Olsen, G. H. Reeves, and H. H. Welsh. 1999. *The strategy and design of the effectiveness monitoring program for the Northwest Forest Plan*. General Technical Report, PNW-GTR-437. U.S. Department of Agriculture, Forest Service, Portland, Ore.

Noon, B. R., K. H. McKelvey, and D. D. Murphy. 1996. Developing an analytical context for multispecies conservation planning. Pp. 43–59 in *The ecological basis of conservation: Heterogeneity, ecosystems, and biodiversity*, edited by S. T. A. Pickett, R. S. Ostefeld, and M. Shachak. New York: Chapman and Hall.

Noss, R. 1990. Indicators for monitoring biodiversity: A hierarchical approach. *Conservation Biology* 4:355–364.

Noss, R. F. 1999. Assessing and monitoring forest biodiversity: a suggested framework and indicators. *Forest Ecology and Management* 115:135–146.

NRC (National Research Council). 1990. *Managing troubled waters: role of marine environmental monitoring*. Washington, D.C.: National Academy Press.

————. 1994a. *Review of EPA's environmental monitoring and assessment program: Forests and estuaries*. Washington, D.C.: National Academy Press.

————. 1994b. *Review of EPA's environmental monitoring and assessment program: Surface waters*. Washington, D.C.: National Academy Press.

————. 1995. *Review of EPA's environmental monitoring and assessment program: Overall evaluation*. Washington, D.C.: National Academy Press.

————. 2000. *Ecological indicators for the nation*. Washington, D.C.: National Academy Press.

Olsen, A. 1992. *The indicator development strategy for the environmental monitoring and assessment program*. EPAA600391023. Environmental Protection Agency. Corvallis, Ore.

Olsen, A. R., J. Sedransky, D. Edwards, C. A. Gotway, W. Liggett, S. Rathbun, K. Reckhow, and L. J. Young. 1999. Statistical issues for monitoring ecological and natural resources in the United States. *Environmental Monitoring and Assessment* 54:1–45.

O'Neill, R. V., D. L. DeAngelis, J. B. Waide, and T. F. H. Allen. 1986. A hierarchical concept of ecosystems. *Monographs in Population Biology* 23:1–272.

Osenberg, C. W., R. J. Schmitt, S. J. Holbrook, K. E. Abu-Saba, and A. R. Flegal. 1994. Detection of environmental impacts: natural variability, effect size, and power analysis. *Ecological Applications* 4:16–30.

Peterman, R. M. 1990. Statistical power analysis can improve fisheries research and management. *Canadian Journal of Fisheries and Aquatic Science* 47:2–15.

Pickett, S. T. A., and P. S. White, eds. 1985. *The ecology of natural disturbance and patch dynamics*. New York: Academic Press.

Possingham, H. P. 1997. State-dependent decision analysis for conservation biology. Pp. 298–304 in *The ecological basis of conservation: Heterogeneity, ecosystems, and biodiversity*, edited by S. T. A. Pickett, R. S. Ostefeld, and M. Shachak. New York: Chapman and Hall.

Possingham, H. P, S. J. Andelman, B. R. Noon, S. Trombulak, and H. R. Pulliam. 2001. Making smart conservation decisions. Pp. 225–244 in *Conservation biology: Research priorities for the next decade*, edited by M. S. Soule and G. H. Orians. Washington, D.C.: Island Press.

Power, M. E., D. Tilman, J. A. Estes, B. A. Menge, W. J. Bond, L. S. Mills, G. Daily, J. C. Castilla, J. Lubchenco, and R. T. Paine. 1996. Challenges in the quest for keystones. *Bioscience* 46:609–620.

Raftery, A. E., G. H. Givens, and J. E. Zeh. 1995. Inference from a deterministic population dynamics model for bowhead whales. *Journal of the American Statistical Association* 90(430): 402–430.

Rapport, D. J., and W. G. Whitford. 1999. How ecosystems respond to stress. *Bioscience* 49:193–203.

Rapport, D. J., W. G. Whitford, and M. Hilden. 1998. Common patterns of ecosystem breakdown under stress. *Environmental Monitoring and Assessment* 51:171–178.

Reeves, G. H., L. E. Benda, K. M. Burnett, P. A. Bisson, and J. R. Sedell. 1995. A disturbance-based ecosystem approach to maintaining and restoring freshwater habitats of evolutionary significant units of anadromous salmonids in the Pacific Northwest. *American Fisheries Society Symposium*. 17:334–349.

ROD (Record of Decision). 1994. *Record of Decision for amendments to Forest Service and Bureau of Land Management planning documents within the range of the northern spotted owl*. U.S. Department of Agriculture, Forest Service and U.S. Department of the Interior, Bureau of Land Management. Portland, Ore.

Sauer, J. R., and S. Droege, eds. 1990. *Survey designs and statistical methods for the estimation of avian population trends*. U.S. Fish and Wildlife Service. Biological Report 90(1). Washington, D.C.

Schmitt, R. J., and C. W. Osenberg, eds. 1996. *Detecting ecological impacts: Concepts and applications in coastal habitats.* New York: Academic Press.

Shrader-Frechette, K. S., and E. D. McCoy. 1993. *Method in ecology: Strategies for conservation.* Cambridge, U.K.: Cambridge University Press.

Skalski, J. R. 1995. Statistical considerations in the design and analysis of environmental damage assessment studies. *Journal of Environmental Management* 43:67–85.

Smith, G. R. 1997. Making decisions in a complex and dynamic world. Pp. 419–435 in *Creating a forestry for the twenty-first century,* edited by K. A. Kohn and J. F. Franklin. Washington, D.C.: Island Press.

Spies, T. A., and J. F. Franklin. 1988. Old-growth and forest dynamics in the Douglas-fir region of western Oregon and Washington. *Natural Areas Journal* 8:190–201.

Stephenson, N. L. 1999. Reference conditions for giant sequoia forest restoration: Structure, process, and precision. *Ecological Applications* 9:1253–1265.

Stewart-Oaten. 1996. Problems in the analysis of ecological monitoring data. Pp. 109–132 in *Detecting ecological impacts: Concepts and applications in coastal habitats* edited by P. J. Schmitt and C. W. Osenberg. San Diego: Academic Press.

Suter, G. 1990. Endpoints for regional ecological assessments. *Environmental Management* 14:9–23.

Suter, G. W. 1993. *Ecological Risk Assessment.* Chelsea, Mich.: Lewis Publishers.

Swetnam, T. W., C. D. Allen, and J. L. Batancourt. 1999. Applied historical ecology: Using the past to manage for the future. *Ecological Applications* 9:1189–1206.

Thompson, W. L., G. W. White, and C. Gowan. 1998. *Monitoring vertebrate populations.* New York: Academic Press.

Thornton, K. W., D. E. Hyatt, and C. B. Chapman, eds. 1993. *Environmental Monitoring and Assessment Program Guide.* EPA/620/R-93/012. U.S. Environmental Protection Agency, Office of Research and Development, Environmental Monitoring and Assessment Program, EMAP Research and Assessment Center, Research Triangle Park, N.C.

Thornton, K. W., G. E. Saul, and D. E. Hyatt. 1994. *Environmental Monitoring and Assessment Program Assessment Framework.* EPA/620/R-94/016, U.S. Environmental Protection Agency, Office of Research and Development, Environmental Monitoring and Assessment Program, EMAP Research and Assessment Center, Research Triangle Park, N.C.

Turner, M. G. 1989. Landscape ecology: The effects of pattern on process. *Annual Review of Ecology and Systematics* 20:171–197.

Turner, M. G, W. H. Romme, R. H. Gardner, and R. V. O'Neill. 1993. A revised concept of landscape equilibrium: Disturbance on scaled landscapes. *Landscape Ecology* 8:213–227.

USDA (U.S. Department of Agriculture), U.S. Forest Service, U.S. Fish and Wildlife Service, National Oceanic and Atmospheric Administration, National Marine Fisheries Service, National Park Service, Bureau of Land Management, and Environmental Protection Agency. 1993. *Forest ecosystem management: An ecological, economic, and social assessment:* Report of the Forest Ecosystem Management Assessment Team. Portland, Ore.

USEPA (U.S. Environmental Protection Agency). 1993. EMAP (Environmental Monitoring and Assessment Program). *Master Glossary.* EPA/630/R-93/013. U.S. Environmental Protection Agency, Research Triangle Park, N.C.

USEPA (U.S. Environmental Protection Agency). 1992. *Framework for ecological risk assessment.* EPA/630/R-92/001. Risk Assessment Forum, U.S. Environmental Protection Agency, Washington, D.C.

Walters, C. J. 1986. *Adaptive management of renewable natural resources.* New York: MacMillan.

White, P. S., J. Harrod, W. H. Romme, and J. Betancourt. 1999. Disturbance and temporal dynamics. Pp. 281–312 in *Ecological stewardship: A common reference for ecosystem management.* Vol. 2. Edited by R. C. Szaro, N. C. Johnson, W. T. Sexton, and A. J. Malk. The Netherlands: Elsevier Science.

White, P. S., and S. T. A. Pickett. 1985. Natural disturbance and patch dynamics: An introduction. Pp. 3–13 in *The ecology of natural disturbance and patch dynamics,* edited by S. T. A. Pickett and P. S. White. London, U.K.: Academic Press.

Wilcox, B. A. 1984. In situ conservation of genetic resources: Determinants of minimum area requirements. Pp. 639–647 in *National Parks: Conservation and Development,* edited by J. A. McNeeley and K. R. Miller. Washington, D.C.: Smithsonian Institution Press.

Zacharias, M. A., and J. C. Roff. 2001. Use of focal species in marine conservation and management: A review and critique. *Aquatic Conservation: Marine and Freshwater Ecosystems* 11:59–76.

Zielinski, W. J., and H. B. Stauffer. 1996. Monitoring *Martes* populations in California: Survey design and power analysis. *Ecological Applications* 6:1254–1267.

CHAPTER 3

Design of an Ecological Monitoring Strategy for the Forest Plan in the Pacific Northwest

Paul L. Ringold, Barry Mulder, Jim Alegria,
Raymond L. Czaplewski, Tim Tolle, and Kelly Burnett

The growing literature on ecosystem management describes an adaptive system in which monitoring measures progress toward goals, increases our knowledge, and improves our plans (e.g., Holling 1978; Walters 1986; Duffus III 1994; Everett et al. 1994; Grumbine 1994; Bormann et al. 1995; Gunderson et al. 1995; Interagency Ecosystem Management Task Force 1995; Montgomery et al. 1995; Christensen et al. 1996; Yaffee et al. 1996). Adaptive management acknowledges that action is necessary or appropriate, although knowledge may be imperfect (Raiffa 1968; Holling 1978; Walters 1986; Everett et al. 1994; Grumbine 1994; USFS and BLM 1994b; Gunderson et al. 1995). Less-than-perfect knowledge is likely to be the rule in ecosystem management, especially when the objectives include the restoration and maintenance of complex ecological patterns and processes over large areas and long periods of time.

We describe two general lessons (adaptive monitoring design, and scale) and three general issues (integration, infrastructure, and individuals) that arose in the early design of an ecological monitoring strategy intended to support ecosystem management under the Northwest Forest Plan. We primarily focus on the former (lessons), but briefly describe the latter (issues), since these management issues are of special importance to the implementation of a monitoring system for ecosystem management. It is not, however, our purpose here to describe the details of our monitoring strategy, since it is described elsewhere (Mulder et al. 1995, 1999; Tolle and Czaplewski 1995).

The Northwest Forest Plan

Enough timber for 300,000 homes was being harvested from federal land in the northwestern United States during an average year in the 1980s (Johnson et al. 1991; USFS and BLM 1994a). By 1994, sales on these lands had halted as a result of more than a dozen lawsuits. The lawsuits centered around concerns over an endangered species, the northern spotted owl, and the old-growth forest upon which is it dependent.

A presidential conference in April 1993 led to the development of the Northwest Forest Plan in 1994. The Plan is designed to provide for a "steady supply of timber sales and nontimber resources that can be sustained over the long term without degrading the health of the forest or other environmental resources" (USFS and BLM 1994b). The Plan amends the nineteen U.S. Forest Service and seven U.S. Bureau of Land Management resource management plans for the 100,000 square kilometers within the range of the northern spotted owl (fig. 3-1). It is implemented under a Memorandum of Understanding that includes the White House Office on Environmental Policy, the Department of Agriculture, five components of the Department of the Interior, the National Marine Fisheries Service, and the Environmental Protection Agency (see Appendix E of USFS and BLM 1994b). Extensive discussions on the Plan have been well developed in a number of widely available sources (Dwyer 1994; Society of American Foresters 1994; Tokar 1994; USFS and BLM 1994a,b; Yaffee 1994; Tuchmann et al. 1996).

A key feature underlying the lawsuits was the perspective that the previous scale of management planning—at the level of nineteen individual national forests and seven individual Bureau of Land Management districts (fig. 3-1)—was inadequate to respond to many ecological concerns. Specific concerns included the northern spotted owl and other species dependent upon a network of old-growth forests (Yaffee 1994) and the cumulative watershed effects on fishes and other organisms dependent upon the aquatic resources of the region.

Based on presidential direction, the Northwest Forest Plan specifies a broad array of regional ecological (table 3-1) objectives. These are divided into terrestrial objectives and Aquatic Conservation Strategy (ACS) objectives (table 3-2). Key spatial reporting or evaluation scales envisioned by the plan are regional and provincial in extent (the region is divided into

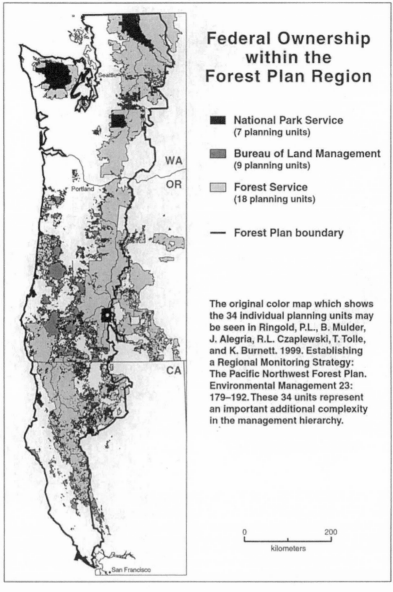

Federal Ownership within the Forest Plan Region

■ National Park Service
(7 planning units)

▣ Bureau of Land Management
(9 planning units)

▢ Forest Service
(18 planning units)

— Forest Plan boundary

The original color map which shows
the 34 individual planning units may
be seen in Ringold, P.L., B. Mulder,
J. Alegria, R.L. Czaplewski, T. Tolle,
and K. Burnett. 1999. Establishing
a Regional Monitoring Strategy:
The Pacific Northwest Forest Plan.
Environmental Management 23:
179–192. These 34 units represent
an important additional complexity
in the management hierarchy.

WA
OR
CA

Seattle
Portland
San Francisco

0   200
kilometers

Figure 3-1. Map of the region. The region is the range of the northern spotted owl in the United States. It includes 99,000 square kilometers of federally administered land embedded in a region containing 230,000 square kilometers of total land. The region is divided into twelve physiographic
provinces to allow for "differentiation between areas of common biological
and physical processes." The text refers to twenty-six planning units; this figure shows twenty-seven Bureau of Land Management and Forest Service
planning units, along with seven National Park Service areas. The different
number of planning units reflects a change in the number from that in existence at the time of the record of decision (USFS and BLM 1994b).

Table 3-1. Hierarchy of goals and monitoring questions.

Level	Source	Statement
1	Presidential	" . . . protect the long-term health of our forests, our wildlife, and our waterways . . . "
2	Cabinet-level goals	Includes: "restore and maintain the ecological health of watersheds and aquatic ecosystems contained within them on public lands"
3	Cabinet-level objectives	Includes nine Aquatic Conservation Strategy Objectives, including: "4. Maintain and restore water quality necessary to support healthy riparian, aquatic, and wetland ecosystems. Water quality must remain within the range that maintains the biological, physical, and chemical integrity of the system and benefits survival, growth, reproduction, and migration of individuals composing aquatic and riparian communities."
4	Interagency monitoring design group general question	Includes: "How many miles/acres of 'waterbodies' have specific chemical and biological characteristics?"
5	Interagency monitoring design group specific monitoring question	Includes: "How many miles (with 10 percent accuracy) of perennial fish-bearing streams within each province and the region contain water whose temperature does not fall above 18 degrees C and below 4 mg/l dissolved oxygen for more than one hour during course of the year?"

twelve ecophysiographic provinces). The spatial grain of the individual features described in these objective ranges from very small—site or local level—to very large—watershed and landscape level. Two examples illustrate this variability in feature size. Terrestrial objective 3 in table 3-2B, "to maintain and enhance late-successional forests as a network of existing old-growth forest ecosystems" describes an inherently large pattern to be maintained or enhanced. In contrast, aquatic conservation strategy (ACS) objective 3, which cites the need to "maintain and restore the physical integrity of the aquatic ecosystem, including shorelines, banks, and bottom configurations," is an example of an objective with inherently

Table 3-2. Forest Plan ecological objectives and their spatial grain. The Forest Plan established two sets of ecological objectives, one for aquatic systems (Table 3-2A) and one for terrestrial systems (Table 3-2B). It is the intention of the plan that these objectives be pursued on a regional and provincial extent. We have added information on the grain of the indicator associated with each objective. *Grain* is the size of the feature that would need to be described or inferred in a regional monitoring program. It reflects the sizes of features that are ecologically meaningful. The challenge here is that both fine- and coarse-grained features are to be described and managed over a large spatial extent. The Aquatic Conservation Strategy Objectives are taken directly from the Record of Decision (USFS and BLM 1994b). The Terrestrial Objectives are adapted from the same source. The information on spatial grain of the indicator was devised by the monitoring design group.

3-2A

Aquatic Conservation Strategy Objectives	*Spatial Grain of Indicator*		
	Site Reach	*Watershed*	*Landscape*
1. Maintain and restore the distribution, diversity, and complexity of watershed- and landscape-scale features to ensure protection of the aquatic systems to which species, populations, and communities are uniquely adapted.		▓	▓
2. Maintain and restore spatial and temporal connectivity within and between watersheds. Lateral, longitudinal, and drainage network connections include floodplains, wetlands, upslope areas, headwater tributaries, and intact refugia. These network connections must provide chemically and physically unobstructed routes to areas critical for fulfilling life-history requirements of aquatic- and riparian-dependent species.		▓	▓
3. Maintain and restore the physical integrity of the aquatic system, including shorelines, banks, and bottom configurations.	▓		
4. Maintain and restore water quality necessary to support healthy riparian, aquatic, and wetland ecosystems. Water quality must remain within the range that maintains the biological, physical, and chemical integrity of the system and benefits survival, growth, reproduction, and migration of individuals composing aquatic and riparian communities.	▓	▓	

continues

Table 3-2. (*Continued*)

3-2A (continued)

Aquatic Conservation Strategy Objectives	Spatial Grain of Indicator		
	Site Reach	Watershed	Landscape
5. Maintain and restore the sediment regime under which aquatic ecosystems evolved. Elements of the sediment regime include the timing, volume, rate, and character of sediment input, storage, and transport.			
6. Maintain and restore in-stream flows sufficient to create and sustain riparian, aquatic, and wetland habitats and to retain patterns of sediment, nutrient, and wood routing. The timing, magnitude, duration, and spatial distribution of peak, high, and low flows must be protected.			
7. Maintain and restore the timing, variability, and duration of floodplain inundation and water table elevation in meadows and wetlands.			
8. Maintain and restore the species composition and structural diversity of plant communities in riparian areas and wetlands to provide adequate summer and winter thermal regulation, nutrient filtering, appropriate rates of surface erosion, bank erosion, and channel migration and to supply amounts and distributions of coarse woody debris sufficient to sustain physical complexity and stability.			
9. Maintain and restore habitat to support well-distributed populations of native plant, invertebrate, and vertebrate riparian-dependent species.			

The spatial grain of the indicator.

Terrestrial Objectives	Spatial Grain of Indicator	
	Stand	Landscape
1. To maintain late-successional and old-growth species habitat and ecosystems		
2. To provide habitat for a variety of organisms associated with both late-successional and younger forests		
3. To maintain and enhance late-successional forests as a network of existing old-growth forest ecosystems		
4. To function as connectivity between late-successional reserves		
5. To perform an important role in maintaining and promoting biodiversity associated with native species		
6. To help ensure that late-successional species diversity will be conserved		
7. To maintain ecological processes, including those natural changes that are essential for the development and maintenance of late-successional and old-growth forest ecosystems		
8. To provide for important ecological functions such as dispersal of organisms, carryover of some species from one stand to the next, and maintenance of ecologically valuable structural components such as down logs, snags, and large trees		
9. To reduce risk to late-successional ecosystems resulting from large-scale disturbances and unacceptable loss of habitat due to large-scale fire, insects, and disease and major human impacts		
10. To accelerate the development of overstocked young plantations into stands with late-successional and old-growth forest characteristics		
11. To provide for desired habitat conditions for at-risk fish stocks		

The spatial grain of the indicator

local focus. This does not imply that fine-scale features are not influenced by those at broader scales, or that broad-scale patterns and processes cannot be inferred from observations of fine-scale features. In fact, many of the features have characteristics at both small and large scales.

Of special interest is the fact that many of the Plan objectives apply to larger features—features not necessarily observable or manageable at the level of the twenty-six individual planning units that comprise the region (fig. 3-1). In summary, the expectation is that both small and large features will be restored and maintained throughout the region and its component provinces. The temporal extent of the Plan, and the individual agency resource plans it amends, is intended to be a century or longer.

Monitoring Design

In the fall of 1994, an interagency monitoring design group (MDG) began to develop a strategy for forest plan monitoring. The ecological monitoring portion of the strategy was to focus on effectiveness monitoring, designed to determine if the objectives of the Plan are being achieved. This assignment was distinct from but bounded by parallel assignments to other groups focusing on implementation monitoring (determines if the steps of the Plan are being implemented), validation monitoring (determines if the assumptions underlying the Plan are correct), and nonecological monitoring.

The MDG worked collectively on a monitoring strategy that included a general approach to ecosystem monitoring and proposals (each prepared by groups of scientists) for each of five priority elements: (1) the northern spotted owl, a threatened species; (2) the marbled murrelet, also threatened; (3) "survey and manage" species—a large group of over four hundred rare species; (4) late-successional/old-growth forests; and (5) riparian/aquatic ecosystems. The resultant strategy (Mulder et al. 1995) includes the application of statistically rigorous sample surveys, modeling, GIS analyses, remote sensing, and synthesis to address the effectiveness monitoring needs of the Plan. The report also discusses (1) integrating monitoring requirements of the five specific resource elements in a cost-effective, synergistic manner, (2) establishing an institutional infrastructure, (3) developing new monitoring methodologies, and (4) establishing intra- and interdisciplinary linkages, as well as interagency linkages.

Our efforts not only resulted in a workable strategy, but also revealed two important lessons. First, we learned to view monitoring design as an adaptive process (Ringold et al. 1996) since the many uncertainties and unknowns make it difficult to adhere to the rigorous standards that define good monitoring practice. Viewing monitoring as an adaptive process is consistent with the prototyping notion of introducing new approaches to ecosystem management (Brunner and Clark 1997). Second, we identified the need to define a monitoring program in the context of the large and multiscalar spatial and temporal elements of the Plan rather than in the context of the single and smaller scales of traditional monitoring and management.

Lesson One: Monitoring Design as an Adaptive Process

An effectiveness monitoring strategy provides the data that allows us to evaluate progress toward the goals of a plan. The goals of the Northwest Forest Plan are hierarchically structured, as shown in table 3-1, to flow from broad presidential and departmental direction (USFS and BLM 1994a, b), to more specific objectives for the Plan, and finally to specific monitoring questions. The hierarchical structure ensures that higher-level goals, which are properly broad and qualitative, are clearly linked to more specific questions around which a technically sound and useful monitoring program can be designed and implemented.

Among other things, a technically sound and usable regional monitoring strategy should define the questions to be addressed in rigorous quantitative terms (Green 1979; Likens 1983; Wolfe et al. 1987; Hicks and Brydges 1994; Noss and Cooperrider 1994; Kondolf 1995; Larsen et al. 1995). Such a definition must also consider the following:

- The population to be monitored (e.g., operational definitions of intermittent streams or old-growth forests).
- The feature of that population to be measured (e.g., temperature, or dissolved oxygen in a stream, temporal connectivity between watersheds, and snag density or connectivity of old-growth forests).
- The temporal and spatial characteristics of the members of the populations, including their distribution across the region, and statistical issues, such as type I and II probabilities, (e.g., Peterman 1990), or required precision and accuracy of the results.
- Measurement methods, which must be regionally consistent so that measurements taken in different places or by different organizations can be accurately combined to describe the characteristics of the region.
- The mechanistic or conceptual linkage between the monitoring question and the broader objectives, to ensure that monitoring focuses on questions of policy and ecological relevance.

These attributes have considerable merit. However, three technical and institutional barriers make it difficult to design a monitoring program with such attributes. The first barrier is vaguely defined objectives. Often, designers of monitoring programs are faced with the task of translating broadly stated policy goals into technically sound, specific questions that monitoring can be used to answer. This is especially true for policy objectives at the national level (see levels 1–3 in table 3-1), which must be refined (but not *redefined*) in a carefully structured, open, and iterative interaction between the scientific community and other participants in the public policy process. This type of objective refinement is analogous to the problems and processes encountered in defining measurement and assessment end points in ecological risk assessment (e.g., Suter 1990; Norton et al. 1992; Risk Assessment Forum 1996).

An example of an appropriate quantitative question is that posed in table 3-1, level 5. Although this particular example provides a technically sound foundation for a monitoring design, the information may not necessarily be useful for management or for tracking elements of an objective. Moreover, a tentative quantitative question may not be logistically parsimonious or consistent with the current state of scientific understanding.

Exacerbating the issue of refining higher-level goals into appropriately specific monitoring questions and then implementing a monitoring program incorporating such questions is the fact that management priorities are not necessarily apparent or well articulated at the outset of the design of a regional monitoring strategy. Priorities are also likely to change over time and they often differ from institution to institution. In addition, management actions or ecological consequences related to one set of questions may overlap with another, requiring that additional effort be made to identify and resolve duplicated information. Underlying our proposal for adaptive monitoring design is the notion that as data sets and their analyses and the costs of their acquisition are made available, and as managers engage in the implementation of the Northwest Forest Plan, more tangible foundations for setting priorities will emerge.

A second barrier to designing a monitoring program that has all the desirable attributes at the outset is nonexistent, inconsistent, or changing methodology. Methods may not be available to measure some important characteristics, or one organization may measure or classify a feature differently than others. Three examples illustrate this problem.

- *Nonexistent Methods.* Riparian forests have considerable ecological importance within this region (e.g., Gregory et al. 1991) and are directly referenced in four of the nine ACS objectives (table 3-2A). Despite their significance, there is not an operational method to characterize these systems consistently throughout the region (Mulder et al. 1995; Smith et al. 1997).

- *Inconsistent Methods.* The inability to combine data from different areas within the region or to compare results from one time period with those from another is a major barrier to meaningfully describe the status and trends of the region. Figure 3-2 illustrates differences in classification schemes for tree size used by terrestrial, riparian, and aquatic resource managers within the region. When regional analyses were developed, they could not take advantage of the finer (and ecologically useful) detail of classification in the larger size classes of Forest Service data sets because the Bureau of Land Management data did not contain those classes. Similarly, because the terrestrial and aquatic classification schemes do not match, linkages between these two components are not as rigorous as would be possible with an integrated classification scheme. This figure illustrates two issues affecting the development of a regional monitoring scheme—loss of information relevant to management associated with merging differing classification schemes, and the need for broad interdisciplinary consideration in developing measurement protocols. For example, the process leading up to the development of the Northwest Forest Plan included a need to analyze vegetation patterns and amounts across the entire range of northern spotted owl. Combining data from categories led to an overly simplified set, resulting in a great loss of information. This problem occurs when one is combining data across ownerships to assess a landscape, such as was described above for the range of the northern spotted owl. It also occurs when one is combining historical records for one place. Take, for example, the case where one finds that the break between two categories is discovered to actually be 50 centimeters instead of 43 centimeters. Without continuous data, the historical record cannot be used to show trends in this size distinction. Second, this figure illustrates the need for broad interdisciplinary consideration in the collection of data. If we wish to relate this year's riparian areas to the coarse woody debris in streams in the future, our classification schemes should be compatible.
- *Changing Methods.* Translating the objectives of the Plan into measurable indicators requires ecological understanding, which may take time to evolve. For example, terrestrial objective 8, listed in table 3-2B, includes providing for "important ecological functions such as dispersal of organisms." Schumaker (1996) showed that nine common indices of habitat pattern did not correlate with mechanistic predictions of dispersal success. He identified a new pattern index, patch cohesion, which provided a better match to this process. Thus, the set of indicators useful for evaluating an objective may change over time as our understanding and capabilities change, even if the policy objectives remain constant.

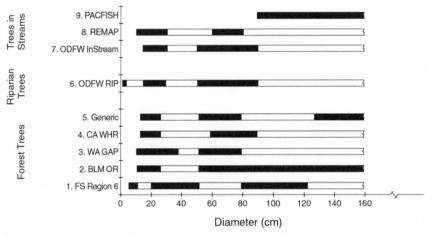

Figure 3-2. Tree classification schemes used by different organizations in the Pacific Northwest. This figure summarizes the classification schemes used to classify tree diameter by several organizations responsible for conducting surveys or inventories of resources in the region. Classification schemes 1 through 5 are for trees in forests; classification scheme 6 is for trees in a forested riparian setting; classification schemes 7 through 9 are for woody debris in streams. Forests of different size classes create habitat for different species, wood in streams forms important habitat for fishes, and wood of different sizes functions and persists differently in streams. Information for schemes 1 through 5 is from John Tepley (personal communication to Tim Tolle); information for schemes 6 and 7 is from the Oregon Department of Forestry 1994; information for scheme 8 is from Hayslip et al. 1994; and information for scheme 9 is from Haughen 1994. In all cases, the upper size class is an open-ended one; the upper limit of 200 centimeters and greater is shown here with an upper limit of 200 centimeters to allow easier graphical comparisons between classification schemes.

The third barrier in monitoring program design is a lack of information needed to estimate the variability of environmental features and measurement error over the relevant temporal and spatial scales. These estimates of variance are necessary if we are to plan the level of effort appropriate for monitoring systems to meet users' quantitative requirement for status and trend information (McRoberts et al. 1994; Larsen et al. 1995; Stapanian et al. 1997).

Exacerbating each of these three barriers is the historic reliance upon the twenty-six federal planning units within the region (fig. 3-1) to individually identify, develop, assess, and manage information.

Recognizing the pervasive nature of the foregoing obstacles, we propose an adaptive approach to monitoring design. This approach iteratively refines the monitoring design as the barriers are reduced or addressed as a result of experience in implementing the monitoring program, assessing its results, and interacting with users. In the course of this adaptive process, monitoring should meet both user and producer needs thusly:

- *User needs* support natural resource managers, regulators, policy makers, and the public in refining the monitoring questions, revising the implementation of the plan, setting monitoring priorities, and understanding the status and trends of the resource being managed.
- *Producer needs* support monitoring system designers in revising the monitoring design by providing a tangible foundation to discussing user needs; by gaining the insights necessary to minimize the most prominent technical barriers; by enhancing the efficiency of the collection, analysis and assessment systems; and by refining the monitoring questions.

The difference between the traditional approach to monitoring design and an adaptive approach is that in the latter, barriers to monitoring design are overcome by adaptively implementing monitoring rather than waiting for new information or designing a system that does not anticipate new information. Two examples from our effort illustrate this adaptive approach. In each case, aspects of all three barriers exist.

The first example addresses "survey and manage" species, the label given to over four hundred rare species, including 338 species of nonvascular plants and four arthropod communities. More detail describing this category can be found in USFS and BLM (1994b). The Northwest Forest Plan establishes four approaches for enhancing the probability of the continued viability of these species. The first approach is to manage sites where each species is known to occur; the second, third, and fourth require the collection of new information on the locations and viability of these populations. However, the distribution, abundance, life histories, and habitats of many of these rare species are poorly understood; in some cases, organisms are available for identification only for short fractions of years. In general, there is a gap between the need for information and the ability of methodologies to provide this information efficiently. In traditional monitoring design, two courses of action could be followed (depicted by "yes" and "no" choices from box 2 in fig. 3-3). First, following the "no" course, we could defer monitoring until distribution and habitat requirements were known and proven protocols existed for identifying each species; following the "yes" course, we could assume that our current knowledge is sufficient for making viability assessments and conduct no additional monitoring. In the first instance, the "no" pathway, the development of any spatially consistent, albeit imperfect, baseline information would be deferred until some future time when methods were perfected. In the second instance, imperfect monitoring would be implemented without recognizing improvements in knowledge that would be derived over time.

We propose a third course in which field sampling efforts would be deployed in an adaptive approach (the provisional paths from box 2 of fig. 3-3). This approach refines habitat-species models (e.g., Scott et al. 1993; White et al. 1997) and thereby the habitat characterization, monitoring, and management efforts as more information becomes available. A key feature of this approach is that it integrates communication, or feedback, between information producers and users, improving knowledge of habitat-species relationships and supporting management decision making; improvements to methodologies halt when the marginal return no longer improves decision making.

A second example of adaptive monitoring design relates to the implementation of a monitoring program to characterize the status of streams throughout a region or a province using site scale indicators of instream condition (see ACS objectives 3–9 in table 3-2A). Numerous protocols could be used to describe the status of the streams in this region (e.g., MacDonald et al. 1991; USFS 1993; Haughen 1994; Hayslip et al. 1994; Oregon Department of Fish and Wildlife 1994). Despite the availability of these protocols, we lack (1) understanding or agreement to determine which, if any, of the existing protocols is most appropriate for use; (2) quantification of the existing qualitative site-level monitoring objectives (i.e., producing a set of monitoring questions of the detail shown in the level 5 of table 3-1); (3) understanding of the existing temporal and spatial characteristics of the population of interest; and (4) understanding of the logistics involved in implementing a regional monitoring strategy. Traditional options are to await the outcome of research to resolve these barriers, to deploy a new monitoring program requiring extensive change in existing programs, or to utilize existing programs designed for disparate purposes as the regional program.

We propose an alternative in the form of an adaptive approach. To respond to these barriers, we recommended a series of modest efforts starting with a comparison of feasible protocols in a subcomponent of the region, for example, within one of the twelve provinces comprising the region. This effort would help to refine our collective understanding of the technical and institutional barriers involved in designing and implementing a regional monitoring program. It would also enable us to work with land managers and others to address necessary quantifications of the existing objectives. This understanding would lead to agreement on an approach for conducting monitoring throughout the region. The

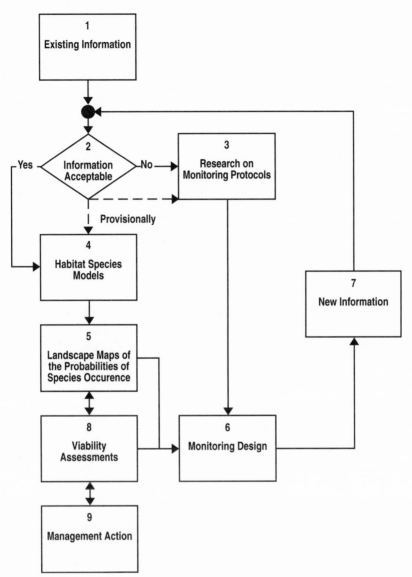

Figure 3-3. Adaptive monitoring design for survey and manage species. An adaptive design provides a mechanism to address multiple resource issues in a cost-effective way. It does this by institutionalizing an iterative feedback loop so that information is used more effectively. One way in which adaptive monitoring design differs from traditional design is by allowing for models or assessments to be provisionally acceptable. In these cases, available information is considered to be superior to no information at all and can be used to support management decision making *and* to assist in the design for collection of additional information. Once data and understanding are considered to be fully acceptable, research (step 3) and monitoring design (step 6) can be deleted, and the development of new information can proceed at an appropriate level of intensity.

implementation of this effort would also build on and supplement existing operational monitoring and research efforts. This adaptive approach is summarized in table 3-3.

The adaptive approach to monitoring design described in these two examples not only addresses the obstacles identified, but also increases the opportunity to link the development of scientific information with resource management needs. This linkage ensures that the design and implementation of the monitoring strategy is responsive to those needs so that institutions and personnel understand one another and work together toward common long-term objectives. Linkage to user needs is one of the ways to increase the probability of long-term support (Levitt 1986; Kotler and Andreason 1996).

An incremental approach to ecological monitoring design is not novel (e.g., Cain and Castro 1959; Gauch 1982; Strayer et al. 1986; Morrison 1994; Stevens 1994; Gunderson et al. 1995; Powell 1995; Steele 1995; Neave et al. 1996). However, the explicit inclusion of both information producers and information users in plan refinement is a new feature. An adaptive approach is also consistent with the notion of a practice-based approach to ecosystem management, as put forth by Brunner and Clark (1997), and with the iterative nature of problem definition in the context of solving complex ecological problems, noted in Clark et al. (1996).

Although an adaptive approach would appear to have many advantages, it is clear there are potential disadvantages. These include concerns that

- Initiating a program with goals that are neither fixed nor clear can easily contribute to the failure of those programs (e.g., Clark et al. 1996).
- The long-term organizational flexibility (in terms of dollars, personnel, or organizational infrastructure) that may be required under an adaptive approach to allow for change in design or reallocation of effort may not be feasible.
- The demands of an adaptive approach on personnel engaged in this type of structured learning process may be high (e.g., Simon 1977; Rosenberger and Lachin 1993) and may require more senior personnel than would a fixed-design monitoring program.
- The reduction of long-term comparability in the data may not be worth the improvements in the design and changes may present significant statistical challenges (Rosenberger and Lachin 1993; Overton and Stehman 1996).

Ecosystem management requires fundamental change (Grumbine 1994); the adaptive approach that we define appears an appropriate way to facilitate this change. Nonetheless, the articulation of these potential

Table 3-3. Example adaptive design for Aquatic Conservation Strategy (ACS) goals dependent upon site-level indicators.

Phase	Input	Output	Result
I. Provincial-scale demonstration and intercomparison	A. ACS objectives B. Multiple candidate protocols	Common understanding of the capabilities of protocols Information on resource status at provincial scale Information on statistical character of candidate indicators at a provincial scale Science/policy dialogue on utility of indicators	Agreement on Phase II protocols Preliminary information on plan effectiveness at a provincial sale Improved quantification of tradeoffs between certainty of answer and cost Improved understanding of how monitoring results will be used
II. Regional-scale demonstration	A. ACS Objectives with quantification from Phase I B. Existing protocols as refined in Phase I	Common understanding of the capabilities of protocols Information on resource status at regional scale Information on statistical character of candidate indicators at a regional scale Science/policy dialogue on utility of indicators	Agreement on operational protocols, the level of certainty they will provide, the associated cost, and how the information will be used
III. Operational monitoring program	A. Quantified monitoring objectives tightly tied to ACS B. Proven protocols, design and reporting methodology	Description of the status and trends in select resources over time and space with known certainty	Useable information on the effectiveness of the plan

pitfalls is an important element that needs to be considered in the development of a long-term plan for adaptive management and monitoring design.

Lesson Two: Accommodating Multiple Scales
SPATIAL SCALE

The Northwest Forest Plan centers on objectives that are regional in extent, while the Aquatic Conservation Strategy calls for restoration and maintenance of features that range in size from small to large (see table 3-2A). The population viability of many species (including threatened and endangered species) is dependent upon the status of small and large features in the region as a whole. In contrast, traditional effectiveness monitoring designs (and the traditional scale of management) have been constructed around questions of smaller spatial extent and small features (MacDonald et al. 1991; Runyon 1995) such as: "Are state water quality standards being met where road prisms are contributing sediment to streams?" or "Are there measurable changes in channel erosion . . . not intended by fish habitat improvement activities?" (Siuslaw National Forest 1990). The Plan makes these questions of smaller extent no less important but adds foci on questions of larger extent and of larger features.

The challenge for monitoring system design arises from the fact that the design for a regional monitoring program is very different from a monitoring program for individual projects within the region or from individual standards and guidelines within a planning unit. The key difference is that the spatial design for a regional program must reflect sample surveys at some level (Stehman and Overton 1994; Stevens 1994; Landers et al. 1995; Paulsen et al. personal communication) or modeling approaches (Schreuder et al. 1993) that combine empirical models and stratification with rigorous sample survey methods. Sample survey techniques, in which members of a population are sampled on a probabilistic basis, have been used in a cost-effective manner to enumerate the characteristics of populations at a national level for four decades or more (Hansen et al. 1953). Inventory, the complete description of the resource, is too costly and not necessary; collection of data at sites selected on a nonprobabilistic basis, or chosen because of the presence of particular project or management practice, is likely to lead to biased estimates (Peterson et al. 1998; Paulsen et al. personal communication). Thus, the design of a regional strategy requires a commitment to sampling at locations selected at random, in a manner that permits description of features of the region with known confidence. Table 3-4 summarizes the differences in the design between small- and larger-scale monitoring programs.

Table 3-4. Different Scales of Monitoring.

Characteristic	Local Monitoring	Regional Monitoring
Purpose	Determine the effects of an individual project	Determine the aggregate effect of management activities on the region
Dominant user of information	Local managers and publics	Applicable to all managers and interests in the region
Limitations	Cannot extrapolate findings to other projects or to the region	Cannot describe the consequences of management at an individual site
Site selection	Project or activity of interest	Selected (on a probability basis) to be representative of the region

Along with the need to focus on the status of the region, there is also clearly a need to respond to questions of smaller spatial extent. The Plan (see especially VIII-14 [FEMAT 1993]) envisions that monitoring will provide information on the effectiveness of hundreds of management prescriptions, formally known as "standards and guidelines" (Alverts et al. 1996), applied to thousands of individual project decisions made every year. An example of such a question is, "What is the effect of this project on this stream?" (a project-scale question). Focusing a design around these individual projects and management decisions would be one way to address questions of smaller extent. Monitoring would, however, be prohibitively expensive and would not provide for inference on the status of the regional objectives. Thus, an important dilemma results from the demands imposed by the two types of questions as highlighted in table 3-4.

How do we reconcile the monitoring design needed to describe the status of the region with the monitoring design needed to describe the effects of a single specific management practice? We have proposed that a multipopulation (or sampling frame) approach be taken. One population would be sampled to describe the status of the region as a whole. The second population would comprise locations known to have been subject to particular types of management actions of interest. In both cases, the specific locations to be sampled would be selected on a probability basis. In the first case, the sample might be drawn from all of the stream reaches within the region. In the second case, the sample might be drawn from all stream reaches within the region on which a particular standard or guideline had been applied or in a location in which there was a special concern. In both populations the sample protocol at the site would be consistent, if not identical. Because monitoring the effectiveness of all the standards and guidelines of the Plan is not feasible, a very small subset would be selected to be the focus of sampling efforts in this second population. We have recommended that a subset be selected in response to the intersection between the standards or guidelines that have highest scientific uncertainty and those with greatest management interest.

TEMPORAL SCALE

Monitoring design must account for the temporal resolution (or grain) and extent of the features to be monitored. This discussion focuses only on temporal extent, not on temporal resolution. A regional design not only reflects large spatial extent, it also reflects long temporal frames. This linkage between broad spatial extent and long temporal periods is consistent with general ecological observations that system size and response time are directly related (e.g., Harris 1980; Forman and Godron 1986; Allen and Hoekstra 1992; Gunderson et al. 1995; Powell 1995). The Plan analyses reflect this; they evaluate alternatives with respect to their likelihood of providing desired habitat conditions and processes for one hundred years and longer (USFS and BLM 1994a; Thomas et al. 1993). This long temporal extent has two implications. The first is that we must design a monitoring approach to be in place for many years. This would seem a relatively straightforward technical task; however, the calls for long-term monitoring are legion, and the data sets are few, even for single sites. Thus, our technical and institutional experience in this area is limited. The second implication is that we have a need to consider past disturbances of the landscape in the design of the monitoring program. Disturbances arise from both natural sources, such as climate or fire (Agee 1991; Brubaker 1991), and anthropogenic sources such as forest harvesting, road building, or stream restoration. Characterization of the past disturbance of the landscape provides insight on the reference conditions that are to be maintained and restored in the ecological objectives of the Plan. It also offers insight into how rapidly those conditions can be achieved, and it supplies a key piece of information in the interpretation of existing patterns.

The ability to describe the history of management actions offers two advantages. First, it can reduce a source of variance in our analysis—the confounding effect that would be introduced by labeling a site undisturbed when it actually has experienced disturbance. Second, *if* space provides a reasonable surrogate for time (Pickett 1989), and *if* land cover or land-use history can be developed, this information can provide considerable insight about the current and future status of the region in a timely manner. We may be able to use this historical insight to make inferences about the consequences of current management actions that take a long time to manifest themselves. To the extent that this can be done successfully, the monitoring program can provide guidance on questions of smaller extent, such as the consequences of types of projects or particular standards or guidelines, in a relatively short period of time.

Other Key Points

The two lessons we have discussed are not the only points we identified in our efforts to develop an ecosystem monitoring strategy. The following three are no less important, but they are more closely related to the process of developing and implementing a strategy than to the resolution of techniques critical to the strategy. These issues reflect more the underlying or inherent lack of understanding or acceptance of monitoring among agencies and managers and what it take to successfully monitor, which may be the most critical barrier to overcome.

We do not claim to have fully identified and resolved the highest-priority issues. We also emphasize that these three issues are highly interdependent.

- *The need to integrate regional programs.* Regional analysis requires the integration of information from multiple sources to address management and science questions. For any particular question, information from multiple sources has to be integrated in a constructive and cost-effective manner, which requires compatibility among multiple sets of data. The significance of this issue with regard to even the simplest questions cannot be underestimated if we are to have regionally compatible and useful data sets for current and future questions. Figure 3-2 provides a simple example of the institutional and ecological barriers to integrating information from multiple sources. The diagram shows that classification schemes for characterizing tree size differ not only among foresters from different organizations, but also among foresters and aquatic ecologists. Thus, integration is a process that cuts across organizations, disciplines, and other boundaries.
- *The need to develop a regional reporting and decision-making capacity while maintaining similar capacities for smaller areas.* The Northwest Forest Plan amends rather than replaces twenty-six existing resource management plans. As a result, regional interagency data management and analysis is needed to support interagency decisions at the planning unit and at larger scales. The challenge for Plan managers is allocating personnel and resources to support regional interagency analysis when infrastructure is not well developed, even at agency and individual planning unit levels (see the twenty-six planning units within the region in fig. 3-1).

- *The need to develop incentives for individuals to work in an interagency setting.* Implementation of the Plan requires that agencies work together to develop common protocols, procedures, and data in pursuit of common regional objectives. Individuals from the agencies must therefore work together toward the common goals that their agencies have an interest in achieving. However, interagency organizations do not manage or supervise personnel. They do require individuals participating in their deliberations to balance the demands and interests of the interagency group against those of their home agency. The potential for conflict can create considerable tension. Although there is a single set of interagency goals, it is likely that there are multiple sets of differing agency goals. If not, an interagency organization would not be necessary. Thus, considerable thought needs to be given to providing incentives to individuals to pursue the goals of interagency organizations.

Conclusion

The design of a monitoring strategy in support of multiagency, multiscale ecosystem management poses unique technical and institutional challenges. Many of these were envisioned in the Northwest Forest Plan. Adaptive monitoring design helps to address these challenges and thus minimizes their potential for inhibiting the development of information in two ways. First, it promotes stronger links between management and monitoring; second, it minimizes risk at the outset of change, as in a prototyping approach proposed for ecosystem management (Brunner and Clark 1997). This linkage is important for ensuring that long-term monitoring enjoys the broad level of support and clarity of mission that will ensure it long-term financial support, technical success, and policy relevance.

ACKNOWLEDGEMENTS
The authors acknowledge the thoughtful comments of Gay Bradshaw, Ila Cote, Carol Eckhardt, Jeroen Gerritsen, Bob Hughes, Dixon Landers, Phil Larsen, Willa Nehlsen, Tony Olsen, Steve Paulsen, Tom Spies, Rich Sumner, and others in developing this analysis. Francie Faure prepared figure 3-1 in her usual unusually gracious and timely manner. Miriam Pendergraft and Susan Christie provided editorial assistance. This document has been subject to the Environmental Protection Agency's peer and administrative review and is approved for publication. This chapter is reprinted from Ringold, P. L., B. Mulder, J. Alegria, R. L. Czaplewski, T. Tolle, and K. Burnett, "Establishing a regional monitoring strategy: The

Pacific Northwest Forest Plan," in *Environmental Management* 23:179–192.

LITERATURE CITED

Agee, J. K. 1991. Fire history of Douglas-fir forests in the Pacific Northwest. Pp. 25–34 in *Wildlife and vegetation of unmanaged Douglas-fir forests*, edited by L. F. Ruggiero, K. B. Aubry, A. B. Carey, and M. H. Huff. U.S. Department of Agriculture, Forest Service, Pacific Northwest Research Station, Portland, Ore.

Allen, T. F. H., and T. W. Hoekstra. 1992. *Toward a unified ecology*. New York: Columbia University Press.

Alverts, B., A. Horton, and B. Stone 1996. *Results of the FY 1996 (pilot year) regional implementation monitoring program*. Research and Monitoring Committee, Regional Ecosystem Office, Portland, Ore.

Bormann, B. T., J. R. Martin, F. H. Wagner, G. Wood, J. Alegria, P. G. Cunningham, M. H. Brookes, P. Friesema, J. Berg, and J. Henshaw. 1999. Adaptive management. Pp. 505–533 in *Ecological stewardship: A common reference for ecosystem management*, edited by N. C. Johnson, A. J. Malk, W. Sexton, and R. Szaro. Amsterdam: Elsevier.

Brubaker, L. B. 1991. Climate change and the origin of old growth Douglas-fir forests in the Puget Sound lowland. Pp. 17–24 in *Wildlife and vegetation of unmanaged Douglas-fir forests*, edited by L. F. Ruggiero, K. B. Aubry, A. B. Carey, and M. H. Huff. U.S. Department of Agriculture, Forest Service, Pacific Northwest Research Station, Portland, Ore.

Brunner, R. D., and T. W. Clark. 1997. A practice-based approach to ecosystem management. *Conservation Biology* 11:45–58.

Cain, S. A., and G. M. d. O. Castro. 1959. *Manual of vegetation analysis*. New York: Harper and Brothers.

Christensen, N. L., A. M. Bartuska, J. H. Brown, S. Carpenter, C. D'Antonio, R. Francis, J. F. Franklin, J. A. MacMahon, R. F. Noss, D. J. Parsons, C. H. Peterson, M. G. Turner, and R. G. Woodmansee. 1996. The report of the Ecological Society of America committee on the scientific basis for ecosystem management. *Ecological Applications* 6(3):665–691.

Clark, T. W., A. P. Curlee, and R. P. Reading. 1996. Crafting effective solutions to the large carnivore conservation problem. *Conservation Biology* 10(4):940–948.

Duffus, J., III. 1994. *Ecosystem management: Additional actions needed to adequately test a promising approach*. GAO/T-RCED-94-308 U.S. General Accounting Office, Washington, D.C.

Dwyer, W. L. 1994. Seattle Audubon Society, et al. v. James Lyons, Assistant Secretary of Agriculture et al. Order on motions for Summary Judgment RE 1994 Forest Plan. U.S. District Court, Western District of Washington at Seattle.

Everett, R., C. Oliver, J. Saveland, P. Hessburg, N. Diaz, and L. Irwin. 1994. Adaptive ecosystem management. Pp. 340–354 in *Ecosystem management: Principles and applications*. Vol. 2. Edited by M. E. Jensen and P. S. Bourgeron. U.S. Department of Agriculture, Forest Service, Pacific Northwest Research Station, Portland, Ore.

Forest Ecosystem Management Assessment Team (FEMAT). 1993. Forest ecosystem management: An ecological, economic, and social assessment. Washington, D.C.: US Department of Agriculture, Forest Service; US Department of Commerce, National Oceanic and Atmospheric Administration, US Department of Interior, Bureau of Land Management, U.S. Fish and Wildlife Service, and National Park Service; and Environmental Protection Agency.

Forman, R. T. T., and M. Godron. 1986. *Landscape ecology*. New York: John Wiley and Sons.

Gauch, H. G. 1982. *Multivariate analysis in community ecology*. New York: Cambridge University Press.

Green, R. H. 1979. *Sampling design and statistical methods for environmental biologists*. New York: Wiley-Interscience.

Gregory, S. V., F. J. Swanson, W. A. McKee, and K. W. Cummins. 1991. An ecosystem perspective of riparian zones. *BioScience* 41:540–551.

Grumbine, R. E. 1994. What is ecosystem management? *Conservation Biology* 1994(1):27–38.

Gunderson, L. H., C. S. Holling, and S. S. Light, eds. 1995. *Barriers and bridges to the renewal of ecosystems and institutions*. New York: Columbia University Press.

Hansen, M. H., W. N. Hurwitz, and W. G. Madow. 1953. *Sample survey methods and theory*. Vol. 1. *Methods and applications*. New York: John Wiley and Sons.

Harris, G. P. 1980. Temporal and spatial scales in phytoplankton ecology: Mechanisms, methods, models, and management. *Canadian Journal of Fisheries and Aquatic Sciences* 37:877–900.

Haughen, G. 1994. *Section 7 fish habitat monitoring protocol for the upper Columbia River Basin*. U.S. Department of Agriculture, Forest Service, Regions 1, 4, and 6, Logan, Utah.

Hayslip, G., D. Klemm, and J. Lazorchak, eds. 1994. *Environmental Monitoring and Assessment Program Surface Waters and Region10 Regional Environmental Monitoring and Assessment Program: 1994 pilot field operations and methods manual for streams on the Coast Range Ecoregion of Oregon and Washington and the Yakima River Basin*. Office of Research and Development, U.S. Environmental Protection Agency, Cincinnati, Ohio.

Hicks, B. B., and T. G. Brydges. 1994. Forum: A strategy for integrated monitoring. *Environmental Management* 18(1):1–12.

Holling, C. S., ed. 1978. *Adaptive environmental assessment and management*. Hoboken, N.J.: John Wiley and Sons.

Interagency Ecosystem Management Task Force. 1995. *The Ecosystem Approach: Healthy ecosystems and sustainable economies*. Vol. 1. *Overview*. Interagency Ecosystem Management Task Force, Washington, D.C.

Johnson, K. N., J. F. Franklin, J. W. Thomas, and J. Gordon. 1991. Scientific Panel on Late-Successional Forest Ecosystems. *Alternatives for the management of late-successional forests of the Pacific Northwest: A report to the U.S. House of Representatives, Committee on Agriculture, Subcommittee on Forests, Family Farms, and Energy, Committee on Merchant Marine and Fisheries, Subcommittee on Fisheries and Wildlife, Conservation, and the Environment*. Washington, D.C.: Government Printing Office.

Kondolf, G. M. 1995. Five elements for effective evaluation of stream restoration. *Restoration Ecology* 3(2):133–136.

Kotler, P., and A. R. Andreason. 1996. *Strategic marketing for nonprofit organizations*. Upper Saddle River, N.J.: Prentice Hall.

Landers, D. H., R. M. Hughes, S. G. Paulsen, D. P. Larsen, and J. M. Omernik. 1998. How can regionalization and survey sampling make limnological research more relevant? *Verhandlungen Internationale Vereinigung für Theoretische und Angewandte Limnologie*, 26:2428–2436.

Larsen, D. P., N. S. Urquhart, and D. L. Kugler. 1995. Regional scale trend monitoring of indicators of trophic condition of lakes. *Water Resources Bulletin* 31(1):1–23.

Levitt, T. 1986. *The marketing imagination*. New York: Free Press.

Likens, G. E. 1983. A priority for ecological research. *Bulletin of the Ecological Society of America* 64:234–243.

MacDonald, L. H., A. W. Smart, and R. C. Wissmar 1991. *Monitoring guidelines to evaluate effects of forestry activities on streams in the Pacific Northwest and Alaska.* EPA/910/9-91-001. Environmental Protection Agency Region X, Water Division and Center for Streamside Studies, University of Washington. Seattle, Wash.

McRoberts, R. E., J. T. Hahn, G. J. Hefty, and J. R. Van Cleve. 1994. Variation in forest inventory field measurements. *Canadian Journal of Forest Research* 24:1766–1770.

Montgomery, D. R., G. E. Grant, and K. Sullivan. 1995. Watershed analysis as a framework for implementing ecosystem management. *Water Resources Bulletin* 31:369–386.

Morrison, M. L. 1994. Resource inventory and monitoring. *Restoration and Management Notes* 12(2):179–183.

Mulder, B., J. Alegria, R. Czaplewski, P. Ringold, and T. Tolle 1995. *Effectiveness monitoring: An interagency program for the Northwest Forest Plan with an emphasis on late-successional forest, northern spotted owl, marbled murrelet, survey and manage, and riparian and aquatic.* Research and Monitoring Committee, Regional Ecosystem Office, Portland, Ore.

Mulder, B. S., B. R. Noon, T. A. Spies, M. G. Raphael, C. J. Palmer, A. R. Olson, G. H. Reeves, and H. H. Welsh. 1999. *The strategy and design of the effectiveness monitoring program for the Northwest Forest Plan.* General Technical Report PNW-GTR-437. U.S. Department of Agriculture, Forest Service, Pacific Northwest Research Station, Portland, Ore.

Neave, H., R. Cunningham, T. Norton, and H. Nix. 1996. Biological inventory for conservation evaluation. 3. Relationships between birds, vegetation and environmental attributes in southern Australia. *Forest Ecology and Management* 85(1–3): 197–218.

Norton, S. B., D. J. Rodier, J. H. Gentile, W. H. Van Der Schalie, W. P. Wood, and M. W. Slimak. 1992. A framework for ecological risk assessment at the Environmental Protection Agency. *Environmental Toxicology and Chemistry* 11: 1663–1672.

Noss, R. F., and A. Y. Cooperrider. 1994. *Saving nature's legacy: Protecting and restoring biodiversity.* Washington, D.C.: Island Press.

Oregon Department of Fish and Wildlife. 1994. *Methods for stream habitat surveys.* Ver. 4.2. Oregon Department of Fish and Wildlife, Corvallis, Ore.

———. 1994. *Water classification and protection rules reference guide.* Oregon Department of Forestry, Salem, Ore.

Overton, W. S., and S. V. Stehman. 1996. Desirable design characteristics for long-term monitoring of ecological variables. *Environmental and Ecological Statistics* 3:349–361.

Peterman, R. M. 1990. Statistical power analysis can improve fisheries research and management. *Canadian Journal of Fisheries and Aquatic Sciences* 47:2–15.

Peterson, S. A., D. P. Larsen, S. G. Paulsen, and N. S. Urquhart. 1998. Regional lake trophic patterns in the northeastern United States: Three approaches. *Environmental Management* 20(5): 789–801.

Pickett, S. T. A. 1989. Space for time substitution as an alternative to long-term studies. Pp. 110–135 in *Long-term studies in ecology*, edited by G. E. Likens. New York: Springer-Verlag.

Powell, T. M. 1995. Physical and biological scales of variability in lakes, estuaries and the coastal ocean. Pp. 119–138 in *Ecological time series*, edited by T. M. Powell and J. H. Steele. New York: Chapman and Hall.

Raiffa, H. 1968. *Decision analysis: Introductory lectures on choices under uncertainty.* Menlo Park, Calif.: Addison-Wesley Publishing.

Ringold, P. L., J. Alegria, R. L. Czaplewski, B. Mulder, T. Tolle and K. Burnett. 1996. Adaptive monitoring for ecosystem management. *Ecological Applications* 6(3):745–747.

Risk Assessment Forum 1996. *Proposed guidelines for ecological risk assessment.* EPA/530/R-95/002B. U.S. Environmental Protection Agency, Washington, D.C.

Rosenberger, W. F., and J. M. Lachin. 1993. The use of response-adaptive designs in clinical trials. *Controlled Clinical Trials* 14(6):471–484.

Runyon, J. 1995. The forest practices monitoring program. *Forest Log* (January–February):8–12.

Schreuder, H. T., T. G. Gregoire, and G. B. Wood. 1993. *Sampling methods for multiresource forest inventory.* New York: John Wiley and Sons.

Schumaker, N. H. 1996. Using landscape indices to predict habitat connectivity. *Ecology* 77(4):1210–1225.

Scott, J. M., F. Davis, B. Csuti, R. Noss, B. Butterfield, C. Groves, H. Anderson, S. Caicco, F. D'Erchia, T. C. Edwards Jr., J. Ulliman, and R. G. Wright. 1993. *Gap analysis: A geographical approach to protection of biological diversity.* Wildlife Monographs 123:1–41.

Simon, R. 1977. Adaptive treatment assignment methods and clinical trials. *Biometrics* 33:743–749.

Siuslaw National Forest 1990. *Land and resource management plan.* Sec. 5, *Monitoring and evaluation.* U.S. Department of Agriculture, Forest Service, Pacific Northwest Region, Portland, Ore.

Smith, J. P., R. E. Gresswell, and J. P. Hayes. 1997. *A research problem analysis in support of the Cooperative Forest Ecosystem Research Center (CFER) program.* Forest and Rangeland Ecosystem Science Center, Biological Resources Division, U.S. Geological Survey, Corvallis, Ore.

Society of American Foresters. 1994. The forest ecosystem management assessment team: Reports and critiques. *Journal of Forestry* 92:12–47.

Stapanian, M. A., S. P. Cline, and D. L. Cassell. 1997. Evaluation of a measurement method for forest vegetation in a large-scale ecological survey. *Environmental Monitoring and Assessment* 45:237–257.

Steele, J. H. 1995. Can ecological concepts span the land and ocean domains? Pp. 5–19 in *Ecological time series,* edited by T. M. Powell and J. H. Steele. New York: Chapman and Hall.

Stehman, S. V., and W. S. Overton. 1994. Environmental sampling and monitoring. Pp. 263–305 in *Handbook of Statistics,* edited by G. P. Patil and C. R. Rao. New York: Elsevier Science.

Stevens, D. L. 1994. Implementation of a national monitoring program. *Journal of Environmental Management* 42:1–29.

Strayer, D., J. S. Glitzenstein, C. G. Jones, J. Kolasa, G. Likens, M. J. McDonnell, G. G. Parker, and S. T. A. Pickett 1986. *Long-term ecological studies: An illustrated account of their design, operation, and importance to ecology.* Occasional publication of the Institute of Ecosystem Studies. No. 2. Millbrook, New York.

Suter, G. W., II. 1990. End points for regional ecological assessments. *Environmental Management* 14(1):9–23.

Thomas, J. W., M. G. Raphael, R. G. Anthony, E. D. Forsman, A. G. Gunderson, R. S. Holthausen, B. G. Marcot, G. H. Reeves, J. R. Sedell, and D. M. Solis. 1993. *The report of the scientific analysis team: Viability assessments and management considerations for species associated with late-successional and old-growth forests of the Pacific Northwest.* U.S. Department of Agriculture, U.S. Forest Service, Portland, Ore.

Tokar, B. 1994. Between the loggers and the owls: The Clinton Northwest Forest Plan. *Ecologist* 24(4):149–153.

Tolle, T., and R. L. Czaplewski 1995. *An interagency monitoring design for the Pacific Northwest Forest Ecosystem Plan: Analysis in support of ecosystem management.* U.S. Department of Agriculture, Forest Service, Ecosystem Management Analysis Center, Fort Collins, Colo.

Tuchmann, E. T., K. P. Connaughton, L. E. Freedman, and C. B. Moriwaki. 1996. *The Northwest Forest Plan: A report to the president and congress.* U.S. Department of Agriculture, Forest Service, Pacific Northwest Research Station, Portland, Ore.

USFS (U.S. Forest Service, Region 6). 1993. *Stream inventory handbook: Level 1 and 2.* Ver. 7. U.S. Department of Agriculture, U.S. Forest Service, Region 6. Portland, Ore.

USFS and BLM (U.S. Department of Agriculture, Forest Service and Department of Interior, Bureau of Land Management) 1994a. *Final supplemental environmental impact statement on management of habitat for late-successional and old-growth forest related species within the range of the northern spotted owl.* Washington, D.C.: USFS and BLM.

————. 1994b. Record of decision for amendments to Forest Service and Bureau Of Land Management planning documents within the range of the northern spotted owl and standards and guidelines for management of habitat for late-successional and old-growth forest related species within the range of the northern spotted owl. Washington, D.C.: USFS and BLM

Walters, C. J. 1986. *Adaptive management of renewable resources.* New York: Macmillan.

White, D., P. G. Minotti, M. J. Barczak, J. C. Sifneos, K. E. Freemark, M. V. Santelmann, C. F. Steinitz, A. R. Kiester, and E. M. Preston. 1997. Assessing risks to biodiversity from future landscape change. *Conservation Biology* 11(1):349–360.

Wolfe, D. A., M. A. Champ, D. A. Flemer, and A. J. Mearns. 1987. Long-term biological data sets: Their role in research, monitoring, and management of estuarine and coastal marine systems. *Estuaries* 10(3):181–193.

Yaffee, S. L. 1994. *The wisdom of the spotted owl: Policy lessons for a new century.* Washington, D.C.: Island Press.

Yaffee, S., A. Phillips, I. Frentz, P. Hardy, S. Maleki, and B. Thorpe. 1996. *Ecosystem management in the United States: An assessment of current experience.* Washington, D.C.: Island Press.

CHAPTER 4

Monitoring for Adaptive Management of the Colorado River Ecosystem in Glen and Grand Canyons

Lawrence E. Stevens and Barry D. Gold

Adaptive ecosystem management (AEM) has become the paradigm for the management of large, complex, human-dominated systems (Holling 1978, 1986; Lee 1993; Gunderson et al. 1995). AEM is particularly appropriate for large river ecosystems, which are often in a degraded condition and are managed by multiple agencies for diverse social goals (Lee 1991, 1995; Johnson et al. 1995; Light et al. 1995; Naiman et al. 1995; Sparks 1995; Stanford et al. 1996; Poff et al. 1997). Some form of AEM is currently used to manage most large North American rivers.

Development of a science-based, long-term AEM program, of which monitoring is an important facet, requires agreement on a supporting administrative process (Holling 1978; Walters 1986; Lee 1993; Stanford and Poole 1996). First, a collaboratively developed vision of goals for the management program is needed (National Research Council 1986). From this vision statement, management goals and objectives must be openly debated, clearly defined, and agreed upon by the stakeholders. Stakeholders should recognize that large uncertainties and risks exist in the management of large ecosystems and that insufficient information usually exists to make an indisputably best decision. Therefore, ecosystem management actions are best approached as carefully formulated scientific experiments, and the resulting information is used to test specific hypotheses about processes, rates of change, and management impacts across the broad spectrum of physical, biological, and cultural resources that characterize large river ecosystems (Stanford and Poole 1996). Ecosystem modeling is a key component in this process and can be used to expose gaps in data and understanding and to evaluate policy options that are trivial or risky given uncertainty about response directions (Stanford and Poole 1996; Walters et al. 2000). Sound and continuing external peer review is needed for project proposals, results, and program directions, and a rigorous information management system is required to archive and present results. Decisions on how, what, when, and where to monitor, and the key scientific elements requiring additional research, are large challenges that require concentrated discussion with experts as part of planning and protocol evaluation (National Research Council 1990; Davis 1993). Also, the results of individual studies must be swiftly integrated into an outreach program that keeps stakeholders abreast of new understanding and hypotheses. In turn, new information and insights need to be used to periodically review management objectives and activities and to revise monitoring (Ringold et al., chap. 3), research, and outreach activities in support of the program.

Although currently emerging as the most appropriate approach for ecosystem management, adaptive management founders when stakeholder values are not clearly stated and agreed upon, or when conflicting social issues (i.e., stakeholder positions, special interests, and inadequate outreach programs) limit scientific components (i.e., monitoring, research, and modeling) that are regarded by the scientific community as necessary to advance understanding (Naiman and Turner 2000). Therefore, successful implementation of a rigorous monitoring and research program for AEM requires extensive and open communication between stakeholders and scientists, as well as thorough external peer review and a commitment by one or more administrative "champions" who thoroughly support the program. The role of science in AEM is not resource valuation or advocacy for specific resources but rather to provide information on the consequences of different management options on resource and process conditions specified by society (Schmidt et al. 1998), as well as advocacy for sound science.

In this chapter, we describe a science-based AEM program for the Colorado River ecosystem affected by Glen Canyon Dam. This large, desert river ecosystem is managed by a public process for economic values related to hydroelectric power, water storage and recreation, and environmental concerns related to wilderness aesthetics, ecosystem health, endangered species, cultural resources, and the Grand Canyon's status as a great landscape park and World Heritage Site. The Glen Canyon Dam Environmental Impact Statement (EIS; U.S. Bureau of Reclamation 1995) was one of the nation's largest EISs, requiring five years to prepare. We describe ecosystem characteristics, the historical and administrative context of the AEM effort, elements of the monitoring and research program, and our accomplishments and challenges to date. Monitoring and research programs are always "works in progress," but we see the Colorado River AEM program as one of the best examples of the application of science for AEM. Both the successes and the setbacks of this program provide useful instruction to the managers and stakeholders of other large ecosystems.

The Colorado River Ecosystem

The Colorado River ecosystem considered by the 1995 Glen Canyon Dam EIS and 1996 Secretary of the Interior Record of Decision encompasses the mainstream river and its floodplain affected by Glen Canyon Dam from Lake Powell to the western-most boundary of Grand Canyon National Park on upper Lake Mead (fig. 4-1). The study area includes Lake Powell (for some characteristics), as well as some aspects of flow, sediment transport, biology, and other aspects of some tributaries. The Colorado River flows 473 kilometers from Glen Canyon Dam (960 meters in elevation) to the Grand Wash Cliffs on upper Lake Mead (full pool is 372.2 meters in elevation) through Sonoran and Mojave Desert terrain, through lower Glen Canyon and all of Grand Canyon. By convention, locations along the Colorado River are designated in river miles from Lees Ferry. The river passes through thirteen bedrock-defined geomorphic reaches, which strongly affect channel geometry and terrestrial productivity (Schmidt and Graf 1990; Stevens et al. 1997a,b). Also, the Paria (kilometer 1) and Little Colorado (kilometer 98) Rivers create three segments with increasing turbidity that strongly influence aquatic ecosystem structure. From a management perspective, this is a large arid-land river that supplies much of the water for the Southwest. It is impounded by a very large dam in its middle reaches, and it has only a few, small tributaries downstream that provide suspended sediment, but little flow (Stevens et al. 1997b). Therefore, the river is not capable of recovering much of its natural character as it moves through the study reach (Ward and Stanford 1983; Schmidt et al. 1998).

The Colorado River ecosystem is structured by geology, geomorphology, past and present floods, dam operations, and tributary sediment inflow in a strongly "bottom-up" fashion. Unlike large unconstrained

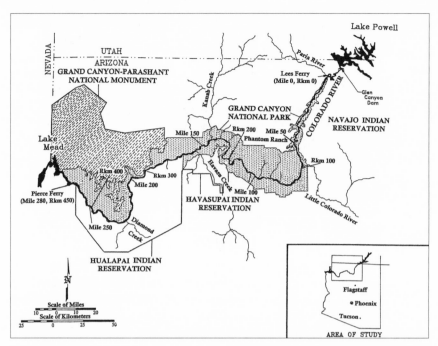

Figure 4-1. The Colorado River ecosystem extends from Lake Powell to the western boundary of Grand Canyon National Park, 447 kilometers downstream from Lees Ferry, Ariz., from which river distances are measured. Four streamflow gauging stations exist along the river (Lees Ferry, above the Little Colorado River, near Phantom Ranch, and at Diamond Creek), and major tributaries are shown.

rivers with broad floodplains, the Colorado River in Glen and Grand Canyons is confined by talus slopes and bedrock to a relatively narrow channel. Debris flows have reached the river at the mouths of more than five hundred tributaries in this system, creating debris-fan constrictions and eddies that control where sand and finer-grain alluvial sediments settle, and dictating the location of sand deposits (Howard and Dolan 1981; Kieffer 1985; Melis 1997; Schmidt 1990; Webb et al. 1989, 1999a). Sand bars in Grand Canyon are spatially fixed by channel geometry and do not move, as they do in unconfined systems such as the Mississippi River (Schmidt and Graf 1990). These debris-fan-eddy complexes are distinctive, repeated geomorphic units separated by relatively uniform runs of channel (Schmidt and Rubin 1995). Sand deposits and debris fans support distinctive vegetation assemblages that, in turn, provide food and habitat for fish and wildlife (Stevens et al. 1995, 1997a,b, 2001).

Administrative Context

The Colorado River ecosystem affected by Glen Canyon Dam has had a lengthy history of multi-agency federal, tribal, and state management (box 4-1). During his initial exploration of the river corridor in 1869–1871 and his tenure as the second director or the U.S. Geological Survey, John Wesley Powell recognized both the aesthetic significance of Grand Canyon and the need for water resources development (Stegner 1954; Rabbit 1975). Potential conflicts between resource exploitation and conservation were recognized in the authorizing legislation for Grand Canyon National Park (1919–1925); however, legislation authorizing water resource exploitation predominated over river-affected resource conservation until the passage of the National Environmental Policy Act (1973), the Endangered Species Act (1973), and the Grand Canyon Protection Act (1992).

The Secretary of the Interior's Record of Decision (ROD) on the operation of Glen Canyon Dam (1996) authorized long-term monitoring and research and experimentation on river ecosystem and hydroeconomic management options in lower Lake Powell, Glen Canyon, and Grand Canyon. The ROD called for formulation of an Adaptive Management Work Group (AMWG) as a Federal Advisory Committee consisting of a diverse group of twenty-seven stakeholders, including Department of the Interior agencies (Bureau of Indian Affairs, Bureau of Reclamation, Fish and Wildlife Service, National Park Service), Western Area Power Administration, Colorado River basin states, Native American tribes, and representatives of power marketing, recreational and environmental advocacy groups (fig. 4-2). The AMWG meets semiannually to discuss dam management, review the progress of scientific activities, develop plans for future activities, and provide recommendations to the Secretary of the Interior on Glen Canyon Dam operations. Scientific studies are coordinated by the Grand Canyon Monitoring and Research Center (GCMRC), and the AMWG is advised by a subcommittee, the Technical Work Group (TWG), which works closely with GCMRC. As a Federal Advisory Committee, all AMWG and TWG meetings are advertised and open to the public.

The GCMRC is a U.S. Geological Survey (USGS) office and is overseen by a Department of the Interior management team composed of representatives from the office of the Assistant Secretary for Water and Science, the USGS, the Bureau of Reclamation, and the National Park Service. The GCMRC is responsible for short-term and longer-term strategic planning, communications with the TWG and AMWG, and staff coordination. GCMRC activities are defined in five-year strategic

Box 4-1 The administrative history of the Colorado River ecosystem (updated from Stevens and Wegner 1995).

1902	Reclamation Act creates the Bureau of Reclamation.
1904	Grand Canyon declared a national game reserve (T. Roosevelt).
1916	National Park Service Organic Act.
1919	Grand Canyon declared a national park, stipulating "reclamation projects" within park boundaries.
1922	The Colorado River Compact allocates the river's water between the upper (Wyoming, Colorado, Utah, and New Mexico) and lower (Arizona, Nevada, and California) basins.
1920–1925	U.S. Geological Survey scouts potential dam sites.
1929–1935	Construction of Hoover Dam creates Lake Mead.
1945	The Mexican Treaty guarantees 1.5 million acre-feet per year of water to Mexico.
1948	The Upper Basin Compact allocates Colorado River water between the upper basin states.
1956	The Colorado River Storage Project (CRSP) Act is passed, authorizing construction of upper basin dams.
1957–1963	Glen Canyon Dam construction; power production starts in 1964.
1967	Humpback chub and Colorado pikeminnow listed as endangered (32 Federal Register 4001).
1970	The National Environmental Policy Act passed.
1973	The Endangered Species Act passed.
1976	The National Park Service coordinates the first ecological inventory of the Grand Canyon (Carothers and Aitchison 1976) and the first sociological studies; last Colorado pikeminnow caught in Grand Canyon at Havasu Creek.

Box 4-1 *(continued)*

1978   U.S. Fish and Wildlife Service Jeopardy Opinion on the operation of Glen Canyon Dam.

1980   Lake Powell fills for the first time; bonytail chub listed as endangered.

1981–1982 Rewind of Glen Canyon Dam turbines; Bureau of Reclamation states that there will be no significant effect on downstream river ecosystem. Secretary of the Interior James Watt orders Bureau of Reclamation Glen Canyon Environmental Studies (GCES) Program to study dam impacts.

1983   Post-dam record 2,724 cubic meters per second flow is released.

1983–1986 Forty GCES studies of dam effects conducted during exceptionally high inflow and dam releases.

1987   National Academy of Sciences (NRC 1987) review of GCES Phase I.

1988   Cooperating agencies conclude that GCES Phase I (U.S. Bureau of Reclamation 1988) showed (1) dam affects river ecosystem, but (2) more data were needed on low and fluctuating flows to determine how to best manage the system.

1989   Secretary of the Interior Manuel Lujan orders an ex post facto EIS on dam operations; initiation of GCES Phase II to support EIS preparation.

1990–1991 Test flows were used to determine effects of individual flow regimes (Patten 1991).

1991   Interim flows (low hourly change in flow) implemented on 1 August to protect river resources while EIS is prepared; interim flows monitoring approved by Reclamation in November 1991; Santa Fe "State of Knowledge" symposium (National Research Council 1991); razorback sucker listed as endangered; Kanab ambersnail proposed for emergency listing.

continues

Box 4-1 *(continued)*

1992   NAS "Delphi Process" symposium in Irvine, Calif., to plan long-term monitoring for the Colorado River corridor. Passage of the Grand Canyon Protection Act provides for a speedy resolution of the EIS and balancing environmental protection with economic benefits. Kanab ambersnail is listed.

1993   U.S. Fish and Wildlife Service proposes listing of southwestern willow flycatcher. It was listed in 1997; critical habitat was designated from Colorado River km 74–116.

1994   U.S. Fish and Wildlife Service Biological Opinion concludes that Glen Canyon Dam still jeopardizes native fish.

1995   Final EIS submitted to Secretary of the Interior Bruce Babbitt, calling for (1) low-flow fluctuations to preserve tributary-derived bed sand, (2) planned flooding to restore higher-elevation sand bars, (3) adaptive management based on (4) long-term monitoring and cooperative, interagency discussion. Establishment of Grand Canyon Monitoring and Research Center.

1996   A beach habitat building flow (BHBF, experimental flood) was conducted from Glen Canyon Dam from 26 March to 2 April. The U.S. Fish and Wildlife Service Biological Opinion on the planned flood restricted take of Kanab ambersnail habitat to less than 10 percent and stipulated that no additional planned high flows be conducted until at least one additional Kanab ambersnail population is discovered or established in Arizona. The Secretary's Record of Decision (ROD) is signed, formalizing the flow regime and adaptive management framework.

1997   GCES is replaced by the Grand Canyon Monitoring and Research Center. Flows (765 cubic meters per second) above the ROD occurred in

Box 4-1 *(continued)*

1997 *(continued)*	February/March and again in mid-summer. The Adaptative Management Work Group (with the Technical Work Group) formally convened as a federal advisory committee. An experimental habitat maintenance flow was conducted in early November.
1998	Beach habitat building flows triggering criteria formalized by AMWG.
1999	Development of AMWG Vision Statement.
2000	The first low-normal inflow year since the ROD. A low steady seasonal flows scenario (spring "spike" of 892 cubic meters per second followed by steady 226 cubic meters per second through the summer, and an early autumn spike) is initiated and monitored.

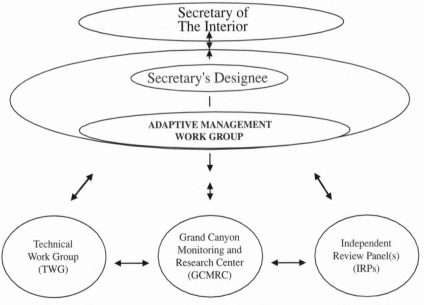

Figure 4-2. Administrative structure of the Glen Canyon Dam Adaptive Management Program.

plans, the second of which will be produced pending completion of the AMWG's strategic plan. GCMRC's staff have expertise in six program areas: physical resources; biological resources, including Lake Powell limnological studies; cultural-socioeconomic resources; information management; and logistical coordination. Each resource area is overseen by a program manager or coordinator, who serves as a senior scientist. Most monitoring and research is contracted externally, except surveying, georeferencing, and Lake Powell limnological studies, which are conducted inhouse for continuity and to reduce costs.

Ecosystem Characteristics and Change through Time

Although logistical access to the Colorado River in the Grand Canyon is challenging, it has become one of the most thoroughly studied large regulated river ecosystems in the world. Much pre-dam and post-dam information has been published on the geology, water quality, and flow of this ecosystem, which has long-term streamflow gauges at Lees Ferry and Phantom Ranch (Andrews 1991). Webb et al. (1999b) and Patten and Stevens (2001) provide the results and syntheses of numerous studies of the 1996 experimental flood. Pre-dam biological information is available from diaries, oblique and aerial photography, a few published studies (e.g., Clover and Jotter 1944), and surviving pre-dam residents and river runners (Webb 1996; Stevens et al. 1997a). Carothers and Brown (1991) and Patten (1991) presented conceptual relationships among physical and biological components and interactions. Walters et al. (2000) quantified many of those relationships with empirical flow, sediment transport, benthic, drift, and fisheries data, and developed a conceptual model that describes physical and biological interactions in the aquatic domain of the ecosystem.

In contrast to the empirically robust literature on flow effects and physical processes in this system, biological inquiries have focused primarily on a few topic areas, particularly benthos, fish, riparian vegetation, and birds (e.g., National Research Council 1991). Attempts to link biological characteristics to flows in a predictive fashion are relatively recent (Walters et al. 2000). No current systemwide vegetation map exists for habitat analysis, and a considerable amount of basic biological inventory is still outstanding, particularly distribution data on riparian fungi, aquatic and terrestrial invertebrates, herpetofaunae, and mammals.

The literature on terrestrial-aquatic interactions and ecological processes indicates that flow regulation by Glen Canyon Dam transformed both the aquatic and terrestrial domains of the ecosystem from an allochthonous, low-moderate diversity, low-productivity condition to an autochthonous, high-diversity, high-productivity system (Turner and Karpiscak 1980; Johnson 1991; Angradi 1994; Stevens et al. 1995, 1997a,b, 1998; Webb 1996; Schmidt et al. 1998; Marzolf et al. 1999; Walters et al. 2000). This system provides an excellent example and test of basic ecological disturbance theory, which predicts that reduced disturbance intensity (reduced flood frequency and magnitude) will increase biodiversity (Connell 1978; Huston 1979; Stevens and Ayers 2002). This system is also strongly influenced by the introduction of nonnative fish, fish parasites, and plant species (Johnson 1991; Minckley 1991; Stevens 1989; Brouder and Hoffnagle 1996; Stevens and Ayers 2002), species that appear to have permanently altered ecosystem structure and function.

The role of impoundment on ecosystem development differs between the aquatic and terrestrial domains in this system. Flow regulation has largely overridden the effects of local geomorphology on the benthos, as variation in mainstream turbidity is now a function of tributary location (particularly the Paria and Little Colorado rivers; Stevens et al. 1997b). In contrast, flow regulation has enhanced the effects of local geomorphology on the floodplain, as riparian landforms now largely control vegetation development and wildlife habitat availability (Stevens et al. 1995, 1997a). Ecological stability from flood control has increased aquatic and terrestrial productivity, diversity, population size, and the predictability of food web interactions in this ecosystem for many species; exceptions include Goodding willow (*Salix gooddingii*; Mast and Waring 1997), an unknown number of aquatic invertebrate taxa (Stevens et al. 1997b), four native and now endangered fish species (Minckley 1991), locally extirpated zebra-tailed lizard (*Callisaurus draconoides*) and muskrat (*Ondatra zibitheca*), declining northern leopard frog (*Rana pipiens*) and southwestern willow flycatcher (*Empidonax trailii extimus*), and probably extinct Colorado river otter (*Lontra canadensis sonora*; Stevens et al. 2001).

Elements of the Monitoring and Research Program

GCMRC is responsible for development and implementation of a long-term, science-based program to monitor and conduct research on the Colorado River ecosystem and to provide that information to the AMWG (Adaptive Management Work Group), as described by the 1996 ROD. This science program involves eight elements: (1) agreement among stakeholders on goals, management objectives, funding, and timing; (2) comprehensive syntheses of existing information on the physical, biological, cultural, and socioeconomic resources affected by dam management; (3) conduct of a competitive, peer-reviewed monitoring and research program; (4) rigorous protocol evaluation; (5) regular, rigorous review of project results, and GCMRC and AMWG program directions; (6) information management and archival; (7) efficient transfer of information to stakeholders, cooperating scientists, the larger scientific community, and the public; and (8) feedback of knowledge to improve and refine future monitoring and research. Accomplishing these elements provides the best long-term approach for adaptive management of this large, dynamic river ecosystem; however, all elements are necessary to the success of the program. In the following section, we describe GCMRC's monitoring and research strategies, and the administrative and scientific challenges inherent in this approach.

Adaptive Ecosystem Management Mission and Goals

The AMWG stakeholders recognize that clearly defined goals and objectives are needed to clarify the process of AEM (adaptive ecosystem management) in the Colorado River ecosystem. Agreement on the group's vision and mission was achieved during an AMWG river trip through Grand Canyon in May 1999 and adopted by the AMWG on 6 July 2000:

> *The Grand Canyon is a homeland for some, sacred to many, and a national treasure for all. In honor of past generations, and on behalf of those of the present and future, we envision an ecosystem where the resources and natural processes are in harmony under a stewardship worthy of the Grand Canyon.*
>
> *We advise the Secretary of the Interior on how best to protect, mitigate adverse impacts to, and improve the integrity of the Colorado River ecosystem affected by Glen Canyon Dam, including natural biological diversity (emphasizing native biodiversity), traditional cultural properties, spiritual values, and cultural, physical, and recreational resources through the operation of Glen Canyon Dam and other means.*
>
> *We do so in keeping with the federal trust responsibilities to Indian tribes, in compliance with applicable federal, state, and tribal laws, including the water delivery obligations of the Law of the River, and with due consideration to the economic value of power resources.*
>
> *This will be accomplished through our long-term partnership utilizing the best available scientific and other information through an adaptive ecosystem management process.*

To achieve this mission, the AMWG has developed, reviewed, and revised a list of goals (box 4-2) and associated management objectives. The

Box 4-2. Adaptive Management Work Group goals for adaptive ecosystem management of the Colorado River in Lake Powell, lower Glen Canyon and Grand Canyon.

- Goal 1. Protect or improve the aquatic food base so that it will support viable populations of desired species at higher trophic levels.
- Goal 2. Maintain or attain viable populations of existing native fish and remove jeopardy from humpback chub and razorback sucker.
- Goal 3. Restore populations of extirpated species, as feasible.
- Goal 4. Maintain a wild reproducing population of rainbow trout above the Paria River to the extent practicable and consistent with the maintenance of viable populations of native fish.
- Goal 5. Establish water temperature, quality, and flow dynamics to achieve GCDAMP ecosystem goals.
- Goal 6. Maintain or attain levels of sediment storage within the main channel and along shorelines to achieve GCD-AMP ecosystem goals.
- Goal 7. Maintain or attain viable populations of Kanab ambersnail.
- Goal 8. Protect the presence of southwestern willow flycatcher and its critical habitat in a manner consistent with riparian ecosystem goals.
- Goal 9. Protect or improve the biotic riparian and spring communities.
- Goal 10. Maintain or improve the quality of recreational experiences for users of the Colorado River ecosystem, within the framework of GCDAMP ecosystem goals.
- Goal 11. Maintain or increase power and energy generation within the framework of GCDAMP ecosystem goals.
- Goal 12. Preserve, protect, manage, and treat cultural resources for the inspiration and benefit of past, present, and future generations.
- Goal 13. Maintain a high-quality monitoring, research, and adaptive management program.

AMWG's present emphasis is on ecosystem restoration toward the river ecosystem's natural condition of flow, sediment transport, and (it is hoped) biological condition, to the extent possible, given the presence and continued operation of the dam. This emphasis has been developed despite substantial theoretical, informational, technological, and societal challenges to the logic behind them (Ward and Stanford 1983; Gore and Shields 1995; Schmidt et al. 1998; Walters et al. 2000; Stevens et al. 2001), and most debates surrounding these goals remain unresolved.

Consensus on management among a large number of stakeholders with widely differing values and missions is rare, and often involves heated debate. The TWG (Technical Work Group) was envisioned in the Glen Canyon Dam EIS to provide technical support to the policy-oriented AMWG. However, the TWG members are typically more oriented toward management than science, and decisions that have large economic or environmental ramifications are made slowly. Trust between managers and scientists, built on open communication and understanding of issues, is required for this AEM effort to succeed, and improvement of the technical expertise of the TWG will help.

Syntheses of Existing Information

In order to achieve the above objectives and proceed with adaptive management of this large regulated river, comprehensive syntheses are required in the several physical, biological, and socioeconomic disciplines. These syntheses are designed to draw together the robust global literature on regulated river management as well as that on the Colorado River. Syntheses on several resource topics have been undertaken, but much information has yet to be compiled, integrated, published, and archived.

Much Colorado River fisheries information has been published, an initial synthesis has been conducted, and compilation of the voluminous data collected from 1970 through 2000 is underway. Syntheses are presently being conducted on the ecology of the lower trophic levels, the trout fishery upstream from the Paria River, and riparian vegetation and habitat. Our goal is to have such syntheses performed by researchers who are independent of the actual data collection projects to avoid conflicts of interest and field researcher biases.

To obtain the "big picture" for short-term scientific planning, one of GCMRC's first contracts was for the development of a conceptual model of the river ecosystem (Walters et al. 2000). This model combines submodels of flow, sediment transport, habitat, food resources, and population dynamics of selected species, particularly trout and endangered humpback chub, and is stepped through numerous short reaches of the river (fig. 4-3). The model successfully demonstrates interactive responses of aquatic resources to dam operations alternatives and economics; however, model development reveals that three decades of fisheries research data have not been compiled or analyzed to provide a baseline for monitoring the alternative impacts of different flow regimes. In fact, these fisheries data may not be useful for monitoring purposes. Also, the model is not calibrated for riparian resources, which are more strongly influenced by reach-based geomorphology and microsite conditions: integrated local landscape modeling will provide a better description of riparian processes.

Several other syntheses have been undertaken. Results of the 1996 experimental flood were presented in a GCMRC-coordinated symposium in 1997 and in three edited formats (Schmidt et al. 1999a; Webb et al. 1999b; Patten and Stevens 2001). Also, a symposium on the history and management of the trophy rainbow trout fishery in the Glen Canyon tailwaters reach was held in spring 2000 and publication of results is

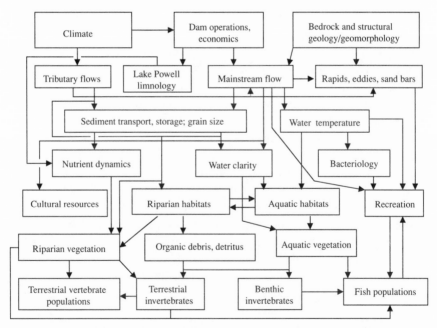

Figure 4-3. Colorado River ecosystem component interactions modeled by Walters et al. (2000). The model couples a river continuum approach (cumulative downstream sediment and water-quality changes related to dam and tributary discharges) with population growth models of key species, particularly fish. Only a few of the many ecosystem interactions are indicated on this simplified diagram.

anticipated. Several Native American tribes have conducted syntheses of cultural issues, ethnology, and traditional ecological knowledge (e.g., Lomaomvaya et al. 2001). River recreational user group preferences and safety issues have been synthesized by Myers et al. (2000) and Stewart et al. (2001).

Competitive, Peer-Reviewed Scientific Monitoring and Research

The collection and analysis of credible, integrated ecosystem information is a primary objective for GCMRC and the AEM program. The present monitoring program involves sampling of numerous resources at different time frames and with varying intensity, depending on the resource in question, and is designed to provide an integrated evaluation of physical, biological, and cultural-socioeconomic resource changes in relation to managed and natural state variables (e.g., climate, flow, channel geometry; table 4-1). Monitoring must take place on a schedule appropriate for the sensitivity of individual resources. For example, the benthos responds to dewatering events as short as one hour (Angradi and Kubly 1994), and monitoring frequency may best be coupled to changes in the flow regime. In contrast, upper riparian zone vegetation is dominated by long-lived tree species (i.e., *Prosopis juliflora, Celtis reticulata*) that only slowly respond to dam operations (Pucherelli 1988; Salzer et al. 1996); a decadal or longer monitoring frequency may be most appropriate for that vegetation. Research is regarded as a necessary element of the program to evaluate monitoring frequency or complex interactions, such as interactions of flow dynamics on competitive relationships between native and nonnative fish.

The GCMRC monitoring program is conducted by competitively awarded contracts. An independent panel of experts on individual topic areas is convened to review the submitted proposals. This process requires periodically assembling and coordinating a group of independent reviewers in each of the major disciplines, processing the results of the reviews, working with the awardees to refine research methods, finalizing the contracts, and coordinating logistics and the receipt of deliverables. Logistical and administrative constraints associated with working in a remote region of a heavily regulated national park may limit the number of new scientists brought into the system. But in time, an open, competitive research approach is likely to attract the best expertise and produce the highest-quality scientific information.

Table 4-1. Annotated bibliographic overview and present monitoring of physical, biological, and cultural resources of the Colorado River ecosystem affected by Glen Canyon Dam.

Resource	Subcategory	References	Monitored components	Monitoring frequency
Climate	General	Sellers et al. 1985	Phantom Ranch temperature and precipitation	Daily minimum and maximum
Physical resources	Geomorphology	Howard and Dolan 1981; Schmidt 1990; Schmidt and Graf 1990; Schmidt and Rubin 1995; Melis 1997; Schmidt et al. 1999a,b; Webb et al. 1989, 1999a,b	Aerial imagery at 0.5-meter contour accuracy, oblique photos, bathymetric mapping	Annual to decadal
	Flow	Kieffer 1985; Andrews 1991; O'Conner et al. 1994; Graf 1995; Wiele and Smith 1996; Smith 1999	Lees Ferry, Little Colorado River, Phantom Ranch and Diamond Creek mainstream gauges; Paria, LCR, Kanab Creek and Havasu Creek tributaries	15-minute intervals to erratic
	Water chemistry	Cole and Kubly 1976; Stanford and Ward 1991; Stevens et al. 1997b; Marzolf et al. 1999		15 minutes to more than annual or sporadic
	Sediment	Andrews 1991; Graf et al. 1991; Melis 1997; Rubin et al. 1998; Hazel et al. 1999; Wiele et al. 1999; Topping et al. 2000a,b	Sandbar erosion at thirty-six sites; GIS mapping from aerial images	Annual to decadal
	Flooding	Webb et al. 1999b; Patten and Stevens 2001	Debris flow fans, geomorphology	Irregular intervals

continues

Table 4-1. (*continued*)

Resource	Subcategory	References	Monitored components	Monitoring frequency
Physical resources (*continued*)	Lake Powell limnology	Potter and Drake 1989; Stanford and Ward 1991; Marzolf et al. 1998; Hueftle and Stevens 2001	Limnology of reservoir's main axis	Seasonal
Aquatic biology	Benthos	Blinn and Cole 1991; Angradi 1994; Blinn et al. 1995; Stevens et al. 1997b; Stevens et al. 1998	Benthic algae and invertebrates in mainstream and tributaries	Seasonal in mainstream, more frequent at LF; annual in tributaries
	Drift	Angradi and Kubly 1994; Shannon et al. 1996; Walters et al. 2000		Seasonally
	Productivity	Marzolf et al. 1999; Walters et al. 2000	Gross and net 1° and 2° productivity	Sporadic
	Fish	Minckley 1991; Walters et al. 2000; see Endangered species (below)	Endangered and nonendangered native and nonnative fish and fish habitats	Monthly to seasonally, life stages and spawning
	Waterbirds	Brown et al. 1989; Brown and Stevens 1997; Stevens et al. 1997a; Brown et al. 1998	Waterfowl, waders, shorebirds, fish-eating raptors	Monthly to seasonally
Terrestrial ecosystem	Vegetation	Carothers et al. 1979; Turner and Karpiscak 1980; Warren et al. 1982; Stevens and Waring 1985; Phillips et al. 1987; Anderson and Ruffner 1988; Pucherelli 1988; Stevens 1989; Johnson 1991; Stevens et al. 1995; Salzer et al. 1996; Mast and Waring 1997; Stevens and Ayers 2002; Stevens et al. 2001	Riparian zone vegetation cover from aerial photography	Annual to decadal at reference sites

continues

	Terrestrial invertebrates	Stevens 1976; Spamer and Bogan 1993	Spot and quantitative collections	Rarely performed
	Herpetofauna	Miller et al. 1982; Warren and Schwalbe 1985; Drost and Sogge 1995	Spot and quantitative collections, observations	Rarely performed
	Birds	Brown et al. 1987, 1994; LaRue et al. 2001	Observations at sentinel sites	Biweekly to seasonal
	Mammals	Ruffner et al. 1978; Suttkus et al. 1978	Spot and quantitative collections, observations	Rarely performed
Endangered species	Kanab ambersnail (Succineidae: *Oxyloma haydeni kanabensis*)	Spamer and Bogan 1993; Meretsky et al. 2000	Systematic sampling at Vasey's Paradise	Monthly to seasonal
	Humpback chub (Cyprinidae: *Gila cypha*)	Minckley 1991; Valdez and Ryel 1997; Valdez and Carothers 1998	Different life stages, spawning and habitat	Monthly to seasonally, life stages and spawning
	Colorado pikeminnow (Cyprinidae: *Ptychocheilus lucius*)	Minckley 1991; Valdez and Carothers 1998	Extirpated—searches continue through fisheries sampling	Monthly to seasonally, life stages and spawning
	Razorback sucker (Catostomidae: *Xyrauchen texanus*)	Minckley 1991; Valdez and Carothers 1998	Extirpated—searches continue through fisheries sampling	Monthly to seasonally, life stages and spawning
	Bonytail chub (Cyprinidae: *Gila elegans*)	Minckley 1991; Valdez and Carothers 1998	Extirpated—searches continue through fisheries sampling	Monthly to seasonally, life stages and spawning

Table 4-1. (*continued*)

Resource	Subcategory	References	Monitored components	Monitoring frequency
Endangered species (*continued*)	Southwestern willow flycatcher (Tyrannidae: *Empidonax trailii extimus*)	Brown 1988; Sogge et al. 1997; Stevens et al. 2001	Migration, nesting success	Biweekly to monthly
	Other species of concern (leopard frogs, bald eagle, peregrine falcon, etc.)	Brown et al. 1989, 1992, 1998; Johnson 1991; Drost and Sogge 1995; Brown and Stevens 1997; Stevens et al. 1997a, 2001	Sampling conducted as part of other studies	Biweekly–decadal
Cultural resources	Archeology	Fairley et al. 1989a,b, 1994; Hereford et al. 1993	Cultural site erosion and associated geomorphology	Semi–annual (biennial for each site)
	Ethnology	Stoffle and Evans 1978; Mast and Waring 1997; Lomaomvaya et al. 2001	Culturally significant species	Annual for each tribe
	Cultural properties	Fairley 1994; Lomaomvaya et al. 2001; Stevens et al. 2001; Balsom 1999	Cutural and historic sites, tribal outreach	Annual for each tribe
	History	Stegner 1954; Hughes 1978; Rusho and Crampton 1981; Lavender 1985; Smith and Crampton 1987; Reilly 1999; Spamer annual bibliographic additions	Historical information	Sporadic
Recreation	Visitor experience and use levels	Shelby et al. 1992; Walters et al. 2000; Stewart et al. 2001	User–day	Daily by National Park Service and AGFD, rarely by GCMRC

120

		References	Focus	Monitoring
	Safety	Brown and Hahn-O'Neil 1988; Myers et al. 1999	Visitor injury and equipment damage	By incident reports, National Park Service
	Environmental impacts	Kearsley et al. 1994	Sandbar area	Annual Adopt-a-Beach Program and Northern Arizona University Sandbar Erosion Program
Economic resources	Hydroelectric power	U.S. Bureau of Reclamation 1995; Harpman 1999; Walters et al. 2000	Power production	Hourly by Bureau of Reclamation
	Water storage	U.S. Bureau of Reclamation 1995; Harpman 1999	Lake Powell and downstream stage	Daily and 15-minute intervals, respectively
	Non-use values	Bishop and Welsh 1992a,b	Cost analysis	None
	Recreation	Bishop et al. 1992a; Stewart et al. 2001	Cost analysis	Sporadic
Integrated ecosystem studies	All resources	Carothers and Aitchison 1976; Carothers and Minckley 1981; Carothers and Brown 1991; Johnson 1991; Angradi 1994; Schmidt et al. 1998	Integration studies	Sporadic

Protocol Evaluation

Technological and conceptual advances, and changing information and management needs are likely to result in evolution of strategies, information needs, and monitoring and research protocols over time. Therefore, GCMRC designed a protocol evaluation program to determine how resources of concern are best monitored and when to update monitoring technology. This program is intended to advise the Adaptive Management Work Group (AMWG) and GCMRC regarding specific monitoring protocols to be used in the ecosystem. Protocol evaluation program workshops and evaluations are being conducted in cooperation with external experts identified through a nationwide scoping and competitive selection process, as well as GCMRC science cooperators, contractors, and the Technical Workshop Group.

Details on the protocol evaluation program results are discussed with the AMWG and reviewed by the science advisory board. Conclusions about the program's process are conveyed through the requests for proposals (RFPs) to contractors. GCMRC plans to conduct pilot monitoring tests, where appropriate, to determine the effectiveness of modified protocols and their applicability to previously collected information, and GCMRC is committed to ensuring that all monitoring data sets are comparable to the greatest extent possible with previously collected information. The Program's evaluation of current and alternative methodologies in all program resource areas is scheduled for completion by the end of fiscal year 2002. Program review workshops have been held on such topics as Lake Powell limnological studies, remote sensing, physical-resource monitoring, terrestrial biology, and cultural resources. Reports on the results of those reviews are presently being evaluated by GCMRC, the Technical Work Group, and the AMWG. GCMRC plans to reinitiate a protocol evaluation program in each resource area on a regular (ca. five-year) basis to maintain the highest-quality long-term monitoring program. Overall, this process will lead to the most credible long-term monitoring program.

Project and Program Review

Scientific monitoring and research is most credible when it is subject to rigorous, independent peer review. The GCMRC program welcomes and incorporates independent peer review at several levels, including review of project proposals and reports, and both GCMRC and AEM program directions.

Review at the project level includes independent peer review of RFPs and project reports. Two RFP review panels have been successfully completed and translated into on-the-ground monitoring and research projects. Review of project reports involves several levels, including review by GCMRC, independent reviewers, and the Technical Work Group and Adaptive Management Work Group. The success of this review process has varied, with some reports proceeding quickly through to peer-reviewed publication and others remaining without review for lengthy periods. More consistency is desirable in this process.

Scientific oversight of the Glen Canyon Environmental Studies Program and the first stages of the present adaptive ecosystem management program were provided by the National Research Council (NRC 1987, 1991, 1996, 1999). The most recent NRC review (1999) recommended numerous changes, including further external review and inclusion of a senior scientist on the GCMRC staff. This most recent review has undergone a lengthy administrative evaluation by the Adaptive Management Work Group.

Program review of the Colorado River AEM process will be overseen by a just-formed Independent Review Panel (IRP; fig. 4-2). The Panel reports to the Secretary of the Interior on the program, including the AMWG's progress and GCMRC's oversight of the monitoring and research program. Until now, the lack of program review has involved added costs, interruption of ecosystem and AEM plan development, and confusion among stakeholders over values and program direction. Two sets of administrative decisions are described below that will benefit from more thorough external program oversight.

The first set of administrative decisions involves flood-triggering criteria approved by the AMWG in 1998, but numerous questions remain regarding the magnitude, duration, frequency, and timing of planned floods (Webb et al. 1999b; Patten et al. 2001). In particular, shorter-duration, higher-magnitude floods have been proposed by sediment transport researchers to increase sand residence time (Schmidt et al. 1999b), whereas, higher-flow, longer-duration flows may be necessary to rejuvenate fish nursery habitats. Stakeholders have been concerned about the effects of high monthly releases during years when planned floods occur, particularly when discharges approach power-plant capacity (940 cubic meters per second). Planned flood "triggering criteria" were developed in 1998 by the Technical Working Group to define the timing of spills. At present, there is an approximately one-in-three chance of a planned flood being triggered between January and June in any year when the reservoir is at its target storage level of 26.51 cubic kilometers by 1 January. Flow-triggering criteria may be useful for stakeholder planning when our understanding of ecosystem responses to high flows is robust; however, definition of triggering criteria is premature at this time because the understanding of how to use high flows has not been resolved, and the triggering criteria do not provide sufficient time for planning scientific activities.

The second set of administrative decisions was made by the U.S. Fish and Wildlife, which concluded, based on information from the early 1990s, that steady flows during the summer months would benefit native fish by increasing backwater nursery habitat area. This view was formalized in the Glen Canyon Dam EIS, which requires low, steady summer flows in five of the next ten low inflow years. Much uncertainty exists concerning the benefits of these low, steady flows on nonnative fish populations; indeed, the impacts of nonnative fish may outweigh the habitat benefits derived from low flows. Such flows promote establishment of nonnative saltcedar (*Tamarix ramosissima*) and other invasive plant species in the lower riparian zone (Stevens et al. 2001). In addition, implementation of summertime low flows may reduce dam revenues by up to 50 percent, because the Western Area Power Administration must purchase power from other power producers, a market that was recently deregulated. A scientific oversight committee would likely recognize these issues before committing the entire ecosystem to potentially harmful treatments. However, without such oversight, the AMWG abided by the U.S. Fish and Wildlife recommendations, implementing summertime steady low flows in 2000. Recent observations by the senior author indicate that extensive saltcedar establishment occurred throughout the river corridor from the year 2000 flows, the largest such establishment pulse of this nonnative plant species in Grand Canyon since at least 1983. A monitoring program has been implemented, but insufficient time series data exist to determine whether the 2000 flows will have a negative, neutral, or positive impact on key native and nonnative fish species (Walters et al. 2000). Recent concern over potential population declines of endangered humpback chubs has stimulated plans for nonnative fish control using winter floods and highly fluctuating flows.

Information Management and Archiving

The NRC has recommended intensive information management and archival in each of its program reviews since 1987; however, only recently has the administrative structure been established for electronic information management. GCMRC inherited a comprehensive collection of hardcopy reports, administrative documents, publications, maps, and photographs from the Glen Canyon Environmental Studies Program and has maintained that library. However, no central repository for data has existed until recently, so that most of the data from hundreds of projects still reside with individual researchers or agencies. The present strategy is to systematically acquire and check those data for accuracy, and then place them in both a geographic information system and a database management system. Georeferencing remains a substantial challenge for researchers in this remote ecosystem, and GCMRC has continued the robust surveying and georeferencing efforts implemented by the Glen Canyon Environmental Studies Program. The task of compiling all historical data is impossible: some early researchers have retired or died, and information from which numerous early reports were produced has been lost. Nevertheless, the large-scale GCMRC information management program will acquire many critical data over time. A continuing problem facing the information management program is that some researchers are reluctant to turn over their data even after it is published, despite the public funding of their work.

Information Transfer

Education of the AMWG stakeholders, the larger scientific community, and the public is necessary to clarify which changes in the ecosystem are dam-related and to what extent the affected resources can be manipulated by altering flow regimes. This education outreach takes place through several mechanisms. GCMRC and the Bureau of Reclamation have assisted the stakeholders in developing electronic administrative communications; however, some tribal entities are still not electronically linked to the information system. GCMRC has initiated several outreach efforts to facilitate information flow and inform the stakeholders and the public about the river ecosystem and its management. It conducted a scientific symposium on the results of the 1996 planned flood in Grand Canyon, assisted in two groups of publications on that experiment (Webb et al. 1999b; Patten and Stevens 2001). In addition, GCMRC convened symposia on Colorado River monitoring and research results in 1999, 2001, and 2002, and a symposium on the Lees Ferry trout fishery in 2000. In response to stakeholder requests, GCMRC provides an online account of current resource conditions in the ecosystem. The State of the Colorado River Ecosystem Report (www.gcmrc.gov) provides information on current physical, aquatic biological, terrestrial biological, cultural and socioeconomic resource conditions over time, and especially related to 1996–1999 flows. This report fulfills part of the requirements of the 1992 Grand Canyon Protection Act, as well as some requirements of the Glen Canyon Dam EIS (1995) and the 1996 Record of Decision.

Improving Future Adaptive Ecosystem Management

The scientific information gathered by GCMRC and presented to the AMWG is designed to improve ecosystem planning, management, and monitoring. This feedback is essential for adaptive ecosystem management as a "learning by doing" process (Walters and Holling 1990). Although the rudiments of this feedback process have been undertaken, several challenges to program improvement remain. Importantly, there is presently no agreement to revisit and update the 1996 Record of Decision. Although substantial individual tests, such as the 1996 and 1997 planned floods and the 2001 test of Biological Opinion flows, can be undertaken as experiments, no administrative mechanism exists to feed the conceptual and management findings and improvements back into the overall AEM program. Modification of the Record of Decision may require another environmental impact statement, a process that the AMWG may need to consider on an infrequent (e.g., decadal) basis. Also, the many embedded, poorly defined and conflicting biases within the Record of Decision and held by various stakeholder and research groups creates a social context that can overwhelm science-based adaptive ecosystem management.

Conclusions

Adaptive management of the Colorado River ecosystem affected by Glen Canyon Dam is a work in progress. GCMRC, which oversees AEM science elements, is presently concluding its fifth year and has made considerable headway in many of the administrative and science areas described above. These efforts and the continuing programmatic development promise to make monitoring and research information readily available to the AMWG and the public to improve future adaptive ecosystem management of the Colorado River ecosystem.

In this chapter, we described eight elements that are required to monitor the progress of adaptive management of the Colorado River ecosystem, including (1) definition of goals and objectives; (2) synthesis of existing information; (3) competitive, peer-reviewed monitoring and research projects, (4) rigorously peer-reviewed protocols; (5) independent review of project and program approaches and results; (6) information management and archival; (7) education of stakeholders, the scientific community, and the public; and (8) feedback of information into the AEM process. GCMRC's overall approach is to monitor river ecosystem resources at a rate and intensity appropriate for the sensitivity of the given resource and to fund research proposals (both solicited and unsolicited proposals) that independent review panels consider worthwhile. This approach is based on the stakeholder's perceived information needs and is also driven by stakeholder demands. The 1999 National Research Council report recommended improved scientific oversight of the overall program, TWG and AMWG proposals, and GCMRC itself to improve the scientific credibility of scientific monitoring. Given the present structure and level of funding, we expect that this program will continue to improve adaptive ecosystem management through time, particularly when all review elements are implemented.

LITERATURE CITED

Anderson, L. S., and G. A. Ruttner. 1988. *Effects of post–Glen Canyon Dam flow regime on the old high water zone plant community along the Colorado River in Grand Canyon.* National Technical Information Service PB 88-183504/AS. Washington, D.C.: United States Government Printing Office.

Andrews, E. D. 1991. Sediment transport in the Colorado River basin. Pp. 54–74 in *Colorado River ecology and dam management.* Washington, D.C.: National Academy Press.

Angradi, T. 1994. Trophic linkages in the lower Colorado River: Multiple stable isotope evidence. *Journal of the North American Benthological Society* 13:479–495.

Angradi, T. R., and D. M. Kubly. 1994. Concentration and transport of particulate organic matter below Glen Canyon Dam on the Colorado River, Arizona. *Journal of the Arizona-Nevada Academy of Sciences* 28:12–22.

Balsom, J. R. 1999. Cultural resources and the Glen Canyon Dam–Colorado River experimental flow of 1996. Pp. 183–194 in *Proceedings of the Fourth Biennial Conference of Research on the Colorado Plateau,* edited by C. van Riper III and M. A. Stuart. U.S. Geological Survey/FRESC Report Series USGSFRESC/COPL/1999/16.

Billingsley, G. H., E. E. Spamer, and D. Menkes. 1997. *Quest for the pillar of gold: The mines and miners of the Grand Canyon.* Grand Canyon: Grand Canyon Association.

Bishop, R. C., and M. P. Welsh. 1992a. Strategy for estimating total value: A case study involving Grand Canyon resources. In *Benefits and costs in natural resources planning, fifth interim report,* edited by R. B. Retting. Corvallis, Ore.: USDA Western Regional Research Publication W-133.

———. 1992b. Existence values in benefit-cost analysis and damage assessment. *Land Economics* 68:405–417.

Blinn, D. W., and G. A. Cole. 1991. Algal and invertebrate biota in the Colorado River: Comparison of pre- and post-dam conditions. Pp. 102–123 in *Colorado River ecology and dam management.* Washington, D.C.: National Academy Press.

Blinn, D. W., J. P. Shannon, L. E. Stevens, and J. P. Carder. 1995. Consequences of fluctuating discharge for lotic communities. *Journal of the North American Benthological Society* 14:233–248.

Brouder, M. J., and T. L. Hoffnagle. 1996. Distribution and prevalence of the Asian tapeworm, *Bothriocephalus acheilognathi,* in the Little Colorado River and tributaries, Grand Canyon, Arizona, including two new host records. *Journal of the Helminthological Society of Washington* 64:219–226.

Brown, B. T. 1988. Breeding ecology of a willow flycatcher population in Grand Canyon, Arizona. *Western Birds* 19:25–33.

Brown, B. T., S. W. Carothers, and R. R. Johnson. 1987. *Grand Canyon birds.* Tucson: University of Arizona Press.

Brown, B. T., S. W. Carothers, R. R. Johnson, M. M. Riffey, and L. E. Stevens. 1994. *Checklist of the birds of the Grand Canyon region.* Grand Canyon: Grand Canyon Natural History Association.

Brown, B. T., R. Mesta, L. E. Stevens, and J. Weisheit. 1989. Changes in winter distribution of bald eagles along the Colorado River in Grand Canyon, Arizona. *Journal of Raptor Research* 23:110–113.

Brown, B. T., G. S. Mills, R. L. Glinski, and S. W. Hoffman. 1992. Density of nesting peregrine falcons in Grand Canyon National Park, Arizona. *Southwestern Naturalist* 37:188–193.

Brown, B. T., and L. E. Stevens. 1997. Winter bald eagle distribution is inversely correlated with human activity along the Colorado River, Arizona. *Journal of Raptor Research* 31:7–10.

Brown, B. T., L. E. Stevens, and T. A. Yates. 1998. Influences of fluctuating river flows on bald eagle foraging behavior. *Condor* 100:745–748.

Brown, C. A., and M. G. Hahn-O'Neil. 1988. *The effects of flows in the Colorado River on reported and observed boating accidents in Grand Canyon.* Washington, D.C.: National Technical Information Service PB88-183553.

Carothers, S. W., and S. W. Aitchison, eds. 1976. *An ecological survey of the riparian zone of the Colorado River between Lees Ferry and the Grand Wash Cliffs, Arizona.* Grand Canyon, Ariz.: National Park Service, Colorado River Research Series Technical Report No.10.

Carothers, S. W, S. W. Aitchison, and R. R. Johnson. 1979. Natural resources, white-water recreation and river management alternatives on the Colorado River, Grand Canyon National Park, Arizona. Pp. 253–260 in *Proceedings of the First Conference on Scientific Research in the National Parks.* Vol. 1. Washington, D.C.: U.S. National Park Service.

Carothers, S. W., and B. T. Brown. 1991. *The Colorado River through Grand Canyon: Natural history and human change.* Tucson: University of Arizona Press.

Carothers, S. W., and C. O. Minckley. 1981. *A survey of the aquatic flora and fauna of the Grand Canyon.* U.S. Water and Power Service, Boulder City, Nev.

Clover, E. U., and L. Jotter. 1944. Floristic studies in the canyon of the Colorado and tributaries. *American Midland Naturalist* 32:591–642.

Cole, G. A., and D. M. Kubly. 1976. *Limnologic studies on the Colorado River from Lees Ferry to Diamond Creek.* Colorado River Research Studies Contribution N. 37, Grand Canyon National Park, Grand Canyon, Ariz.

Connell, J. H. 1978. Diversity in tropical rain forests and coral reefs. *Science* 199:1302–1310.

Davis, G. E. 1993. Design elements of monitoring programs: The necessary ingredients for success. *Environmental Monitoring and Assessment* 26:99–105.

Drost, C. A., and M. K. Sogge. 1995. Preliminary survey of leopard frogs in Glen Canyon National Recreation Area. Pp. 239–254 in *Proceedings of the Second Biennial Conference on Research in Colorado Plateau National Parks,* edited by C. van Riper III. National Park Service Transactions and Proceedings Series NPS/NRNAU/NRTP-95-11.

Fairley, H. C. 1989a. Prehistory. Pp. 85–152 in *Man, models and management: An overview of the archeology of the Arizona Strip and the management of its cultural resources,* edited by J. H. Altschul and H. C. Fairley. U.S. Forest Service and U.S. Bureau of Land Management Report 1989-0-673-212/3028. Washington, D.C.: Government Printing Office.

————. 1989b. History. Pp. 153–218 in *Man, models and management: An overview of the archeology of the Arizona strip and the management of its cultural resources*, edited by J. H. Altschul and H. C. Fairley. U.S. Forest Service and U.S. Bureau of Land Management Report 1989-0-673-212/3028. Washington, D.C.: Government Printing Office.

Fairley, H. C., ed. 1994. *The Grand Canyon River Corridor Survey Project: Archaeological survey along the Colorado River between Glen Canyon Dam and Separation Rapid*. U.S. Bureau of Reclamation Glen Canyon Environmental Studies Program Cooperative Agreement No. 9AA-40-07920. Flagstaff, Ariz.

Gore, J. A., and F. D. Shields Jr. 1995. Can large rivers be restored? *BioScience* 45:142–152.

Graf, J. B. 1995. Measured and predicted velocity and longitudinal dispersion at steady and unsteady flow, Colorado River, Glen Canyon Dam to Lake Mead. *Water Resources Bulletin* 31:265–281.

Graf, J. B., R. H. Webb, and R. Hereford. 1991. Relation of sediment load and flood-plain formation to climatic variability, Paria River drainage basin, Utah and Arizona. *Geological Society of America Bulletin* 103:1405–1415.

Gunderson, L. H., C. S. Holling, and S. S. Light. 1995. Barriers and bridges to learning in a turbulent human ecology. Pp. 461–488 in *Barriers and bridges to the renewal of ecosystems and institutions*, edited by L. H. Gunderson, C. S. Holling, and S. S. Light. New York: Columbia University Press.

Harpman, D. A. 1999. The economic cost of the 1996 controlled flood. Pp. 351–357 in *The controlled flood in Grand Canyon. Geophysical Monograph 110*, edited by R. H. Webb, J. C. Schmidt, G. R. Marzolf, and R. A. Valdez. Washington, D.C.: American Geophysical Union.

Hazel, J. E., Jr., M. Kaplinski, R. Parnell, M. Manone, and A. Dale. 1999. Topographic and bathymetric changers at thirty-three long-term study sites. *American Geophysical Union Geophysical Monograph* 110:161–183.

Hereford, R., H. C. Fairley, K. S. Thompson, and J. R. Balsom. 1993. *Surficial geology, geomorphology, and erosion of archaeological sites along the Colorado River, eastern Grand Canyon National Park, Arizona*. U.S. Geological Survey Open-File Report 93-517. Washington, D.C.: U.S. Government Printing Office.

Holling, C. S. 1978. *Adaptive environmental assessment and management*. London: John Wiley and Sons.

Howard, A. D., and R. Dolan. 1981. Geomorphology of the Colorado River in the Grand Canyon. *Journal of Geology* 89:269–297.

Hueftle, S., and L. E. Stevens. 2001. Experimental flood effects on the limnology of Lake Powell Reservoir, southwestern USA. *Ecological Applications* 11:635–643.

Hughes, J D. 1978. *In the house of stone and light*. Grand Canyon: Grand Canyon Natural History Association.

Huston, M. 1979. A general hypothesis of species diversity. *American Naturalist* 113:81–101.

Johnson, B. L., W. B. Richardson, and T. J. Naimo. 1995. Past, present, and future concepts in large river ecology: How rivers function and how human activities influence river processes. *BioScience* 45:134–141.

Johnson, R. R. 1991. Historic changes in vegetation along the Colorado River in the Grand Canyon. Pp. 178–206 in *Colorado River ecology and dam management*, edited by National Research Council. Washington, D.C.: National Academy Press.

Kieffer, S. W. 1985. The 1983 hydraulic jump in Crystal Rapid: Implications for river-running and geomorphic evolution in the Grand Canyon. *Journal of Geology* 93:385–406.

LaRue, C. T., J. R. Spence, L. L. Dickson, L. E. Stevens, and N. L. Brown. 2001. Recent bird records from the Grand Canyon region, 1975–2000. *Western Birds* 32:101–118.

Lavender, D. 1985. *River runners of the Grand Canyon*. Grand Canyon: Grand Canyon Natural History Association.

Lee, K. N. 1991. Unconventional power: Energy efficiency and environmental rehabilitation under the Northwest Power Act. *Annual Review of Energy and the Environment* 16:337–364.

——— 1993. *Compass and gyroscope: Integrating science and politics for the environment*. Washington, D.C.: Island Press.

———. 1995. Deliberately seeking sustainability in the Columbia River basin. Pp. 214–238 in *Barriers and bridges to the renewal of ecosystems and institutions*, edited by L. H. Gunderson, C. S. Holling, and S. S. Light. New York: Columbia University Press.

Light, S. S., L. H. Gunderson, and C. S. Holling. 1995. The Everglades: Evolution of management in a turbulent ecosystem. Pp. 103–168 in *Barriers and bridges to the renewal of ecosystems and institutions*, edited by L. H. Gunderson, C. S. Holling, and S. S. Light. New York: Columbia University Press.

Lomaomvaya, M., T. J. Ferguson, and M. Yeatts. 2001. *Öngtuvqava Sakwtala: Hopi ethnobotany in the Grand Canyon*. Kykotsmovi, Ariz.: Hopi Cultural Preservation Office.

Marzolf, G. R., C. J. Bowser, R. Hart, D. W. Stephens, and W. S. Verhieu. 1999. Photosynthetic and respiratory processes: An open stream approach. *American Geophysical Union Geophysical Monograph* 110:205–215.

Marzolf, G. R., R. J. Hart, and D. W. Stephens. 1998. *Depth profiles of temperature, specific conductance, and oxygen concentration in Lake Powell, Arizona-Utah, 1992–1995*. U.S. Geological Survey Open-file Report 97-835.

Mast, J. N., and G. L. Waring. 1997. Dendrochronological analysis of Goodding willows in Grand Canyon National Park. Pp. 115–128 in *Proceedings of the Third Biennial Conference of Research on the Colorado Plateau*, edited by C. van Riper III and E. T. Deshler. National Park Service Transactions and Proceedings Series NPS/NRNAU/NRTP-97/12.

Melis, T. S. 1997. Geomorphology of debris flows and alluvial fans in Grand Canyon National Park and their influences on the Colorado River below Glen Canyon Dam, Arizona. Ph.D. diss., University of Arizona, Tucson.

Minckley, W. 1991. Native fishes of the Grand Canyon region: An obituary? Pp. 124–177 in *Colorado River Ecology and Dam Management*. Washington, D.C.: National Academy Press.

Miller, D. M., R. A. Young, T. W. Gatlin, and J. A. Richardson. 1982. *Amphibians and reptiles of the Grand Canyon*. Grand Canyon Natural History Association Monograph No. 4, Grand Canyon.

Myers, T., C. Becker, and L.E. Stevens. 1999. *Fateful journey: Injury and death on the Colorado River in Grand Canyon*. Flagstaff: Red Lake Books.

Naiman, R. J., J. J. Magnuson, D. M. McKnight, and J. A. Stanford. 1995. *The freshwater imperative: A research agenda*. Washington, D.C.: Island Press.

Naiman, R. J., and M. G. Turner. 2000. A future perspective on North America's freshwater ecosystems. *Ecological Applications* 10:958–970.

National Research Council. 1986. *Ecological knowledge and environmental problem-solving: concepts and case studies*. Washington, D.C.: National Academy Press.

———. 1987. *River and dam management: A review of the Bureau of Reclamation's Glen Canyon environmental studies*. Washington, D.C.: National Academy Press.

———. 1990. *Managing troubled waters: the role of marine environmental monitoring*. Washington, D.C.: National Academy Press.

———. 1991. *Colorado River ecology and dam management*. Washington, D.C.: National Academy Press.

——. 1996. *River resource management in the Grand Canyon.* Washington, D.C.: National Academy Press.

——. 1999. *Downstream: adaptive management of Glen Canyon Dam and the Colorado River ecosystem.* Washington, D.C.: National Academy Press.

O'Conner, J. E., L. L. Ely, E. E. Wohl, L. E. Stevens, T. S. Melis, V. S. Kale, and V. R. Baker. 1994. A 4,500-year record of large floods on the Colorado River in the Grand Canyon, Arizona. *Journal of Geology* 102:1–9.

Patten, D. T. 1991. Glen Canyon environmental studies research program: Past, present and future. Pp. 239–251 in *Colorado River ecology and dam management.* Washington, D.C.: National Academy Press.

Patten, D. T., and L. E. Stevens. 2001. Restoration of the Colorado River ecosystem using planned flooding. *Ecological Applications* 11:633–634.

Patten, D. T., D. A. Harpman, M. I. Voita, and T. J. Randle. 2001. A managed flood on the Colorado River: Background, design and implementation. *Ecological Applications* 11:635–643.

Phillips, B. G., A. M. Phillips III, and M. A. Schmidt-Bernzott. 1987. *Annotated checklist of vascular plants of Grand Canyon National Park.* Monogram No. 7. Grand Canyon: Grand Canyon Natural History Association.

Poff, N. L., K. D. Allan, M. B. Bain, J. R. Karr, K. L. Prestegaard, B. D. Richter, R. E. Sparks, and J. C. Stromberg. 1997. The natural flow regime: A paradigm for river conservation and restoration. *BioScience* 47:769–784.

Potter, L. D., and C. L. Drake, 1989. *Lake Powell: Virgin flow to dynamo.* Albuquerque: University of New Mexico Press.

Pucherelli, M. 1988. *Evaluation of riparian vegetation trends in the Grand Canyon using multitemporal remote sensing techniques.* U.S. Department of the Interior, Bureau of Reclamation Glen Canyon Environmental Studies Report No. 18, NTIS No. PB88-183488.

Rabbit, M. D. 1975. *A brief history of the U.S. Geological Survey.* Washington, D.C.: U.S. Government Printing Office.

Reilly, P.T. 1999. *Lee's Ferry: From Mormon crossing to national park.* Logan: Utah State University Press.

Rubin, D. M., J. M. Nelson, and D. J. Topping. 1998. Relation of inversely graded deposits to suspended-sediment grain-size evolution during the 1996 flood experiment in Grand Canyon. *Geology* 26:99–102.

Ruffner, G. A., N. J. Czaplewski, and S. W. Carothers. 1978. Distribution of natural history of some mammals from the inner gorge of the Grand Canyon, Arizona. *Journal of the Arizona–Nevada Academy of Science* 13:85–91.

Rusho, W. L., and C. G. Crampton. 1981. *Desert river crossing: Historic Lee's Ferry on the Colorado River.* Santa Barbara, Calif.: Peregrine Smith.

Salzer, M. W., V. A. S. McCord, L. E. Stevens, and R. H. Webb. 1996. The dendrochronology of *Celtis reticulata* in the Grand Canyon: Assessing the impact of regulated river flow on tree growth. Pp. 273–281 in *Tree rings, environment and humanity,* edited by J. S. Dean, D. M. Meko, and T. W. Swetnam. Published by Radicarbon.

Schmidt, J. C. 1990. Recirculating flow and sedimentation in the Colorado River in Grand Canyon, Arizona. *Journal of Geology* 98:709–724.

——. 1999. Summary and synthesis of geomorphic studies conducted during the 1996 controlled flood in Grand Canyon. *American Geophysical Union Geophysical Monograph* 110: 329–341.

Schmidt, J. C., E. D. Andrews, D. L. Wegner, D. T. Patten, G. R. Marzolf, and T. O. Moody. 1999a. Origins of the 1996 controlled flood in Grand Canyon. *American Geophysical Union Geophysical Monograph* 110:23–36.

Schmidt, J. C., and J. B. Graf. 1990. *Aggradation and degradation of alluvial-sand deposits, 1965 to 1986, Colorado River, Grand Canyon National Park, Arizona.* U.S. Geological Survey Professional Paper 1493. Washington, D.C.: U.S. Government Printing Office.

Schmidt, J. C., P. E. Grams, and M. F. Leschin. 1999b, Variation in the magnitude and style of deposition and erosion in three long (8–12 km) reaches as determined by photographic analysis. *American Geophysical Union Geophysical Monograph* 110:185–203.

Schmidt, J. C., and D. M. Rubin. 1995. Regulated streamflow, fine-grained deposits, and effective discharge in canyons with abundant debris fans. Pp. 177–195 in *Natural and anthropogenic influences in fluvial geomorphology* edited by J. E. Costa, A. J. Miller, K. W. Potter, and P. R. Wilcock. American Geophysical Union Geophysical Monograph 89.

Schmidt, J. C., R. H. Webb, R. A. Valdez, G. R. Marzolf, and L. E. Stevens. 1998. Science and values in river restoration in the Grand Canyon. *BioScience* 48:735–747.

Sellers, W. D., R. H. Hill, and M. Sanderson-Rae. 1985. *Arizona climate: The first hundred years.* Tucson: University of Arizona Press.

Shannon, J. P., D. W. Blinn, P. Benenati, and K. P. Wilson. 1996. Organic drift in a regulated desert river. *Canadian Journal of Fisheries and Aquatic Science* 53:1360–1369.

Shelby, B., T. C. Brown, and R. Baumgartner. 1992. Effects of streamflows on river trips on the Colorado River in Grand Canyon, Arizona. *Rivers* 3:191–201.

Smith, D. L., and C. G. Crampton. 1987. *The Colorado River survey: Robert R. Stanton and the Denver, Colorado River & Pacific Railroad.* Salt Lake City: Howe Brothers.

Smith, J. D. 1999. Flow and suspended-sediment transport in the Colorado River near National Canyon. *American Geophysical Union Geophysical Monograph* 110:99–115.

Sogge, M. K., T. J. Tibbitts, and J. R. Petterson. 1997. Status and breeding ecology of the southwestern willow flycatcher in the Grand Canyon. *Western Birds* 28:142–157.

Spamer, E. E., and A. E. Bogan. 1993. Mollusca of the Grand Canyon and vicinity, Arizona: New and revised data on diversity and distributions, with notes on Pleistocene-Holocene mollusks of the Grand Canyon. *Proceedings of the Academy of Natural Sciences of Philadelphia* 144:21–68.

Sparks, R. E. 1995. Need for ecosystem management of large rivers and their floodplains. *BioScience* 45:168–182.

Stanford, J. A., and G. C. Poole. 1996. A protocol for ecosystem management. *Ecological Applications* 6:741–744.

Stanford, J. A., and J. V. Ward. 1991. Limnology of Lake Powell and the chemistry of the Colorado River. Pp. 75–101 in *Colorado River ecology and dam management*, edited by National Research Council. Washington, D.C.: National Academy Press.

Stanford, J. A., J. V. Ward, W. J. Liss, C. A. Frissell, R. N. Williams, J. A. Lichatowish, and C. C. Coutant. 1996. A general protocol for restoration of regulated rivers. *Regulated Rivers: Research and Management* 12:391–413.

Stegner, W. 1954. *Beyond the hundredth meridian.* Boston: Houghton Mifflin.

Stevens, L. E. 1976. An insect inventory of Grand Canyon. Pp. 141–145 and appendix in *An ecological survey of the riparian zone of the Colorado River between Lees Ferry and the Grand Wash Cliffs, Arizona*, edited by S. W. Carothers and S. W. Aitchison. National Park Service, Colorado River Research Series Technical Report No.10. Grand Canyon, Ariz.

———. 1989. Mechanisms of riparian plant community organization and succession in the Grand Canyon, Arizona. Ph.D. diss. Northern Arizona University, Flagstaff.

Stevens, L. E., and T. J. Ayers. 2002. The biodiversity and distribution of alien vascular plant and animals in the Grand Canyon region. Pp. 241–265 in *Exotic species in the Sonoran Desert*, edited by B. Tellman. Tucson: University of Arizona Press.

Stevens, L. E., T. J. Ayers, J. B. Bennett, K. Christensen, M. J. C. Kearsley, V. J. Meretsky, A. M. Phillips III, R. A. Parnell, J. Spence, M. K. Sogge, A. E. Springer, and D. L. Wegner. 2001. Planned flooding and Colorado River riparian trade-offs downstream from Glen Canyon Dam, Arizona. *Ecological Applications* 11:701–710.

Stevens, L. E., K. A. Buck, B. T. Brown, and N. C. Kline. 1997a. Dam and geomorphological influences on Colorado River waterbird distribution, Grand Canyon, Arizona, USA. *Regulated Rivers: Research and Management* 13:151–169.

Stevens, L. E., J. C. Schmidt, T. J. Ayers, and B. T. Brown. 1995. Flow regulation, geomorphology, and Colorado River marsh development in the Grand Canyon, Arizona. *Ecological Applications* 5:1025–1039.

Stevens, L. E., J. P. Shannon, and D. W. Blinn. 1997b. Benthic ecology of the Colorado River in Grand Canyon: Dam and geomorphic influences. *Regulated Rivers: Research and Management* 13:129–149.

Stevens, L. E., J. E. Sublette, and J. P. Shannon. 1998. Chironomidae (Diptera) of the Colorado River in Grand Canyon, Arizona, USA, II: Distribution and phenology. *Great Basin Naturalist* 58:147–155.

Stevens, L. E., and G. L. Waring. 1985. Effects of prolonged flooding on riparian vegetation in Grand Canyon. Pp. 81–86 in *Riparian ecosystems and their management: Reconciling conflicting uses*, edited by R. R. Johnson, C. D. Ziebell, D. F. Patten, P. F. Ffolliott, and R. H. Hamre. General Technical Report GTR-RM-120. U.S. Department of Agriculture, U.S. Forest Service, Fort Collins, Colo.

Stevens, L. E., and D. L. Wegner. 1995. Changes on the Colorado River: Operating Glen Canyon Dam for environmental criteria. Pages 65–74 in *Proceedings of a national symposium on using ecological restoration to meet Clean Water Act goals*, edited by R. J. Kirschner. Chicago: Northeastern Illinois Planning Commission.

Stewart, W., K. Larkin, R. Manning, D. Cole, and J. Taylor. 2001. *Preferences of recreation user groups of the Colorado River in Grand Canyon*. Flagstaff, Colo.: U.S. Geological Survey Grand Canyon Monitoring and Research Center Report.

Stoffle, R. W., and M. J. Evans. 1978. *Kaibab Paiute history: The early years*. Fredonia, Ariz.: Kaibab Paiute Tribe.

Suttkus, R. D., G. H. Clemmer, and C. Jones. 1978. *Mammals of the riparian region of the Colorado River in the Grand Canyon area of Arizona*. Occasional papers of the Tulane University Museum of Natural History, No. 2, Belle Chasse, La.

Topping, D. J., D. M. Rubin, J. M. Nelson, P. J. III, Kinzel, and I. C. Corson. 2000a. Colorado River sediment transport 2: Systematic bed-elevation and grain-size effects of sand supply limitation. *Water Resources Research* 36:543–570.

Topping, D. J., D. M. Rubin, and L. E. Vierra Jr. 2000b. Colorado River sediment transport 1: Natural sediment supply limitation and the influence of Glen Canyon Dam. *Water Resources Research* 36:515–542.

Turner, R. M., and M. M. Karpiscak. 1980. *Recent vegetation changes along the Colorado River between Glen Canyon Dam and Lake Mead, Arizona*. U.S. Geological Survey Professional Paper 1132.

U.S. Bureau of Reclamation. 1995. *Operation of Glen Canyon Dam, Colorado River Storage Project, Arizona, Final Environmental Impact Statement*. Washington, D.C.: U.S. Government Printing Office.

U.S. Fish and Wildlife Service. 1997. Endangered and threatened wildlife and plants: Final determination of critical habitat for the southwestern willow flycatcher; correction. *Federal Register* 62:44228.

Valdez, R. A., and R. J. Ryel. 1997. Life history and ecology of the humpback chub in the Colorado River in Grand Canyon, Arizona. Pp. 3–31 in *Proceedings of the Third Biennial Conference of Research on the Colorado Plateau*, edited by C. VanRiper III and E. T. Deshler. National Park Service Trans Proc. Serv NPS/NRNAU/NRTP 97/12.

Walters, C. J. 1986. *Adaptive management of renewable resources*. New York: Macmillan.

Walters, C. J., and C. S. Holling. 1990. Large-scale management experiments and learning by doing. *Ecology* 71:2060–2068.

Walters, C. J. Korman, L. E. Stevens, and B. D. Gold. 2000. Ecosystem modeling for evaluation of adaptive management policies in the Grand Canyon. *Conservation Ecology* 4:1. [Online at www.consecol.org/vol4/iss2/art1]

Ward, J. V., and J. A. Stanford. 1983. The serial discontinuity concept of lotic ecosystems. Pp. 29–42 in *Ecology of river systems*, edited by T. D. Fontaine and S. M. Bartell. Dordrecht: Dr. W. Junk Publishers.

Warren, P. L., K. L. Reichhardt, D. A. Mouat, B. T. Brown, and R. R. Johnson. 1982. *Vegetation of Grand Canyon National Park*. U.S. National Park Service Cooperative Studies Unit Technical Report Number 9, Tucson, Ariz.

Warren, P. L., and C.R. Schwalbe. 1985. Herpetofauna in riparian habitats along the Colorado River in Grand Canyon. Pp. 347–354 in *Riparian ecosystems and their management: Reconciling conflicting uses*, technical coordinators, R. R. Johnson, C. D. Ziebell, D. R. Patton, P. F. Ffolliott, and R. H. Hamre. First North American Riparian Conference, April 1985, Tucson, Ariz. General Technical Report GTR-RM-120. U.S. Department of Agriculture, Forest Service. Washington, D.C.: U.S. Government Printing Office.

Webb, R. H. 1996. *Grand Canyon: A century of change*. Tucson: University of Arizona Press.

Webb, R. H., T. S. Melis, P. G. Griffiths, J. G. Elliott, T. E. Cerling, R. J. Poreda, T. W. Wise, and J. E. Pizzuto. 1999a. *Lava Falls Rapid in Grand Canyon: Effects of late Holocene debris flows on the Colorado River*. U.S. Geological Survey Professional Paper 1591.

Webb, R. H., P. T. Pringle, and G. R. Rink. 1989. *Debris flows from tributaries of the Colorado River, Grand Canyon National Park, Arizona*. U.S. Geological Survey Professional Paper 1492.

Webb, R. H., J. C. Schmidt, G. R. Marzolf, and R. A. Valdez, eds. 1999b. The controlled flood in Grand Canyon. *American Geophysical Union Geophysical Monograph* 110, Washington, D.C.

Wiele, S. M., E. D. Andrews, and E. R. Griffin. 1999. The effect of sand concentration on depositional rate, magnitude, and location in the Colorado River below the Little Colorado River. *American Geophysical Union Geophysical Monograph* 110:131–145.

Wiele, S. M., and J. D. Smith. 1996. A reach-averaged model of diurnal discharge wave propagation down the Colorado River through the Grand Canyon. *American Geophysical Union Water Resources Research* 32:1375–1386.

CHAPTER 5

Science Strategy for a Regional Ecosystem Monitoring and Assessment Program: The Florida Everglades Example

John C. Ogden, Steven M. Davis, and Laura A. Brandt

Regional ecosystem restoration programs require a high level of scientific support if the policy and management decisions that are crucial to the long-term success of such programs are to be effectively "science-based." Given the large and complex scales and goals of regional restoration programs, new and more-structured strategies for meeting the science objectives must be developed. The objective of this strategy must be to develop effective means for organizing and applying both current and future understanding of natural and human systems to the planning, evaluation, and assessment phases of the restoration plan.

For "science" to have a strong, guiding role in restoration processes such as that for the Everglades, it must answer a number of questions that are essential to the successful design and implementation of a systemwide restoration program. These questions are:

- How are broadly stated restoration goals and objectives converted into specific, measurable targets?
- What constitutes success?
- How are immense amounts of existing technical and scientific information converted into formats that effectively support restoration planning?

- What are the measures, from among thousands of potential indicators of natural and human systems, that should be used to assess the success of the program?
- How do we minimize the uncertainties that are inherent in all natural and human systems?
- How do we assess and improve the plan during the extended implementation period?
- How do we attract a broad range of scientists and institutions/agencies to participate in an integrated science application process in support of the Everglades restoration plan?

These questions are not simple to answer in cases where

- The temporal and spatial scales of restoration are large, as with regional-scale ecosystem initiatives
- Knowledge of ecosystem parameters and relationships is incomplete
- Uncertainty regarding ecosystem responses is a major consideration
- Existing information is widely scattered in place and time
- Focused efforts are required to identify areas of scientific agreement and disagreement
- Responsibility for managing and protecting the natural resources is divided among many different agencies and jurisdictions

This chapter describes an "applied science strategy" developed by teams of south Florida scientists and resource specialists to address the above questions. The intent of the strategy is to provide the high level of scientific support required for planning, implementing, and assessing a systemwide Comprehensive Everglades Restoration Plan (CERP). Key components of the applied science strategy are a set of conceptual ecological models for the major natural physiographic regions in south Florida, restoration performance measures, an integrated, systemwide monitoring and assessment plan, and an adaptive assessment protocol. In this chapter we provide a brief description of the need and purpose of the Comprehensive Everglades Restoration Plan and describe the context and role for each of these components of the science strategy. One of the conceptual models (that of the Everglades ridge and slough system) is used to illustrate the application of the strategy.

The Comprehensive Everglades Restoration Plan

The greater Everglades landscape prior to human impacts was an expansive and hydrologically interconnected mosaic of freshwater marshes, tree island hammocks, cypress swamps, pinelands, mangrove swamps, lakes and streams, and broad estuarine lagoons (Gunderson and Loftus 1993; Browder and Ogden 1999). The true Everglades portion of this system covered an area of about 1.2 million hectares in a total south Florida wetland basin of 3.6 million hectares. The physical and biological characteristics of the greater Everglades that combined to define this ecosystem included the region's large spatial extent, strong seasonal and annual variation in rainfall and in surface water volumes and flow patterns, naturally low nutrient freshwater marshes and highly productive estuaries, and the great abundance of large vertebrates such as alligators and wading birds (Davis and Ogden 1994; Holling et al. 1994; Ogden et al. 1999).

The pristine condition has been greatly altered, primarily since the 1940s, by a rapidly growing and expanding human population in south Florida. A massive "Central and South Florida Project" (C&SF Project) was authorized by U.S. Congress in the late 1940s with three objectives (Light and Dineen 1994). These were (1) to ensure an adequate supply of water for the growing human systems in south Florida, (2) to provide adequate water for the natural systems, and (3) to greatly improve and expand flood protection for the urban and agricultural components of the region. Construction of approximately 1,000 miles of canals and levees and 150 water control structures occurred primarily between the late 1940s and the 1970s. A major strategy of the C&SF Project was to meet the water supply needs of the region by storing a portion of the summer rains in Lake Okeechobee and in large Everglades impoundments ("Water Conservation Areas"), and draining much of the "surplus" rainfall through canals out to sea. Ridding the region of this water served to greatly reduce flooding in cities and farms.

It is now widely acknowledged that the C&SF Project largely achieved two of its three objectives during the period from the 1950s through the 1980s. Higher priorities for the human systems, and only a primitive understanding of the ecology of the greater Everglades, were factors that contributed to the deterioration of the natural systems as C&SF projects were completed.

By the late 1980s the list of ecological ills in the greater Everglades was long enough to capture the concern of the region and the world (Science Subgroup 1993; U.S. Army Corps of Engineers 1994). Prominent on the list were

- Fifty percent reduction in the spatial extent of Everglades marshes and a 90 percent reduction in numbers of nesting wading birds
- Die-offs by tens of thousands of hectares of seagrass beds in Florida Bay (mainly *Thallassia*)
- Loss of over half of the hardwood tree islands in the Everglades marshes

- A rapid expansion of dense stands of cattails into more than 10,000 hectares of former sawgrass and slough marshes
- Over sixty federally listed species in south Florida
- Predictions of a total collapse in the ecological "health" of several major components of the greater Everglades system, including Lake Okeechobee, extensive marl prairies, and several estuaries

In addition, the urban and agricultural communities began to realize that a project designed in the 1940s and 1950s to meet the water supply and flood protection needs of 2 million people was increasingly failing to meet the needs of 6 million people in the 1990s. In response to this "crisis," federal and state authorities during the 1990s created a new, comprehensive plan for Everglades restoration.

The Comprehensive Everglades Restoration Plan (CERP) is the product of a study authorized by U.S. Congress in 1992 "to reexamine the C&SF Project to determine the feasibility of structural or operational modifications to the project essential to the restoration of the Everglades and the south Florida ecosystem, while providing for other water-related needs such as urban and agricultural water supply and flood protection in those areas served by the project" (U.S. Army Corps of Engineers and South Florida Water Management District 1999). The Comprehensive Plan created by this C&SF Project "restudy" was approved and authorized by Congress in the Water Resources Development Act (WRDA) of 2000. The Comprehensive Everglades Restoration Plan consists of sixty-eight discrete projects to be constructed, implemented, and monitored over a thirty-five-year period, beginning in 2001. Responsibility for funding, implementing, and assessing the plan is shared equally by federal government (U.S. Army Corps of Engineers) and the state of Florida (South Florida Water Management District).

The primary objectives of the Comprehensive Plan are to substantially increase the amount of water (rainfall) that can be retained in the south Florida system; to distribute water throughout the natural system in a way that reestablishes ecologically desirable water quality and water-quantity patterns (depth, distribution, and flow) in the freshwater wetlands and desirable salinity regimes in the estuaries; and to maintain or improve water supply and flood protection in the urban and agricultural systems. Two key requirements specified in the plan are that it (1) "will be based on the best available science" and (2) "must be a flexible plan that is based on the concept of adaptive assessment—recognizing that modifications will be made in the future based on new information" (U.S. Army Corps of Engineers and South Florida Water Management District 1999).

The Water Resources Development Act (WRDA) of 2000 specifically authorizes an Adaptive Assessment and Monitoring Program to be conducted throughout the implementation of the Comprehensive Plan. WRDA 2000 requires that programmatic regulations be developed "to ensure that new information resulting from changed or unforeseen circumstances, new scientific or technical information or information that is developed through the principles of adaptive management contained in the Plan" is incorporated as the plan is implemented. At this point, we can only assume that monitoring and assessment will continue beyond the thirty-five-year implementation of the plan.

Success and Uncertainty

Arguably, the overarching question that must be answered as a means of creating a framework for addressing all other restoration questions is, "What is restoration?" The uncertainties of answering this key question are illustrated in figure 5-1. The greater Everglades originally was a large, interconnected hydrological system. For this diagram, just as for the Everglades restoration plan, the assumption is made that maximum progress toward ecosystem restoration will depend on how successfully the plan achieves hydrological restoration in most areas of the remaining natural system. Thus, in the diagram, the current ecological condition of the Everglades is assumed to be directly linked to the current hydrological patterns (left side of the figure), and the fully restored ecosystem (i.e., pre-drainage condition, right side) is assumed to be closely associated with the pre-drainage hydrological patterns. The Comprehensive Plan is designed to shift the current hydrological patterns as far as possible in the direction of the pre-drainage pattern. As stated above, it is assumed that by doing so, the current ecological condition will also shift in the direction of the pre-drainage condition.

Several important points are prompted by figure 5-1. Human-induced changes in the landscapes of south Florida have occurred on such large scales that full hydrological restoration is no longer an option. Thus, it is accepted that full ecological restoration (i.e., a return to the pre-drainage ecological conditions in the greater Everglades) also is not possible. The restoration plan will shift the hydrological system to some point (or range of points) somewhere along the spectrum between the current condition and the pre-drainage condition. Current understandings of the Everglades strongly support the expectation that with this improvement in hydrological patterns, the ecological conditions in the Everglades will also show substantial improvements.

But current understandings of the Everglades are not sufficient to tell us how far along the ecological restoration spectrum the natural system

What is Restoration?

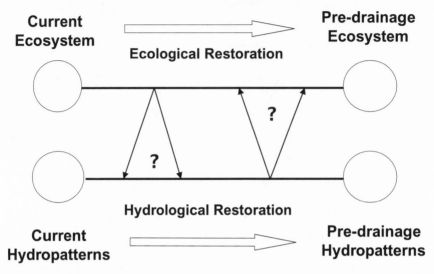

Figure 5-1. Diagram illustrating the uncertainties of the linkages between hydrological and ecological restoration in the Florida Everglades. Current understanding of Everglades ecology is not sufficient to allow for predictions of the ecological conditions that would be associated with any point or range of points along the hydrological line.

will shift with any given hydrological shift. In such a hydrologically dynamic system as the Everglades, it is presumed that any given point along the hydrological spectrum will support a range of ecological responses (depending on time scales), just as any given ecological condition may be supported by some range of hydrological patterns.

A final point needs to be made before moving to a description of the Everglades monitoring and assessment plan and how it was created. Little is actually known of the pre-drainage Everglades. William B. Robertson Jr., an early biologist at Everglades National Park, once wisely quipped that the pre-drainage Everglades is "a lost world, one that we will never know as well as we would like" (Davis and Ogden 1994). Almost everything that we do know about the Everglades ecosystem, and about the animals and plants that occur in this system, has been learned from studies and monitoring programs that were conducted in a substantially altered Everglades.

The Applied Science Strategy

In order to have a process and framework for answering and dealing with the challenges associated with the questions listed above, teams of south Florida agency scientists and resource specialists have created an "applied science strategy" (Ogden and Davis 1999). The overall purpose of the strategy is to effectively apply current and future scientific understandings of natural and human systems to the planning, evaluation, and assessment phases of the Comprehensive Everglades Restoration Plan.

The applied science strategy used in the Everglades restoration program is modeled after the "ecological risk assessment" process of the U.S. Environmental Protection Agency (EPA 1992; Gentile 1996; Harwell et al. 1999). Risk assessments have been used by the EPA as a guideline for conducting ecological evaluations of proposed management actions in both North America and Europe. A similar approach was used by the Man and the Biosphere Human Dominated Systems program to define sustainability goals and identify ecological end points for a series of restoration scenarios for south Florida (Harwell and Long 1992; Harwell et al. 1996, 1999). Many of the same components and steps incorporated into the applied science strategy for the current Everglades program, including conceptual models, were also described in Margoluis and Salafsky (1998).

Gentile (1996) suggested that risk assessment and recovery of ecological systems "can be viewed as opposite sides of the same coin." Risk assessment "is the process of determining the probability (with associated uncertainty) of a particular event occurring as a result of the action of a specific agent or stressor." Recovery of ecological systems "can be viewed as the process of determining the probability (with associated uncertainty) of a particular event occurring (e.g., recovery to a . . . ecologically desired sustainable state) as a result of mitigating the action of a specific agent (e.g., canals, berms) or stressor (e.g., phosphorous)." Gentile (1996) listed three principal functions of the risk assessment process: (1) identification of potential causal relationships between stressors and effects, (2) selection of end points (attributes), indicators, and success criteria, and (3) development of a scale-dependent conceptual model that describes the interrelationships between multiple stressor pathways and multiple ecological receptors.

Gentile (1996) summarized the importance and appropriateness of the risk assessment and conceptual model process for setting ecological targets and conducting ecological evaluations in regional ecosystem restoration and management programs. The points were (1) that there is a large body of peer-reviewed literature that describes these methods and processes for conducting ecological assessments, (2) that this literature represents an internationally accepted framework for structuring assessments, and (3) that considerable effort has been devoted during recent years to formalizing the process of selecting and classifying the end points, indicators, and metrics used in assessments.

To be successful in south Florida, the applied science strategy must

- Serve as a catalyst for promoting consensus among scientists, resource specialists, and managers regarding the nature of the major resource issues and the probable routes for resolving these issues.
- Present a scientifically reviewed sequence of steps for converting research, modeling, and monitoring results into a parsimonious set of performance measures, and evaluation and assessment protocols.
- Be a process that can contribute to the objectives and needs of both the scientific and management components of the regional restoration programs.
- Be a process that can identify and reduce the major areas of technical uncertainty pertaining to the ways that these systems will respond during and following implementation of the restoration plan.
- Provide the quantitative basis for organizing and applying an adaptive assessment protocol during the implementation of the Comprehensive Plan.

Specifically, the applied science strategy must lead to the identification of a discrete set of physical and biological performance measures and restoration targets. These measures and targets can be used (through modeling) to decide the most favorable design and operation of the restoration plan and (through monitoring) to measure how well the restoration plan is meeting its targets. The performance measures and monitoring program are the essential ingredients in an adaptive assessment protocol for measuring actual system responses and recommending improvements in the restoration plan as new information and interpretations become available.

The components and linkages in the science strategy are shown in figure 5-2. The figure shows how scientific research and application (solid boxes) provide the linkage between restoration goals and the Comprehensive Plan (dotted boxes). Four essential components of the science strategy are landscape-scaled conceptual ecological models developed and refined by south Florida scientists, a set of systemwide restoration performance measures, a systemwide monitoring and assessment program, and an adaptive assessment protocol.

Applied Science Strategy

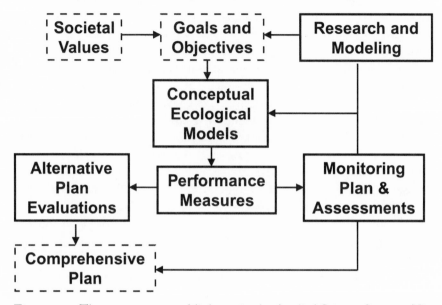

Figure 5-2. The components and linkages in the Applied Science Strategy. The strategy was organized by teams of south Florida scientists to provide more effective scientific support to the Everglades restoration program.

Conceptual Ecological Models and Performance Measures

The heart of the applied science strategy is a set of conceptual ecological models. The development and application of conceptual models provide benefits to the scientists who create them and to the managers and public who use or learn from them to guide or influence resource policy (Gentile 1996; Ogden et al. 1997; Science Coordination Team 1997). The process of creating and reviewing conceptual models benefits scientific endeavors in several ways, including establishing a forum for open, multidisciplinary exchanges of ideas and information pertaining to complex ecological issues; developing scientific consensus regarding current understandings of ecosystems; creating working hypotheses that guide both research and monitoring; better defining areas of scientific uncertainty; and providing a framework for continuing refinement of models as new information becomes available.

For managers and the public, the models serve to demystify the science; provide a means for converting broad policy-level goals into specific, measurable objectives; provide a visual description of the rationale for the prevailing hypotheses and management priorities; reduce the complexity and dimensionality of the problems; separate essential from nonessential information; and provide a tool for improved communication.

These simple, nonquantitative models are an effective means for converting broadly stated restoration goals into more specific measures of restoration success. The first step in this conversion occurs by using the models as a framework for identifying and developing a consensus regarding the set of causal hypotheses that best explain the effects that the major anthropogenic stressors have had on the wetland ecosystems in south Florida (Noon chap. 2). The models link each of the stressors on the ecosystem with one or more biological attributes that are considered to be the best indicators of the effects of the stressors. Each of the linkages in the models represents a working hypothesis based on current knowledge of the system to explain the ecological effects from each of the stressors.

The models are important as a means for reducing a potentially large number of stressors, working hypotheses, and attributes to a parsimonious set of key components and linkages. From this comparatively small set of components, performance measures are created. These measures, derived from the model stressors and attributes, describe the physical, biological, and ecological conditions that will be used to define a successful restoration program.

The attributes that are selected must provide measures of responses at appropriate temporal and spatial scales and from different taxonomic and hierarchical levels in ecological systems. They must be measurable and their historical patterns, relationships, and functions well enough understood that responses can be correctly determined and interpreted. Attributes should be selected from both the category of key ecological conditions (e.g., diversity and functionality) and the category of societal priorities (e.g., endangered species and high-quality sport fishing). Finally, consideration must be given to the most appropriate number of attributes for each model. The desirable trade-off is to have a sufficient number to adequately reflect the major system responses while not having more than is necessary for this purpose or that could be sustained during a long-term monitoring program.

The rationale for having performance measures and targets for each stressor is that the stressors are known or hypothesized to be the immediate causes of the ecological problems in each physiographic region. A successful restoration program must eliminate the unnatural stressors acting on the natural systems. A performance measure describes the stressor, how the stressor should be measured, and how the stressor must change in order to neutralize its adverse effects.

Similarly, restoration performance measures are developed for each attribute in the models. The attributes are selected as the biological or ecological parameters that are the best indicators of responses in the natural systems to the adverse effects of the stressors. The hypotheses used to construct the conceptual models link each attribute to the stressor(s) that are most responsible for unnatural change in that attribute. If the hypotheses are correct, neutralizing the stressors will result in a predictable, positive response by the attribute. Each attribute-based performance measure identifies the parameter of that attribute that should be measured and how that parameter is predicted to change once the effects of the stressor are removed.

The performance measures developed from the models can be used in two ways to support the Comprehensive Plan. The measures define the desired performance of the restoration plan, or, in other words, each measure includes a measurable target or end point. They are used during the planning phases of the restoration program as "evaluation" tools to predict the performance of a proposed plan or to compare the performance of alternative plans. The combination of structural and operational components most likely to achieve the desired objectives of the plan is determined by how well a plan is predicted (by simulation modeling) to meet each of the restoration targets set by the performance measures. In this planning role, the measures not only are used to determine which plan is most likely to be successful, but also are used to influence the selection of components as efforts are made to evaluate combinations of features.

The second use of the performance measures is as "assessment" tools. The measures determine the content and design of a systemwide, ecological monitoring program. The overall purpose of the systemwide monitoring program is to "assess" how a representative selection of elements of the natural system actually respond following the implementation of each iteration of the restoration plan. Because the overall design of the restoration plan is strongly influenced by the selection of components that collectively are predicted to achieve a suite of regional restoration expectations, the monitoring program must measure the same set of parameters. Comparisons between the predicted responses by a set of stressors and attributes and the actual responses by the same stressors and attributes provide a basis for making revisions to the working hypotheses and the conceptual models and for structuring an adaptive assessment strategy throughout the implementation of the plan.

The conceptual models also serve to set research priorities in support of the Comprehensive Plan. The specific target for ongoing and future research is to improve understandings of ecological linkages as a basis for improving the ability of the science teams to interpret responses detected by the monitoring program (adaptive assessment). The conceptual models created to support the Everglades restoration program are based on a large number of working hypotheses that express existing levels of understanding about the ecological linkages between stressors and attributes in the natural systems (Ogden et al. 1999). Although the design of the Comprehensive Plan and the selection of the plan's performance measures are based on an assumption that the hypotheses are correct, in fact levels of certainty in the hypotheses vary depending on the strength of existing databases.

To deal with the differences in levels of certainty among the conceptual model hypotheses, the conceptual model teams ranked each hypothesis as having a high, moderate, or low level of certainty. Levels of certainty were based on existing databases and the collective opinions of the teams regarding the strength of each hypothesis. Using these criteria, 173 working hypotheses were ranked for levels of certainty. The initial ranking resulted in seventy-one hypotheses listed as having a high level of certainty, fifty-three as having a moderate level of certainty, and forty-nine as having a low level of certainty. As part of this process, each hypothesis was converted into one or more research questions that focused on the information needed to strengthen the hypothesis. The highest research priorities in support of the Comprehensive Everglades monitoring and assessment plan are those that strengthen those hypotheses where levels of certainty were lowest. The link between the hypotheses and the research questions is illustrated in the ridge and slough model, described in the section below.

A set of nine conceptual ecological models was created for the major wetland physiographic regions or subregions of the greater Everglades system in south Florida (fig. 5-3). The models are for Lake Okeechobee, Caloosahatchee estuary, St. Lucie estuary and Indian River Lagoon, central Everglades ridge and slough region, southern Everglades marl prairies, Big Cypress swamp, southern Everglades mangrove estuaries, Florida Bay, and Biscayne Bay. The Everglades ridge and slough model is used later in this chapter to illustrate the organization, content, and application of the models in the Everglades restoration program.

The initial step in the development of the conceptual ecological models was to use informal technical workshops of scientists and resource specialists to develop a set of working hypotheses that best explained the major anthropogenically driven alterations known to have occurred at

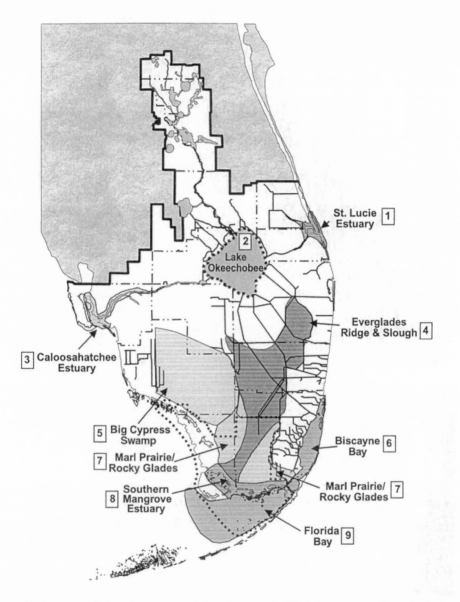

Figure 5-3. Subregions covered by nine south Florida conceptual ecological models: (1) St. Lucie Estuary, (2) Lake Okeechobee, (3) Caloosahatchee Estuary, (4) Everglades Ridge and Slough, (5) Big Cypress Swamp, (6) Biscayne Bay, (7) Marl Prairies/Rocky Glades (8) Southern Mangrove Estuary, and (9) Florida Bay. The waters surrounding the Florida Keys, to the south a. ¹ east of Florida Bay, comprise the Florida Keys National Marine Sanctuary (see chap. 10 for more information).

147

the scale of each physiographic region. The working hypotheses were collected from published literature and from the discussions during the workshops. The workshop participants created lists of stressors, ecological effects, and attributes (indicators) for each of the model regions. The objective was to identify the physical and biological components and linkages in each region that best characterized the changes explained by the hypotheses. Preparers (model leads) used the hypotheses and lists of components to lay out an initial draft of the model, and to prepare a supporting narrative to explain the organization of the model and present the supporting science for the hypotheses.

Ultimately, each model included external drivers, ecological stressors, ecological effects, attributes, and measures. The result was a graphic showing the model linkages and a narrative containing sections that describe the dynamics and problems of the region; the nature of the external drivers and ecological stressors; the characteristics of the ecological changes caused by the stressors; the selection of the appropriate attributes and performance measures; descriptions of the key linkages between stressors and attributes, and levels of certainty for each of the key hypotheses; priority research questions derived from the ranking of hypotheses; recommended performance measures; and literature cited.

Comprehensive Everglades Monitoring and Assessment Plan

The conceptual ecological models and associated performance measures provide the basis for the development of an Everglades monitoring and assessment plan (U.S. Army Corps of Engineers and South Florida Water Management District 2001). The overall purpose of the Everglades monitoring and assessment plan is to identify the performance measures and the specific parameters for each of the measures that should be monitored and assessed in order to determine (and enhance) the success of the Comprehensive Plan. The monitoring and assessment plan describes the specific set of physical and biological performance measures, the geographic and ecologic locations where each should be monitored, and the restoration targets that have been established for each. The performance measures, ecological linkages, and research questions that have been developed from the set of nine conceptual ecological models are the basis for the natural system parameters and targets in the Everglades monitoring and assessment plan. In order to meet its overall purpose, the monitoring and assessment plan has four specific objectives:

1. Establish base-line conditions and variability for each of the measured parameters.

2. Determine the status and trends among the performance measures.
3. Detect unexpected responses for components or measures of the ecosystem that have not been specifically identified as performance measures of the Comprehensive Plan.
4. Support cause-and-effect scientific investigations designed to increase ecosystem understanding, particularly for dealing with unanticipated results.

The monitoring and assessment plan was prepared by a multi-agency team of scientists and resource specialists known as REstoration COordination and VERification (RECOVER). The RECOVER team has the responsibility of conducting and coordinating the evaluation and adaptive assessment tasks, and other elements of the applied science strategy, throughout the full implementation of the Comprehensive Everglades Restoration Plan. RECOVER is a programmatic team established to maintain the systemwide perspective of the Comprehensive Plan and to identify opportunities to improve the overall performance of the plan relative to the restoration targets set by the performance measures.

The Everglades monitoring and assessment plan and its periodic revisions contain the most current versions of the conceptual ecological models and performance measures. In addition to the model diagrams, the monitoring and assessment plan presents the working hypotheses, the research questions in support of assessments of system responses, and the scientific documentation for the models and hypotheses. The plan also includes a documentation sheet for each of the approximately 150 performance measures in the plan at the time that this chapter is being written. The performance measure documentation sheets provide a summary description of each performance measure and restoration target, explain the link(s) between the measure and one or more conceptual models, and then identify the appropriate uses of the measure for plan evaluations (simulation modeling) and for tracking system responses (monitoring).

Over time, the number of performance measures will change, mostly by becoming smaller. The RECOVER team agrees with the views of others (e.g., NRC 2000) that the number of systemwide performance measures must be reduced to less than one hundred if the monitoring program for the Comprehensive Plan is to be sustained over several decades. Additionally, the plan must focus on key restoration expectations and be designed to increase our certainty about the ecological linkages and responses.

A major reason why the original number of measures was high is that they were derived from the stressors and attributes in nine different conceptual models. The large spatial extent of the greater Everglades basin, the distinct physiographic regions within this basin, and the varying scales of stressor effects were all good reasons for creating so many conceptual models. Revisions in the number of Comprehensive Plan performance measures will occur through continuing reviews, including the reevaluation and synthesis of the working hypotheses and conceptual models, and the refinement of the evaluation criteria and priorities used to screen the performance measures.

Adaptive Assessment

Perhaps the ultimate objective of the Everglades monitoring plan is that it must produce the information necessary for sustaining a long-term adaptive assessment protocol. Adaptive assessment offers the Comprehensive Plan teams (RECOVER) with their best opportunities for confronting and reducing the uncertainties that exist about how the natural and human systems in south Florida will respond to a long-term restoration program (AAT 2000). Uncertainties are inevitable in that the restoration teams are dealing with systems that are highly complex and not adequately understood. The current levels of uncertainty have prompted ongoing debates among south Florida scientists regarding the comparative strengths of a range of ecosystem variables and linkages as components of the conceptual models and sources of performance measures. Although these debates have for the most part been essential steps in focusing and prioritizing future research and modeling and for refining the conceptual models and performance measures, they also reaffirm that our current levels of knowledge are inadequate for predicting system responses (primarily at medium to large scales). Adaptive assessment provides the means for continually reducing the levels of uncertainty by learning from system responses and using this new information to refine the design of the restoration plan (Reynolds and Reeves, chap. 7; Ringold et al, chap. 3).

Monitoring data collected for each performance measure are analyzed in formats that are relevant to the restoration targets and used to guide refinements or improvements in the design or operation of the Comprehensive Plan where changes are required. These analyses feed three feedback loops in the adaptive assessment protocol. The simplest feedback occurs when system responses during the implementation of the plan are considered to be consistent with the predictions and objectives of the plan. In this situation, the design of the plan remains unchanged, and the implementation continues as previously scheduled. The alternative feedback occurs when unexpected responses are detected. This event triggers new modeling to identify corrective changes in the restoration plan.

These changes could be comparatively simple, such as changes in operational rules or in the sequence of implementation of future project components of the plan. Or the changes could be complex, such as redesign of the plan on a scale requiring reauthorization by Congress.

The third feedback loop is a learning loop whereby improved understandings of the natural and human systems originating from monitoring, research, and modeling are incorporated. Improved knowledge can lead to refinements in conceptual models and performance measures, to changes in the design of the monitoring and assessment plan, and to revisions in the design or operation of the restoration plan itself. Annual assessments by the RECOVER teams can prompt recommendations for these improvements.

Overall, adaptive assessment provides a process for learning and incorporating new information into the evaluation and assessment phases of the restoration plan. Adaptive assessment is valuable in that it treats all responses, expected or not, as learning opportunities. An unexpected response does not represent a failure for the restoration plan if it can be used to improve our understanding of a complex system so that corrective actions can be made. Implementation of the restoration plan is a "test" of the accuracy of the working hypotheses used to organize the conceptual models and to support the performance measures. Adjustments in the models and measures based on information coming from the monitoring and assessment program, and from supporting research and simulation modeling, is the best route to increased levels of certainty as future iterations of the restoration plan are implemented.

An Example Using the Everglades Ridge and Slough Conceptual Model

The content, organization, and application of the conceptual ecological models in planning, evaluating, and assessing the Comprehensive Everglades Restoration Plan can be illustrated by reviewing the Everglades ridge and slough model (fig. 5-4). The ridge and slough model was created during a series of interagency workshops attended by over fifty agency, university, and nongovernment scientists and resource specialists between October 1996 and July 1997 (Ogden and Davis 1999). The ridge and slough model was intensively reviewed and revised during a similar series of workshops between July and October 2000 (U.S. Army Corps of Engineers and South Florida Water Management District 2001).

In its pre-drainage condition, the ridge and slough portion of the Everglades was an expansive, low-nutrient, freshwater marsh characterized by moderate to deep organic soils and alternating sawgrass "ridges"

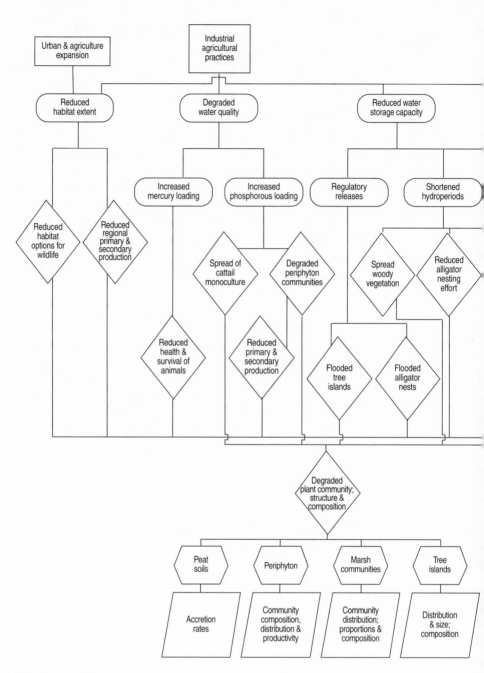

Figure 5-4. The Everglades ridge and slough conceptual ecological model. Each south Florida conceptual model is reviewed and revised at one- to two-year intervals by teams of regional scientists. Performance measures and the regional monitoring program for Everglades restoration are derived from the working hypotheses that structure these models.

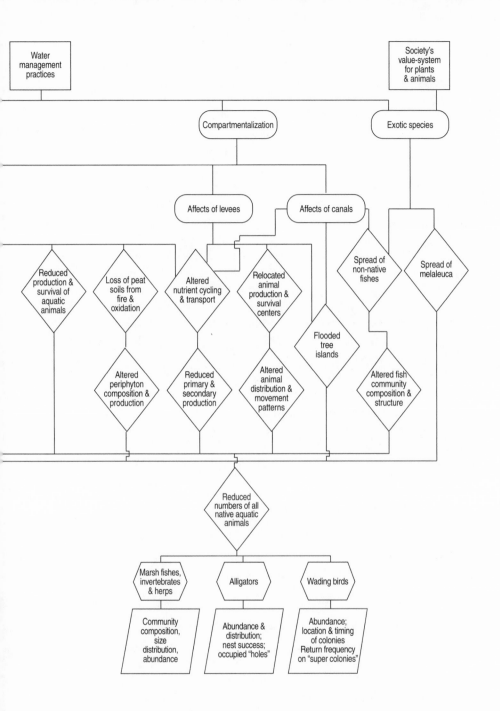

153

and more open-water "sloughs." The ridge and slough systems were dominated by sawgrass (*Cladium*), water lily, and "wet prairie" (*Eleochris*,
Rhynchospora) communities. Although surface water depths rarely ex-
ceeded 1 meter, and annual rainfall was strongly seasonal (80 percent of
annual rainfall occurred during a summer rainy season, May–October),
the deeper central sloughs remained flooded throughout most years. The
subtle north-to-south gradient in the ridge and slough system caused surface water to move slowly and directionally on a broad front as "sheet
flow" through the marshes. A spatially variable mosaic of discrete hardwood hammock, bay head, and willow tree islands occurred throughout,
ranging from widely spaced in the central Everglades to more densely distributed in the northern, Loxahatchee Basin and southern Shark River
Slough. The ridge and slough system was the principal center for primary
and secondary production and interannual survival of aquatic organisms
in the pre-drainage, freshwater wetlands of southern Florida.

The major known changes in the ridge and slough Everglades have
been caused by modern land-use practices and by drainage and water
management programs in south Florida. The effects from these external
drivers include reduced spatial extent, compartmentalization and reduced
sheet flow, loss of organic soils, altered water depth, distribution and flow
patterns, altered water and soil chemistry, and the introduction of exotic
species. Eleven working hypotheses developed by the ridge and slough
model team to explain the major ecological changes caused by these stressors are as follows:

- *Periphyton (Algal) Community hypothesis.* It is hypothesized that diatom
  and green algae mats have higher food value for fish and invertebrate
  browsers than do blue-green algae mats and that shifts in water chemistry (increased phosphorous) and hydropatterns (reduced duration of
  flooding and altered depth patterns) have resulted in an increased presence of blue-green algae and a reduction in the food value of the mats
  in the Everglades food webs.
- *Marsh Community hypothesis no. 1.* Compartmentalization (reduced
  water flow volumes; changes in water depth–distribution patterns) has
  caused the loss of discrete ridge and slough communities (invasion of
  sloughs by sawgrass) and has reduced levels of primary and secondary
  production (disrupted patterns of nutrient dynamics and transport).
- *Marsh Community hypothesis no. 2.* Overdrained marshes and the resulting loss of organic soils through oxidation and increased intensity
  of fires have substantially altered marsh contours, flow patterns, productivity, and plant community composition and spatial patterns (including loss of tree islands and invasions of herbaceous communities by
  woody species).

- *Marsh Community hypothesis no. 3.* Although it is known that increased phosphorous loading in the Everglades has caused substantial conversions of marsh communities to pure stands of cattails, it is only hypothesized that reductions in phosphorous inputs will lead to reduced soil phosphorous levels or reductions in the spatial extent of cattails.
- *Tree Island hypothesis.* The loss and degradation of tree islands in the ridge and slough Everglades has largely been caused by water management practices that have exaggerated the frequency and magnitude of unnatural high-water events, have resulted in increased severity of fires during unnatural drought events, and have increased the frequency of desiccation events in tree island soils.
- *Fish and Aquatic Invertebrates hypothesis no. 1.* The characteristics of aquatic animal communities in Everglades marshes are determined along hydropattern and nutrient/productivity gradients (i.e., hydropatterns alone do not determine aquatic animal community characteristics) and that regionally scaled changes in surface-water flow volumes and water depth–distribution patterns have reduced production and survival rates, altered the community composition for all species in this category, and altered "prey availability" patterns for larger, foraging animals such as wading birds.
- *Fish and Aquatic Invertebrates hypothesis no. 2.* The pre-drainage Everglades was dominated by a crayfish-based food web, which explains the former abundance of white ibis and otter. The abundance and distribution of crayfish have been greatly altered by shortened hydroperiods and lowered dry-season groundwater levels.
- *Fish and Aquatic Invertebrates hypothesis no. 3.* It is hypothesized that large spatial extent of the pre-drainage ridge and slough Everglades was a key factor in the total, systemwide production of small aquatic organisms, and that the approximately 50 percent reduction in the extent of this system has substantially reduced the numbers of "prey" and the foraging-habitat options for wading birds and other large aquatic vertebrates.
- *American Alligator hypothesis.* Densities of alligators and of alligator nests in the pre-drainage Everglades basin were low in the ridge and slough system but relatively high in the adjacent marl prairies and freshwater mangrove swamps. This was because the pre-drainage ridge and slough was too deeply flooded to be ideal alligator habitat, but the pre-drainage marl prairies and mangrove swamps were relatively favorable alligator habitats because of more favorable water depth and salinity characteristics.
- *Wading Bird hypothesis no. 1.* The largest wading-bird nesting colonies in the historical Everglades were located along the ecotone between

the southern ridge and slough system and the downstream estuaries because of elevated secondary production in this transition zone. These conditions have been lost due to reductions in the volume and flow rates of freshwater into the southern Everglades estuaries.

• *Wading Bird hypothesis no. 2.* The former "super nesting colonies" were responses by wading birds to spikes in the abundance of small aquatic prey because of the role that major, natural droughts had on organizing the release and transport of nutrients and on the numbers of the larger predatory fishes in the marshes. Changes in the frequency, duration, and extent of droughts have undermined this historical nesting pattern.

These hypotheses suggest components and linkages for the ridge and slough conceptual model (fig. 5-4). The model team selected five key stressors for the ridge and slough model: reduced habitat extent, degraded water quality, reduced water storage capacity, compartmentalization, and exotic species. The team also selected seven key attributes that were linked to stressors by the hypotheses. These seven ridge and slough attributes, and the restoration performance measure derived from each, are as follows:

• *Peat soil accretion rates.* The recovery of a peat-accreting system throughout the ridge and slough system is the short-term restoration objective. Included in this target is a minimization of anthropogenic enrichment of soils and recovery of peat accretion in portions of the sloughs where overdrainage has converted soil production from peats to marls. The long-term target is the recovery of pre-drainage soil depth and microtopography contour patterns.

• *Periphyton community composition and distribution.* The restoration target for periphyton is to recover the presence of, and increase the cover and extent of, the algal mat that is associated with the long-hydroperiod, low-nutrient indicator, *Utricularia purporea*. Measures of species composition can be used to reflect improvements in water and soil chemistry and in hydroperiod and depth parameters.

• *Marsh community composition and distribution.* The short-term target for cattails and exotic plants is to minimize further invasions of these species into the ridge and slough communities. The long-term targets are to (1) recover, at a systemwide scale, the expected mosaics and spatial and temporal dynamics of the marsh communities, as should occur with natural patterns of hydrology, fire, and soils; (2) decrease the proportion and extent of sawgrass where it has invaded slough and wet prairie communities; (3) reestablish spatial patterns of ridge and slough communities by restoring pre-drainage flow directionality and volumes; and (4) eliminate large, monotypic stands of cattails.

- *Tree island numbers and composition.* The three targets include (1) prevention of any further net loss in the extent and quality of tree islands (determined by landscape distribution and spatial extent, vegetation community composition, habitat value and ecological function, and geomorphology—including accretion, shape, elevation profile, and orientation); (2) recovery of lost tree-island habitat, where natural elevation rises are preserved and natural patterns of succession would be measured; and (3) recovery of a more natural mosaic of tree-island types across the landscape (e.g., hardwood hammocks, bay heads, willow heads).

- *Marsh fishes and invertebrates, and their abundance and composition.* The four targets include

  (1) *Animal density.* Multiyear patterns of fish dynamics (including multiyear increases in marsh fish numbers and biomass in response to uninterrupted, multiyear hydroperiods) should be increased to levels comparable to those of Shark Slough following multiyear hydroperiods. Densities of the pig frog, *Rana grylio*, should be increased in areas of the ridge and slough system where hydroperiods have been unnaturally reduced. Densities of the apple snail, *Pomacea*, should also be increased.

  (2) *Size distribution.* In response to lengthened hydroperiods, the size range in biomass and length distributions of marsh fishes should be increased by increasing the frequency of larger fishes. Current low frequencies of exotic fishes in the interior of the ridge and slough system should be maintained.

  (3) *Relative abundance.* In response to lengthened hydroperiods, the relative abundance of centrarchids and chubsuckers should be increased. In addition, the frequency of *Procambarus fallax*, a crayfish species typical of long hydroperiod marshes, should be increased in relation to *P. alleni*, a species characteristic of shorter hydroperiod marshes.

  (4) *Contaminants.* Reduce levels of mercury and other toxins in marsh fishes.

- *Alligator abundance, distribution, and nesting success.* The six targets include

  (1) A shift in densities of nests in Everglades National Park. Higher densities would move from the center to the edges and downstream portions of Shark River Slough.

(2) Uniform density and distribution of alligators, occupied holes, and nests in the Everglades Water Conservation Areas.

(3) Improvement in the overall "health" of alligators in Shark River Slough and the central and southern Water Conservation Areas. This would include meeting baselines for contaminant loadings.

(4) Increased alligator reproductive rates in Shark River Slough and the central and southern Water Conservation Areas. This would be measured by the frequency of alligator nesting, the frequency of hatching, and the survival of hatchlings.

(5) Decreased incidence of nest flooding. Currently, these are caused by wet season regulatory releases and other water management practices in the Shark River Slough and Water Conservation Areas.

(6) An increase in the proportion of juvenile size classes in the total Everglades alligator population.

- *Wading bird abundance, location and timing of nesting*. The five restoration targets include

  (1) Increased total numbers of wading birds nesting in and adjacent to the ridge and slough system (target numbers have been proposed in Ogden et al. [1996]).

  (2) A substantially increased proportion of the total number of nesting birds that nest in the area of the southern Everglades marsh-mangrove ecotone (i.e., the former center of abundance by nesting wading birds in the Everglades).

  (3) A return of super nesting colonies, with a minimum return frequency of one to two years per decade. To recover their historical size, super colonies should occur in the headwaters ecotone at the lower Shark River Slough.

  (4) Meet threshold criteria for "safe" loadings of contaminants in wading birds.

  (5) Improved foraging habitat for wading bird nesting colonies and super colonies. Specifically, habitats should represent a range of climatological conditions, marsh production rates, and availability of prey and should provide for maximum survival rates of organisms during dry periods.

In setting research priorities based on the working hypotheses from the ridge and slough conceptual model, twenty-three hypotheses and discrete components of hypotheses were ranked for levels of certainty. Those with low levels of certainty were used to identify the priority research needs as a prerequisite for obtaining the information needed to strengthen or revise these hypotheses. Seven of these were categorized as having low levels of certainty. From these seven, the model teams identified sixteen priority research questions loosely grouped into wetland landscape and vegetation patterns, and wetland productivity and aquatic food webs. These two sets of questions are described below.

Wetland Landscape and Vegetation Patterns

The first set of research questions deal with landscape and community vegetation patterns and the physical and chemical factors that determine these patterns. The focus of these questions is on determining how water, fire, and nutrients interact in creating and maintaining the characteristic vegetation mosaics in Everglades sloughs.

- How do hydrology and fire interact to sustain the spatial coverage, directionality, and coexistence of sawgrass ridges, sloughs, and tree islands that characterized pre-drainage landscapes?
- What are the respective roles of hydrologic variables and fire as causal factors that sustain discrete, directional slough systems and prevent encroachment by sawgrass?
- What are the respective roles of hydrologic variables and fire as causal factors that sustain viable tree islands and prevent drowning under high-water conditions and burn-out under low-water conditions?
- In nutrient-enriched areas of the ridge and slough landscape, how can current trends of cattail expansion be reversed once phosphorous inputs are reduced?
- In overdrained, non-enriched areas of the ridge and slough landscape, how can elevated phosphorous concentrations in the subsided peat be sequestered in forms that are unavailable for cattail growth upon the recovery of more natural hydropatterns?
- How are the composition and structure of native plant communities altered when water with higher-than-normal mineral content for surface waters enters wetland systems from canals, aquifer storage and recovery, and in-ground reservoirs?

Wetland Productivity and Aquatic Food Webs

The second set of research questions deal with aquatic organisms in the Everglades wetlands and the factors that influence the diversity, abundance, and survival of these animals. Much of the focus of these research questions is on the food webs and production patterns that once supported the tremendous abundance of animals that characterized the predrainage Everglades.

- How are populations of marsh fishes and other aquatic fauna related to the spatial cover, primary productivity, and taxonomic composition of periphyton mats?
- Are populations of marsh fishes and other aquatic fauna regulated by food webs generated from primary production of periphyton mats?
- Will the restoration of periphyton mat spatial cover, primary productivity, and taxonomic composition yield densities of aquatic fauna above those predicted in response to hydroperiod restoration alone?
- What is the functional importance of crayfish, grass shrimp, and hylid and ranid frogs as intermediate trophic levels and prey bases for higher consumers in freshwater wetlands?
- Are populations of crayfish, grass shrimp, and frogs regulated by the same factors regulating populations of marsh fishes, including hydroperiod, dry season refugia, and primary productivity?
- Is a functional Everglades ecosystem dominated by a crayfish-based food chain, as suggested by the former abundance of white ibis and otter?
- How are crayfish population density and taxonomic composition related to annual and interannual patterns in hydroperiod and depth, including drought?
- Are white ibis reproduction and super colony formation related to crayfish populations, as influenced by interannual hydropatterns?
- What triggers the formation of super colonies of white ibis and other wading birds following major droughts?
- What are the functional contributions and interrelationships of pulsed nutrient releases, reduction in numbers of predatory fishes, and spikes in abundance of small aquatic prey in supporting super colony formation?

Conclusions

The multiyear experience of designing and applying a systemwide strategy for providing scientific support for the Comprehensive Everglades Restoration Plan has inevitably provided improved insights into how an effective science-management linkage can best be achieved. Following are eight "lessons" that have become the focal points for ongoing reviews of the science strategy.

- *Lesson 1.* Where the temporal and spatial scales of regional restoration programs are large, and where the complexity of ecosystems is such that predictions of responses will be difficult, the creation of a successful process for maximizing and sustaining effective application of existing scientific knowledge is as important as the more traditionally recognized need for maintaining a commitment to continuing research and monitoring. Successful organization and application of science at such scales requires new and more focused efforts to achieve successful science support for management and policy in these situations. For the science application process to be both effective and sustainable, it must consist of a sequence of actions for organizing and focusing scientific and resource information and opinion that is broadly supported by the scientific and management communities.

- *Lesson 2.* Just as the restoration plan itself must be adaptable, the science application process must be adaptable, both in organization and content, in response to on-going evaluations of the application process, as well as in response to the acquisition of new information about the natural and human systems affected by the plan. The conceptual models provide a sound scientific basis for organizing the application of science in support of restoration, in that they provide a focus for on-going review of all working hypotheses and performance measures, and they are well designed for refinement as new information becomes available and hypotheses are improved.

- *Lesson 3.* There must be clear linkages between the working hypotheses and conceptual models and the selection of the restoration performance measures, objectives and design of the restoration program, content and design of the monitoring and assessment program, and important research questions that are the basis for improving interpretation of system responses. The working hypotheses and conceptual models should precede the formulation of restoration objectives and point to the design of the restoration plan for achieving these objectives. Solutions to resource problems revealed in the conceptual models (i.e., the stressors) should be clearly apparent in the restoration plan design. The key performance measure indicators for tracking how well the plan resolves resource problems should be addressed in the monitoring and assessment plan.

- *Lesson 4.* The design and application of a science application strategy must be a broadly inclusive undertaking, incorporating the knowledge and professional opinions of the widest possible array of scientific and natural resource talents. All relevant agency and institutional participation in the science application process should be strongly encouraged so that the restoration program has consistent access to the best available science and as a basis for creating a consensus among "stakeholders" as to the objectives of the restoration plan and the means for designing and assessing the plan. Ultimately, the goal would be a single, systemwide set of performance measures and monitoring plans that would serve as a basis for measuring the progress and success of the plan.

- *Lesson 5.* A common set of working hypotheses should be used to design the conceptual models and identify the major management and research questions to be addressed by simulation models supporting the restoration process. To the extent possible, the working hypotheses used to identify the best set of performance measures for the restoration plan should be evaluated ("tested") by simulation models.
- *Lesson 6.* A set of subregional conceptual models and a total system conceptual model should be developed as a framework for identifying and linking issues at different scales to ensure that all major restoration issues and working hypotheses are identified. Looking at the systems at different spatial scales (e.g., at subregional and total system scales) helps to organize and synthesize important hypotheses, set priorities for research questions, and select the most parsimonious set of monitoring parameters needed to adequately measure systemwide and subregional responses to the restoration plan.
- *Lesson 7.* The purpose of the monitoring and assessment plan is to effect change in the design and application of the restoration plan where needed and to provide a basis for measuring the success of the plan.
- *Lesson 8.* The applied science strategy can serve as a model for the organization and application of science in complex ecosystems elsewhere. A key element in its successful application is the new role played by scientists. Senior scientists and resource specialists with strong credentials in regional ecological studies and resource management programs will be responsible for designing and implementing the strategy. The assumption is that these scientists will eventually lead multidisciplinary science teams that plan, evaluate, and assess restoration programs and will serve as scientific representatives in forums also attended by managers and policy makers.

LITERATURE CITED

AAT (Adaptive Assessment Team). 2000. *An adaptive assessment strategy for the Comprehensive Everglades Restoration Plan.* March. Internal report of RECOVER.

Browder, J. A., and J. C. Ogden. 1999. The natural south Florida system II: Predrainage ecology. *Urban Ecosystems* 3:245–277.

Davis, S. M., and J. C. Ogden. 1994. Towards ecosystem restoration. Pp. 769–796 in *Everglades: The ecosystem and its restoration*, edited by S. M. Davis and J. C. Ogden. Delray Beach, Fla.: St. Lucie Press.

EPA (Environmental Protection Agency). 1992. *Framework for ecological risk assessment.* EPA/630/R-92/001. Washington, D.C.: Environmental Protection Agency.

Gentile, J. H. 1996. *Workshop on "South Florida ecological sustainability criteria."* Final Report. University of Miami, Center for Marine and Environmental Analysis, Rosenstiel School of Marine and Atmospheric Science, Miami, Fla.

Gunderson, L. H., and W. F. Loftus. 1993. The Everglades. Pp. 191–255 in *Biodiversity of the southeastern United States/Lowland terrestrial communities*, edited by W. H. Martin, S. G. Boyce, and A. C. Echternacht. New York: John Wiley and Sons.

Harwell, M. A., J. H. Gentile, A. Bartuska, C. C. Harwell, V. Myers, J. Obeysekera, J. C. Ogden, and S. Tosini. 1999. A science-based strategy for ecological restoration in south Florida. *Urban Ecosystems* 3(3/4):201–222.

Harwell, M. A., and J. F. Long. 1992. U.S. M.A.B. Human-Dominated Systems Directorate workshop on ecological endpoints and sustainability goals. University of Miami, Rosenstiel School of Marine and Atmospheric Science, Miami, Fla.

Harwell, M., J. F. Long, A. Bartuska, J. H. Gentile, C. C. Harwell, V. Myers, and J. C. Ogden. 1996. Ecosystem management to achieve ecological sustainability: The case of south Florida. *Environmental Management*. 20:497–521.

Holling, C. S., L. H. Gunderson, and C. J. Walters. 1994. The structure and dynamics of the Everglades system: Guidelines for ecosystem restoration. Pp. 741–756 in *Everglades: The ecosystem and its restoration*, edited by S. M. Davis and J. C. Ogden. Delray Beach, Fla.: St. Lucie Press.

Light, S. S., and J. W. Dineen. 1994. Water control in the Everglades: A historical perspective. Pp. 47–84 in *Everglades: The ecosystem and its restoration*, edited by S. M. Davis and J. C. Ogden. Delray Beach, Fla.: St. Lucie Press.

Margoluis, R., and N. Salafsky. 1998. *Measures of success.* Washington, D.C.: Island Press.

National Research Council. 2000. *Ecological indicators for the nation.* Washington, D.C.: National Academy of Sciences Press.

Ogden, J. C., S. M. Davis, D. Rudnick, and L. Gulick. 1997. *Natural systems team report to the Southern Everglades Restoration Alliance.* July 1997. West Palm Beach, Fla.: South Florida Water Management District.

Obeysehera, J., J. A. Browder, L. Horning, and M. A. Harwell. 1999. The natural South Florida system I: Climate, geology, and hydrology. *Urban Ecosystems* 3(3/4):223–244.

Ogden, J. C., G. T. Bancroft, and P. C. Frederick. 1996. *Everglades restoration success indices: Reestablishment of healthy wading bird populations.* South Florida Ecosystem Restoration Working Group, Science Subgroup. Miami, Fla.: Florida International University.

Ogden, J. C., J. A. Browder, J. H. Gentile, L. H. Gunderson, R. Fennama, and J. Wang. 1999. Environmental management scenarios: Ecological implications. *Urban Ecosystems* 3(3/4):279–303.

Ogden, J. C., and S. M. Davis. 1999. *The use of conceptual ecological landscape models as planning tools for the south Florida ecosystems restoration programs.* Internal report. South Florida Water Management District. West Palm Beach, Fla.

Science Coordination Team. 1997. *Integrated science plan.* Report to the South Florida Ecosystem Restoration Task Force and Working Group. Office of the Executive Director, SFERTF, Florida International University, Miami, Fla.

Science Subgroup. 1993. *Federal objectives for the south Florida restoration.* Science Subgroup of the South Florida Management and Coordination Working Group, South Florida Ecosystem Restoration Task Force, Miami, Fla.

U.S. Army Corps of Engineers. 1994. *Central and Southern Florida Project Comprehensive Review Study Reconnaissance Report.* Jacksonville District, Jacksonville, Fla.

U.S. Army Corps of Engineers and South Florida Water Management District. 1999. *Central and South Florida project comprehensive review study.* Final integrated feasibility report and programmatic environmental impact statement. Jacksonville and West Palm Beach, Fla.

———. 2001. Monitoring and assessment plan. Comprehensive Everglades Restoration Plan. 29 March. Jacksonville and West Palm Beach, Fla.

PART III

Information Management and Modeling for Monitoring Programs

CHAPTER 6

The Use of Models for a Multiscaled Ecological Monitoring System

Donald L. DeAngelis, Louis J. Gross, E. Jane Comiskey,
Wolf M. Mooij, M. Philip Nott, and Sarah Bellmund

Monitoring is the periodic measurement of some attribute of a system over a period of time in a given location or set of locations. It may serve such general purposes as detecting changes in the environment or in renewable natural resources, and it often serves specific conservation and restoration issues. In particular, monitoring may be employed if anthropogenic changes in a region are suspected of negatively affecting an ecological system. Similarly, if managers initiate a conservation or restoration plan, they usually complement it with a monitoring program to determine if the plan is providing the expected benefits to the ecosystem or target components (e.g., populations). In either case, the usual practice is to define a set of *performance measures* (which we will also refer to as *target variables*) to represent the health or condition of the system. These performance measures may span a wide possible range, from underlying abiotic conditions, to the sizes and viabilities of key populations, or more abstract indices such as species richness. Monitoring to evaluate how well a plan is working is often referred to as *effectiveness monitoring* (Noss and Cooperrider 1994).

The process of monitoring involves a number of steps and begins with the need for a clear recognition of why the monitoring needs to be done

(Peck 1998). For example, general goals of the monitoring may be to assess the population trends of an endangered species or to determine whether the composition of a biological community is changing. Once identified, these general goals must be converted into specific objectives, such as what, how, when, and where observations will be made. By defining specific monitoring objectives, modeling (both conceptual and quantitative) can play an important role in this process.

Modeling should be integrated into a monitoring effort for a number of reasons, including the following:

- It is seldom possible to monitor the target variables of interest directly. Usually some indicator variables, which are more easily monitored, are measured. These indicators can be related to the target variables through models.
- Every management plan for conservation or restoration rests on assumptions or hypotheses concerning the causal connections between the operation of the plan and desired effects for the variables of interest, such as the population of an endangered species. A conceptual or quantitative model formalizes these assumptions and hypotheses.
- Modeling enables scientists to project the changes of the target variables into the future, which could possibly signal the need for adjustments in the management plan.

This set of model functions is of a general nature and applies to almost any monitoring program, although our focus here will be on species populations. The important thing is to have a scheme that efficiently integrates these functions into the program. In this chapter, we explore how this might be done.

Three main sections follow this introduction. In the first, we describe in some detail the three functions of modeling outlined above. We emphasize that some of the effort of monitoring must be used to test models, which are in turn used to provide basic support for monitoring. This is an iterative process in which both model accuracy and monitoring efficiency improve together through time. We express our belief that spatially explicit dynamic landscape models, based on geographic information system (GIS) data, and including spatially explicit population individual-based population (SEIB) models, form a useful basis for incorporating the functions listed above. These models are used both to formalize our understanding of the connections between management decisions and natural populations and to project future changes in populations. Finally, we indicate that statistical analysis can be seamlessly integrated into the modeling to allow confidence intervals to be calculated. The second section illustrates this strategy with some examples of models used in the

Everglades. The final section outlines some broad implications and suggestions for integrating modeling with monitoring.

The Functions of Modeling in Monitoring Programs

Modeling has several important uses as part of a monitoring program. First, it can be used to establish relationships between quantities of interest and more easily monitored quantities, which can then be used as surrogates or indicators. Second, it can help scientists and managers interpret and understand monitoring data. Finally, modeling can be used with monitoring data to help make predictions.

The Need for Indicators

A scientist monitoring a bird population might walk through the woods or fields, listening for singing birds and counting the distinct individuals heard. In itself, this is a pure act of counting, but to extract useful information from the exercise, such as an estimate of the population density, the scientist employs a model. The model must at least incorporate such factors as: How far can a singing bird be heard? How frequently on average does an individual sing, given the time of day, weather, and other environmental factors, such that it might be heard? What members of the population are likely to be singing? How is that number likely to be related to the density of the whole population? These factors may be only a mental model in the investigator's head, but they could also be expressed in mathematical form, or even incorporated into a computer code that simulates the monitoring process. From the model, one could estimate the population density and use statistics to calculate the confidence in the estimate.

Another scientist, monitoring the same or a different population, might take a different tack and, instead of counting animals, measure the areas of habitat in a region, ranked by suitability for the species in question. Again, this is a pure counting exercise until the scientist uses it to try to estimate the potential population size; then a model must be employed. The model may be a simple regression between numbers of animals that can be maintained in habitat of a given quality level. The model could also be more complex, taking into account, for example, barriers that might prevent individuals of the population from reaching certain areas of possible suitable habitat, or the presence of disturbances that might make it impossible for the population to fully exploit a given area of habitat. Again, the model used could be just a mental one, it could be a very detailed computer simulation, or it could be something in between.

These are only two examples, but they illustrate the general fact that virtually all monitoring is indirect to some degree. In the vast majority of cases, there is simply no way to reliably count all individuals of a species in nature. A model has to be used to relate what can be observed, whether it is the number of singing birds, patches of habitat, or any other indicator, to estimate the quantities of interest, or target variables—population size, in this case. Is this situation unique to ecology? In fact, it is commonplace in all investigations of nature. The ability to find relationships between one quantity and another is one of the basic aims of science in general. Pursuit of this goal leads to the development of models of the relationships, which allows the monitoring of surrogate measures, as practiced in many fields. For example, the hypothesized warming of the oceans can now be monitored by measuring the time it takes sound to travel through wide expanses of ocean from a sonar transmitter to a receiver. A model relates this measurement to average ocean temperature. Possible volcanic activity is forecast by monitoring the growth of bulges on the side of a volcano. This can by done using global satellite–based positioning systems (GPS). The operations of these sensors are themselves dependent on theory and models; that is, GPS systems are possible because models can predict the position of satellites to within centimeters. In short, monitoring and models are inextricably interwoven in scientific data gathering of all types.

Development of a monitoring strategy must include the measurement of surrogates or indicators, if it is not possible or practical to measure the performance measures directly. In particular, it is best to adopt several, or a "cluster," of such indicators. By adopting several indicators one can form a more reliable estimate of the performance indicator. Also, some indicators may provide more rapid warning of changes in the performance measure, while others, although slower, are better at indicating long-term trends. Development of conceptual and/or quantitative models of a population is indispensable to choosing good surrogates. Modeling is needed to link the performance measure with the cluster of indicators. For example, a model of the life cycles of the members of a population may reveal which population process is most critical to the growth of a particular population: reproductive success in a given year, survival of young adults, survival of older, highly fecund adults, or some other characteristic. It may also reveal which, if any, resources are of critical importance to the population. One or more of these then might be useful indicators to monitor. Modeling may also reveal that other aspects of a population, such as spatial pattern of population density or age structure, may be good early indicators of population trends and could thus be useful indicators of population size, change, or viability.

In a discussion of indicators useful for monitoring biodiversity, Noss and Cooperrider (1994, table 9.3) provided a useful classification of the types of indicators, dividing them into four levels (genetic, population/species, community-ecosystem, and regional landscape) and three categories (composition, structure, and process) (Noss and Cooperrider termed this last category *function*). We can adapt their classification, with little change, to classify the various indicators of the well-being of a species population. Applied to a population, composition includes such measures as absolute or relative abundance, biomass, and percent cover (this last is suitable for vegetation or sessile animals like corals). Structure can include such measures as age- and size-structure, sex ratio, and spatial distribution. Process can include measures like reproductive success, survival, recruitment, and movement. Indicators at other levels are also useful for monitoring the health of a population. For example, for the genetic level, allelic diversity would be an example of a "composition" indicator. At the community-ecosystem level, vegetation structure suitable for nesting would be a useful indicator for a bird species, and production rates for prey would be an indicator in the "process" category. At the regional or landscape level, the pattern of patchiness of suitable habitat would be an indicator in the "structure" category. In a similar way, all of the elements of the 3-x-4 array can be filled out. The result is a matrix of indicators. Tables 6-1, 6-2, 6-3, and 6-5 show matrices of indicator variables that are being used or could be used for certain species in the Everglades. Although the performance measure for these species populations would usually be population size, rate of change of population size, or longer-term population viability, these cannot always be directly measured. A monitoring program must choose one or more indicators, which allow one to construct and verify models, from which population size, change, or viability might be estimated. These tables will be discussed in more detail later.

How well a particular variable works as an indicator depends on how well we understand its relationship to the performance measure. Modeling allows us to hypothesize and test these relationships and thus provides a basis for allowing these surrogates to be used as indicators. Modeling, therefore, can make it possible for monitoring to be more flexible, by relating total population size to other measurable quantities.

The Need for Understanding

Monitoring is of little use if the results are not well understood. Monitoring may reveal trends in populations and other variables of interest. However, by itself it does not provide a means for interpreting those trends or confirming that the trend is a result of the specific actions taken in a conservation or restoration plan. Monitoring data could even be misleading, showing initial favorable changes in a target population that are not sustained over a longer period. This is a common situation for ecological systems for two primary reasons. First, there is inherent variability in most populations. Populations change significantly through time without any human intervention, which can confound a monitoring study that is looking for effects of management actions. In the management of a large region, there is no way to use controls that would help one assess whether a plan applied to the region was responsible for any observed changes. Second, ecological systems are full of nonlinearities and feedbacks that can act in unexpected ways. This is well known in such fields as agricultural pest control, where chemical control may work to control pests for a short time but then suffer unintended reversals when natural predators of the pests are decimated or the pests develop resistance.

To integrate natural population variability and system feedbacks to achieve effective management, models must be developed and used in parallel with monitoring programs. Such models should incorporate the best understanding of the ecological system, including all relevant relationships between target and environmental variables, and should be calibrated and tested using data from the system prior to the implementation of the plan. In this way, the model should be able to represent the natural variability of the system and its possible feedback responses.

Models used to help interpret monitoring should have a sufficient level of mechanistic detail to be useful. Suppose that a population is being monitored after implementation of a plan for increased protection of that population. The model should be able to help discriminate, for example, whether the increased protection of hatchlings or adults was having the most effect on the population response. Thus, an age- or stage-structure model would be essential. Similarly, if management plans are expected to affect a population differentially in important ways across a region, then the model should contain descriptive and predictive spatial capabilities.

The Need for Prediction

Models are often employed before any management changes are put into operation to attempt to predict how well the planned changes will work, or to predict the relative effectiveness of several different alternatives. (We use the term *prediction* in a loose sense here. It is more correct to say *projection*, as absolute prediction is impossible given the uncertainty about future weather and other uncontrollable conditions.) The use of such models should be continued, even after the plan has been implemented, in combination with monitoring. Analogously, meteorologists combine modeling and monitoring simultaneously in tracking such phenomena as hurricanes. Data collected from planes and satellites monitoring the storm are fed into models to try to project its path for the next few hours or days, after which the models are again updated and corrected where necessary. Although ecological monitoring operates on longer time scales than the tracking of a single hurricane, the process of combining modeling and monitoring is similar.

Predictions or projections using a model are a means for estimating how well a conservation or restoration plan will work over some future time horizon, and models can help discern trends early in the application of the plan. This is something that monitoring data alone cannot do. Even if the data are good and sophisticated statistics are used, it may be difficult to discern trends due to a particular influence, because of the variability of natural systems. Particularly when the size of a long-lived population is monitored, it may take many years for a trend in population size to be recognized, and this recognition may come too late to make changes in management to help a threatened population. In addition, obtaining estimates of total population size on a regular basis, for example, may be impractical.

Types of Models to Meet Monitoring Needs

Models are important in enabling indicators to be effectively used in monitoring, increasing the understanding of monitoring results, and making projections into the future. Thus, it is relevant to discuss how models should be integrated into the design of monitoring programs.

Models are routinely applied in ecology to allow the use of indicators for target variables that are difficult to measure. They may be fairly simple. For example, phosphorus concentration in lakes is often a reliable indicator of primary production (Schindler 1977), or an index combining leaf area index and duration of growing season may be a good indicator of gross primary production of a forest (Kira 1975). Somewhat more complex models help estimate the rate of feeding of piscivorous fish in a lake, which is difficult to observe directly. One indirect method is to use stomach contents of the fish, the prey that are still being digested, as an indicator. To estimate the rate of predation, one can use a model that relates the degree of digestion to the time since ingestion. DeAngelis et al. (1984) show that, when such a model of digestion rate is integrated into a stochastic model simulating predation, both the average rate of successful predation and the distribution of intervals between prey captures can be calculated, given a large enough sample. Another way to estimate mean predation rate combines measurements of growth rate of the fish with a bioenergetics model that relates fish growth to intake of prey biomass (Carpenter and Kitchell 1984). Bioenergetics models may also be applied to populations in some cases to estimate survival and future population sizes. For example, because fish require at least some minimum level of stored lipids for survival through winter in temperate lakes, knowledge of the size distribution of first-year fishes at the end of the growing season may be used, along with bioenergetics modeling, to provide an indication of the number fish that will survive the winter.

All of these are important instances in which models can either extend the inferences drawn from monitoring or make it possible to substitute an easily measured indicator for a target that is difficult to measure. However, in general, to provide the means to understand and make useful predictions of populations on landscapes, more ambitious models are needed. Can models be woven into the fabric of ecological monitoring as they are in the physical sciences? Two key areas of technical advance are moving ecology in that direction.

The first advance is the increasing power of remote sensing and the use of GIS (geographic information systems) in ecology. It is now possible to characterize vegetation, topography, and many other key features of a landscape or region, on a scale of resolution of meters or finer. This has profound implications for monitoring a species population if the remotely sensed variables can act as indicators of the population. Whether these variables can act as indicators ultimately depends on having good quantitative information concerning the species' natural history; that is, its habitat preferences and aversions, space and resource needs, and other characteristics that must be learned from field research. We present such examples later.

The second advance is the increasing computational ease of performing simulations. As discussed above, individual-based simulations of growth, based on bioenergetics models, are used to infer predation rates based on growth rates of predators. However, they can be used in even more imaginative ways to facilitate and improve the interpretation of monitoring. An important example is the monitoring of population size, which often depends on use of an indicator, such as the singing of territory-holding male birds. A spatially explicit individual-based model (SEIB) incorporating the monitoring process can greatly improve the estimation of population size (discussed below). Individual-based modeling is also highly compatible with statistical approaches to determining confidence. Variation is incorporated into individuals in Monte Carlo fashion, and replicate simulations can produce statistical information on the population (Levin 1992; DeAngelis and Gross 1992; Mooij and Boersma 1996).

Models can only play the role of providing efficient surrogates in monitoring programs if they have some degree of validity. For this reason, not only is effectiveness monitoring needed, to assess how well a plan is working, but also "validation" monitoring (Noss and Cooperrider 1994) is needed as well, to develop and test assumptions about relationships that can eventually begin to simplify the number and complexity of indicators monitored.

Examples from the Everglades

The greater Everglades region has been the focus of a recent study, the Central and Southern Florida Project Comprehensive Review Study, with the objective of selecting a plan to restore some of the natural characteristics of this system. The most important features of the plan are changes in the structures and procedures by which water is regulated in central and southern Florida. A set of performance criteria was used to evaluate and compare a series of prospective plans, starting with those implemented in 1997. The performance measures numbered in the hundreds, most of which related to specific requirements for amounts, distribution, and timing of water flows.

In addition to these abiotic performance measures, biotic performance measures, such as population size or viability, of a number of important "flagship" species of the Everglades were also chosen. These species include the suite of wading birds (including the endangered wood stork, *Mycteria Americana*), the Cape Sable seaside sparrow (*Ammodramus maritima mirabilis*), the snail kite (*Rostrhamus sociabilis plumbeus*), the Florida panther (*Puma concolor coryi*), the American alligator (*Alligator mississippiensis*), and the American crocodile (*Crocodylus acutus*). This set of species has a broad and diverse range of habitat requirements. Therefore, models capable of dealing with diverse performance measures provide a strong evaluation tool for the restoration plans.

Individual models for several of these species populations have been developed by the Biological Resources Division of the U.S. Geological Survey (Across Trophic Level System Simulation Program; see DeAngelis et al. 1998, 2001) to help compare the performance of these species under current water regulation conditions and under various projected plans. Four of these models—those for the wood stork, the Cape Sable seaside sparrow, the snail kite, and the Florida panther—are used to illustrate interactive strategies of monitoring and modeling. The models are all spatially explicit, and some are individual-based models, describing populations by modeling them as collections of interacting individuals on the landscape. These models have the purpose of predicting population trends over the long term under different project scenarios. In doing so, they serve to both help understand the relationship of these species to environmental conditions and relate the target variables to indicators.

Some aspects of the modeling/monitoring strategies are discussed below for these species. Although the target variable in each of these cases is population abundance, change, or viability, the other attributes categorized under "structure" and "process" at the population, ecosystem, and landscape levels are also crucial indicators connected with the well-being of these populations (tables 6-1 to 6-3 and table 6-5). Modeling helps relate these variables to each other. The following five examples demonstrate ways in which modeling is used to identify appropriate indicators for monitoring and to interpret monitoring data.

Modeling Underlying Ecosystem Variables

Underlying ecosystem variables, such as abiotic conditions and their landscape distributions as well as their relationships to populations, can be modeled and used as indicators to supplement or replace the monitoring of the populations themselves.

For instance, models were developed for the Everglades that focus on nesting habitat conditions as the primary indicator. A spatially explicit species index model (SESI model; Curnutt et al. 2000) has been developed for the Cape Sable seaside sparrow, an endangered species restricted to the southern Everglades, to predict the breeding potential within suitable breeding habitat across the Everglades landscape on an annual time step. The model uses a GIS-based vegetation map and projections of water levels from a hydrologic model and fine-scale topographic data. Hydrology and vegetation are strong indicators of reproductive success in a given year. In recent years, a detailed knowledge of the sparrow's life cycle, the types of vegetation used in breeding and the types avoided, breeding behavior and phenology, and the effects of water depth on nesting has accumulated (Pimm et al. 1995; Nott et al. 1998; Nott 1998). It is observed that sparrows, which nest in *Muhlenbergia filipes* grass and sparse sawgrass (*Cladium jamaicense*), generally will not start nesting until water depths have receded to less than 5 centimeters. Once nesting has started, a rise of water depths to about 15 centimeters during the breeding season is assumed to flood the nest. Besides knowledge of sparrow behavior, the other essential elements for modeling are accurate knowledge of landscape vegetation, topography, and water levels through time. Fine-scale vegetation and topography data have been collected in the sparrow's range and, using this fine-scale topography, output from the South Florida Water Management Model for the southern Florida landscape is adapted to provide water depths on a daily basis on 500-x-500-meter cells or pixels. Both the vegetation and hydrologic factors are combined in an index that can be given a value for every spatial cell in the Cape Sable seaside sparrow's habitat. Thus, this model relates breeding success directly to two environmental indicators: vegetation and annual patterns of water levels. The model's predictive output is provided in the form of maps in which areas of landscape are color coded according to their different projected potentials for breeding success under given rainfall and water regulation scenarios. In model validation tests, the map output agrees well with data on numbers of nests and their spatial locations, providing evidence that landscape knowledge of vegetation and topography, along with data or predictions of hydrologic conditions, are good indicators of the Cape Sable seaside sparrow's reproductive potential.

Modeling Population and Monitoring Processes

Both the population and the monitoring process can be modeled to relate observations of singing males to actual population size. Like many bird species, the Cape Sable seaside sparrow is inconspicuous. In practical terms, these birds can only be counted through their behavioral characteristics, that is, by detection of singing males (table 6-1). During the breeding season, the sparrow population is surveyed by counting the number of singing males heard at each of the vertices of a 1-kilometer grid. In the past, the singing males have been assumed to be a constant fraction of the total sparrow population size and thus a simple indicator of the population. However, the number of singing males may be influenced by factors other than the size of the population. In particular, males may either not be present at a site or not be singing if the water depth at a site is more than a few centimeters. Water depth is a systematic factor that may vary considerably from year to year. Hence, counts of singing

Table 6-1. Indicator variables for monitoring the Cape Sable seaside sparrow (*Ammodramus maritima miribilis*).

Level	Composition	Structure	Process
Population	Abundance	Numbers of singing males (spatially explicit pattern)	Nesting effort and success
Ecosystem		Local temporally changing water depths, local vegetation types	Changes in local vegetation type
Landscape		Landscape scale patterns of water depth and vegetation, connectivity of reproductive habitat	Changes in landscape patterns of hydrology and vegetation

males can be biased and the estimates of total population size will consequently be biased as well.

This problem is addressed by applying a spatially explicit individual-based (SEIB) model of the Cape Sable seaside sparrow (SIMSPAR; Nott 1998; Elderd and Nott in review). The model is built on the same landscape and uses the same hydrologic data or model as the SESI model described above (Curnutt et al. 2000). However, SIMSPAR is a demographic model, which simulates the behavior of each individual Cape Sable seaside sparrow in the population during the breeding season. Nesting is the critical part of the life cycle affected by water (table 6-1). Water levels in sparrow breeding habitat must decline nearly to ground-surface level during the dry season before nesting behavior will begin. Therefore, the model predicts when male sparrows are likely to be singing.

SIMSPAR not only simulates the locations, breeding activities, and movements of each sparrow, but it also simulates the monitoring process itself, which is performed in the field by helicopter flights to a preestablished grid of sites. This model version of monitoring is called a "virtual helicopter survey," simulating precisely the temporal pattern of sampling of the actual survey. Water depths through time at each observation site are incorporated into the model. Thus, when the virtual helicopter lands at a site at which water depth is high, there are no singing male sparrows at that site, and the model survey records no sparrows, just as the field survey would. The virtual helicopter survey was shown to produce spatial and temporal patterns of observed singing males similar to the actual survey. But unlike the field survey, which can only observe the singing males, the SIMSPAR simulation also keeps track of the total model population. Thus, it internally estimates the correction factor that allows one to get an unbiased estimate of the sparrow population based on more-easily monitored water levels. SIMSPAR was applied to the so-called "western" subpopulation of Cape Sable sparrows, northwest of Shark Slough in Everglades National Park. It demonstrated that, although both the actual and virtual helicopter surveys showed both negative and positive fluctuations in the number of singing males from one year to the next between 1992 and 1998, the total western subpopulation of sparrows was very probably declining monotonically during that period.

Modeling Underlying Abiotic Conditions and Spatial Distributions of Species

The snail kite is an extreme specialist, feeding almost exclusively on apple snails, *Pomacea paludosa*, so it can be affected by fluctuations in the availability of this food item (table 6-2). Monitoring apple snails is then a potential indicator of resource availability and thus population status of snail kites (Darby et al. 1997). However, apple snail densities are difficult to monitor, so water depths and vegetation type appropriate to apple snails in known breeding habitat areas are taken to be the important indicators of breeding potential in particular local sites.

Because the snail kite is a nomadic species in which individuals have been described as moving large distances in a period of a month (Bennetts et al. 1998), it is not possible to understand monitoring data on snail kite numbers or to predict the dynamics of the population without taking movements into account. The range of snail kites in central and southern Florida is large enough that there is considerable spatial variation in rainfall during given years. Some parts of this region may experience drought while others have average or high rainfall. Thus, the availability of apple snail prey may vary spatially and temporally in complex ways. The snail kite moves in response to such conditions. How then does one use this information, along with hydrologic indicators, to interpret monitoring data and estimate snail kite population well-being? A SEIB model, EVERKITE, has been developed for the snail kite that relates movement through the life cycle and breeding success in various sites to status of the overall population (Mooij et al. 2002). The model follows individual birds through their lives, with emphasis on their ability to move from one site to another in response to deteriorating conditions in their present locations. The aim of the model is to relate the snail kite population to changes in water depths in the habitat sites within its range.

The monitoring of snail kites has been and is currently being carried

Table 6-2. Indicator variables for monitoring the snail kite (*Rostrhamus sociabilis plumbeus*).

Level	Composition	Structure	Process
Population	Abundance	Numbers of nests; spatial distribution of nests among wetland sites	Reproductive effort and success, timing of nesting, movement between wetlands
Ecosystem		Local temporally changing water depths, local vegetation types, local densities of apple snails, local availability of tree islands for nesting	Local degradation of habitat through prolonged flooding
Landscape		Regional patterns of wetland habitats and droughts in central and south Florida	Losses of wetland areas

out on at least fourteen major wetland sites across central and southern Florida. Much of the available monitoring data on snail kite numbers are from the annual snail kite count that was started in 1970. There are severe problems with using those annual count data as estimates of the total snail kite population, for at least two reasons. First, the number of snail kites recorded is biased by the number of years of experience of the person doing the counting (Bennetts et al. 1999). Second, the count includes only those birds in the fourteen major wetland sites, whereas during years of severe drought many snail kites are known to move to areas other than those sites. Therefore, the data appear at first to be useless for all but the crudest analyses of population trends. However, the application of modeling techniques has allowed important information on the movement patterns of snail kites to be extracted from the annual count data; this information can be incorporated into EVERKITE, since it is a spatially explicit model. The links between modeling and monitoring are provided by an understanding of the rules that govern snail kite movements. Such rules have been hypothesized from other studies on the snail kite over the past decade (Bennetts and Kitchens 1997) and are incorporated in EVERKITE. These will be tested through the studies and monitoring now being carried out on the snail kite.

The use of spatial pattern data from monitoring data in developing EVERKITE is a variation on the "pattern-oriented modeling" approach (Grimm et al. 1999). Such modeling techniques make use of the fact that spatial patterns and other patterns of the snail kite populations are manifestations of the underlying processes in the population. With statistical techniques, the relationships between the patterns and processes can be found and predictive models developed. Thus, modeling may greatly amplify the usefulness of monitoring data.

Monitoring Indicators within a Population: Nest Initiations

There are a number of behavioral aspects of wading birds that are useful targets for monitoring. One aspect is simply where the wading birds have chosen to locate their colonies. Another behavioral aspect is the date of initiation of nesting. Wading birds are believed to initiate nesting behavior in the breeding season after they have been able to forage successfully for a number of days. The date of initiation of nesting in a breeding colony appears to be a good indicator of the ultimate reproductive success of that colony in a given year. This is documented by empirical data on wading bird reproduction in the southern Everglades (Ogden 1994) showing that wood stork colonies initiated in November, December, and January have a much higher probability of success than those initiated later in the dry season (table 6-3). The reason for this relationship is that wood storks and other wading birds depend on decreasing water depths to concentrate prey during the nesting season. When the rainy season starts, usually sometime around May, water depths in the freshwater marshes will start to increase again, diluting prey. Thus, wading bird pairs that start nesting too late in the dry season may not be able have the approximately sixteen weeks of good foraging needed to raise a brood to the fledgling stage.

A wood stork model developed by Wolff (Wolff 1994; Fleming et al. 1994) simulates colonies of wood storks on a realistic, spatially explicit landscape in which temporally and spatially varying water depths and biomass densities of fish are modeled on five-day time steps and 500-by-500-meter cells, each with its own elevation, hydroperiod, and prey dynamics. The purpose of the model is to project how successful colonies in particular locations will be in producing fledglings given the hydrologic and prey dynamics of the area surrounding the sites. The foraging success of wading birds depends on both the depths of water in areas available to foraging and the concentration of fish in the right size classes (table 6-3).

The wood stork model predicts a strong and robust relationship between the date during the dry season on which wading bird nesting behavior starts and the success of the breeding colony in producing fledglings. Table 6-4 presents a number of simulation results in which both the date of nest initiation and the number of fledglings produced are shown. This result strengthens the possibility for use of nesting initiation date as an indicator for monitoring reproductive success.

Table 6-3. Indicator variables for monitoring the wood stork (*Mycteria americana*).

Level	Composition	Structure	Process
Population	Abundance	Numbers of nesting colonies formed, spatial distribution of colonies across south Florida	Timing of start of nesting behavior, nest abandonment, nesting success
Ecosystem		Local temporally changing water depths, local concentrations of fish	Local changes in hydroperiod and fish densities
Landscape		Regional patterns of elevations and connectivity of landscape allowing drying fronts to form during the breeding season in south Florida	Changes in regional hydrology, changes in fish production across region

Table 6-4. Results of computer simulations (means and distributions for fifty simulations) relating date of simulated wood stork colony formation (date measured from start of dry season in simulation) to percentage of pairs that nest and the number of fledglings produced (from Fleming et al. 1994).

Day of colony formation (SD)	Percentage of pairs nesting (SD)	Number of fledglings (SD)
10 (2)	100	742 (7)
12 (6)	93 (2)	743 (3)
23 (6)	92 (3)	621 (49)
53 (34)	66 (4)	154 (33)
75 (4)	62 (7)	22 (21)
89 (18)	37 (2)	4 (2)

Note: The date of colony formation was a function of food availability. In simulations with poor food availability early in the dry season, colony formation was delayed.

Analyzing and Interpreting Monitoring Data

To be useful, monitoring data must be analyzed and interpreted. Experience is showing that the exercise of developing a simulation model from monitoring data, which requires active engagement with the data, benefits both monitoring databases and models that are interactive with such data. This is evident in the case of a SEIB model, SIMPDEL (Comiskey et al. 1994), developed to simulate movement, growth, reproduction, and mortality of Florida panthers and white-tailed deer (an indicator for panthers; see table 6-5) in the Greater Everglades region. The endangered Florida panther, once found throughout the southeastern United States, survives now only in southern Florida, where the population has been closely monitored since 1981. SIMPDEL operates on a daily time step, although within this time step, deer and panther movements are simulated, taking account of local water conditions, forage, and prey availability. Spatially, the model makes use of vegetation data to calculate forage availability on a 100-meter scale but tracks deer and panther locations on the daily time step at a 500-meter scale. The objective of the model has been to determine the potential long-term impacts (e.g., over thirty years or more) of spatially explicit modifications in habitat (particularly hydrology) on the panther population. The spatiotemporal dynamics of habitat coupled with small population size and long-distance movements of panthers, implies that a spatially explicit individual-based modeling approach offered the best opportunity to utilize the extensive available empirical data on panthers to produce a model that tracks the effects of alternative hydrologic scenarios.

Individual-based modeling has proved to be useful in raising important questions not addressed by traditional population models. For example,

Table 6-5. Indicator variables for monitoring the Florida panther (*Puma concolor coryi*).

Level	Composition	Structure	Process
Population	Abundance	Age structure, sex ratio	Percentage of females having reproductive success, kitten/ juvenile mortality, highway deaths, movement patterns
Ecosystem		Local temporally changing water depths, bedding and denning sites, white-tailed deer density	Local changes in hydroperiod, vegetation, and deer densities
Landscape		Barriers to movement, connectedness of habitat corridors, pattern of stalking cover, road patter	Changes in regional hydrology, landscape changes

simpler population-level models generally discard excess reproduction after carrying capacity of the modeled environment has been reached. However, individual-based models must posit specific mechanisms by which the carrying capacity limits population growth. In the case of panthers, three likely mechanisms are suggested by monitoring data and field biologists: (1) intraspecific aggression, (2) kitten mortality, and (3) increased dispersal out of crowded areas. Because models require information about trends in these factors, monitoring attention can be focused on the mechanisms limiting population density.

The effort to parameterize SIMPDEL is aided by software called PANTRACK (Comiskey et al. 2002). PANTRACK is designed to analyze radiotelemetry monitoring data and display observations over a variety of background maps. Subsets of monitoring data, by time periods and/or by individuals, allow movement patterns to be analyzed and interpreted in terms of such environmental factors as habitat variability and seasonality, and such structural factors of the panther population as age, gender, reproductive status, and causes of mortality. PANTRACK helps wildlife biologists translate raw location data into information about key animal behaviors, such as territoriality and movement patterns, breeding, aggression, and other social interactions, predation, dispersal patterns, and habitat use or avoidance. This information can be used to expand what is known about the individual animal being tracked, as well as to develop behavioral rules for predictive models.

The results of parameter estimations for SIMPDEL and exploration of the model have led to better understanding of the Florida panther population and to new recommendations for improved monitoring. An important result has been a revision in thinking about the habitat needs of the Florida panther. Previous analyses (e.g., Kerkhoff et al. 2000) have contended that the Florida panther is primarily a forest species and that habitats with less than 20 percent forest cover are unsuitable for panthers. Parameter estimation for SIMPDEL required examining a larger set of radio-tracking data through more refined analyses, and the results indicate a far broader range of habitat use than the earlier claims (Comiskey et al. 2002). Several recommendations for monitoring have also been made based on aspects of the Florida panther demography indicated by SIMPDEL. These include the recommendation that young panthers, both male and female, be radio-collared before dispersal from the natal range, especially those located near the northern boundary of their occupied habitat in southern Florida. A critical question for defining expansion potential of the population is how female dispersers will be affected as carrying capacity of environment for the panther population is reached. Preliminary results of SIMPDEL simulations indicate that dispersal behaviors play an important role in the ability of panthers to colonize new areas and recolonize habitats that fall vacant. Simulations that assume females will disperse farther from their natal ranges as population pressure increases (as shown to occur in other subspecies of *Puma concolor*) show greater persistence of the population over time and quicker recovery from downturns. Monitoring to document dispersal behavior as the population expands is important to achieve accurate predictive results from models.

Conclusion

In order to comprehend and deal with the increasing threats to the world's biotic resources, effective monitoring programs are desperately needed. But monitoring is usually expensive. This is particularly true for monitoring populations, where labor-intensive observations must be carried out. In order to provide adequate monitoring directed at conservation and restoration activities and at studying changes in the natural system and renewable resources in general, ways of both reducing the costs of monitoring and increasing its effectiveness must be used. One approach, which we have described as indispensable in any large-scale monitoring program, is modeling. In this, we echo the National Science and Technology Council (1997), which wrote: "Models provide a tool to bridge that gap between the requirements of assessment and policy analyses, and the logistical and financial constraints associated with not being able to measure everything everywhere. The information provided by monitoring must be analyzed and interpreted through the use of models. Models range in complexity from simple descriptive statistics to complex, process-based computer simulations that can be used to predict future conditions."

We have stressed three major functions of modeling. First, models make possible the measurement of indicators as surrogates for target variables that are difficult to measure. Second, they aid in the understanding of the relationships between various causal factors in the environment and population changes, thus making it easier to interpret monitoring observations. Third, models can make projections that can indicate when and where modeling should be focused and that can be tested against monitoring data to improve the models.

We can go beyond these general observations to make specific suggestions for coupling modeling and monitoring. First, before management changes are implemented, baseline monitoring should be done both to establish initial conditions and to help determine what factors drive the system. Monitoring at this stage should include efforts to improve estimates of parameters that affect different stages of the species' life cycle: predation rates, dispersal behavior, mating systems, and reproductive success. From this information a first cut of a model or models can be developed. After the plan is initiated, monitoring will be commenced to assess the effectiveness of the plan and to evaluate and improve the models. Primary consideration will be given to statistical power during the early phases of monitoring.

As the models improve, it may be possible to reduce the cost of monitoring. Sensitivity analyses on the models will indicate when and where there might be significant changes in the population that need to be monitored as opposed to times and places where the population should remain stable. This will lead to a better focus of monitoring and an overall reduction in effort.

The models should be spatially explicit. In fact, we argue that virtually all models used in land management should be spatially explicit. This is primarily because conservation and restoration always takes place on a landscape or region where monitoring information can be expected to vary across the region. A spatially explicit model will also reveal the variability of the population across years at different spatial sites. The model may indicate high variability in some sites, suggesting that the population may be undergoing change in some direction, which could be critical for the management plan. Monitoring should be concentrated in such sites rather than in sites where little change is expected to occur.

Spatially explicit models are necessary also because monitoring is spatially explicit. One potential use of modeling is the simulation of the monitoring itself. The example above of the Cape Sable seaside sparrow illustrated that even though monitoring alone showed that the sparrow population was fluctuating in numbers between 1992 and 1997, the model revealed that this was an effect of the times when monitoring was done and that the population was probably steadily declining during that period.

Efforts should be made to collect data on as many environmental variables and abiotic stressors as possible, even if they have not been definitely linked causally to the population, if they can be obtained inexpensively. Remote sensing and other technical developments may make it simple to collect certain types of data, such as leaf area indices, that can serve as an indicator variable for habitat. Finally, modeling can play a key role in adaptive management. Results of model projections into the future may show a need for changes in management before this is indicated by monitoring data alone.

There should be no illusion that modeling is a panacea in monitoring programs. Experience has shown that useful models can take much time and money to develop. However, as is the case in all other fields of science, the development of models must be viewed as an integral element in the monitoring and managing of ecological systems.

ACKNOWLEDGMENTS

This research was supported in significant part by the Department of Interior's Critical Ecosystems Studies Initiative, a special funding initiative for Everglades restoration administered by the National Park Service, and in part by the United States Geological Survey's Florida Caribbean Science Center. This is publication 3058 NI00-KNAW Netherlands Institute of Ecology

LITERATURE CITED

Bennetts, R. E., and W. M. Kitchens. 1997. *The demography and movements of snail kites in Florida*. Technical Report 56. U.S. Gainesville, Fla.: Geological Survey Biological Resources Division and Florida Cooperative Fish and Wildlife Research Unit, Gainesville, Fla.

Bennetts, R. E., W. M. Kitchens, and D. L. DeAngelis. 1998. Recovery of the snail kite in Florida: Beyond a reductionist paradigm. Pp. 486–501 in *Transactions of the 63rd North American Wildlife and Natural Resources Conference*.

Bennetts, R. E., W. A. Link, J. R. Sauer, and P. W. Sykes. 1999. Factors influencing counts in an annual survey of snail kites in Florida. *Auk* 116:316–323.

Carpenter, S. R., and J. F. Kitchell. 1984. Consumer control of lake productivity. *BioScience* 38:764–769.

Comiskey, E. J., O. L. Bass Jr., L. J. Gross, R. T. McBride, and R. A. Salinas. 2002. *Panthers and forests in south Florida: An ecological perspective. Conservation Ecology* 6(18).

Comiskey, E. J., L. J. Gross, D. M. Fleming, M. A. Huston, O. L. Bass Jr., H. K. Luh, and Y. Wu. 1994. A spatially explicit individual-based simulation model for Florida panther and white-tailed deer in the Everglades and Big Cypress landscapes. Pp. 494–503 in *Proceedings of the Florida Panther Conference*, edited by D. Jordan. U.S. Fish and Wildlife Service, Fort Meyers, Fla.

Curnutt, J. L., J. Comiskey, M. P. Nott, and L. J. Gross. 2000. Landscape-level spatially-explicit species index models for Everglades restoration. *Ecological Applications* 10(6):1849–1860.

Darby, P. C., P. L. Valentine-Darby, R. E. Bennetts, J. D. Croop, H. F. Percival, and W. M. Kitchens. 1997. *Ecological studies of apple snails* (Pomacea paludosa Say). Special Publication SJ98-SP6. Gainesville, Fla.: Florida Cooperative Fish and Wildlife Research Unit, Gainesville, Fla.

DeAngelis, D. L., S. M. Adams, J. E. Breck, and L. J. Gross. 1984. A stochastic predation model: Application to largemouth bass observations. *Ecological Modelling* 24:25–41.

DeAngelis, D. L., S. Bellmund, W. M. Mooij, M. P. Nott, E. J. Comiskey, L. J. Gross, M. A. Huston, and W. F. Wolff. 2001. Modeling ecosystem and population dynamics on the South Florida hydroscape. In *Linkages between ecosystems in the South Florida hydroscape*, edited by J. W. Porter and K. G. Porter. Delray Beach, Fla.: CRS Press.

DeAngelis, D. L., and L. J. Gross, eds. 1992. *Individual-based models and approaches in ecology*. New York: Chapman and Hall.

DeAngelis, D. L., L. J. Gross, M. A. Huston, W. F. Wolff, D. M. Fleming, E. J. Comiskey, and S. Sylvester. 1998. Landscape modeling for Everglades ecosystem restoration. *Ecosystems* 1:64–75.

Elderd, B., and M. P. Nott. In review. *Changing landscapes, changing demography: An individual-based model for the endangered Cape Sable seaside sparrow* (Ammodramus maritimus mirabilis).

Fleming, D. M., W. F. Wolff, and D. L. DeAngelis. 1994. Importance of landscape heterogeneity to wood storks in the Everglades. *Environmental Management* 18:743–757.

Grimm, V., K. Frank, F. Jeltsch, R. Brandl, J. Uchmanski, and C. Wissel. 1996. Pattern-oriented modelling in population ecology. *Science of the Total Environment* 183:151–166.

Kerkhoff, A. J., B. T. Milne, and D. S. Maehr. 2000. Toward a panther-centered view of forests of South Florida. *Conservation Ecology* 4(1):1. [Online at www.consecol.org/vol4/iss1/art1]

Kira, T. 1975. Primary production in forests. Pp. 5–40 in *Photosynthesis and productivity in different environments*, edited by J. P. Cooper. Cambridge, U.K.: Cambridge University Press.

Levin, S. A. 1992. The problem of pattern and scale in ecology. *Ecology* 73:1943–1967.

Mooij, W. M., R. E. Bennetts, W. M. Kitchens, D. L. DeAngelis. 2002. Exploring the effect of drought extent and interval on the Florida snail kite: Interplay between spatial and temporal scales. *Ecological Modelling*. 149:25–39.

Mooij, W. M., and M. Boersma. 1996. An object-oriented simulation framework for individual-based simulation (OSIRIS): Daphnia population dynamics as an example. *Ecological Modelling* 93:139–153.

National Science and Technology Council. 1997. *Integrating the nation's environmental monitoring and research networks and programs: A proposed framework*. Office of the White House, Washington, D.C.: The Environmental Monitoring Team of the Committee on Environment and Natural Resources, Office of the White House, Washington, D.C.

Noss, R. F., and A. Y. Cooperrider. 1994. *Saving nature's legacy: Protecting and restoring biodiversity.* Washington, D.C.: Island Press.

Nott, M. P. 1998. Effects of abiotic factors on population dynamics of the Cape Sable seaside sparrow and continental patterns of herpetological species richness: An appropriately scaled landscape approach. Ph.D. diss. University of Tennessee.

Nott, M. P., O. L. Bass Jr., D. M. Fleming, S. E. Killefer, N. Fraley, L. Manne, J. L. Curnutt, T. M. Brooks, R. Powell, and S. L. Pimm. 1998. Water levels, rapid vegetation changes, and the endangered Cape Sable seaside sparrow. *Animal Conservation* 1:23–32.

Ogden, J. C. 1994. A comparison of wading bird nesting colony dynamics (1931–1946 and 1974–1989) as an indication of ecosystem conditions in the southern Everglades. Pp. 533–570 in *Everglades: The ecosystem and its restoration*, edited by J. C. Ogden and S. M. Davis. Delray Beach, Fla.: St. Lucie Press.

Peck, S. 1998. *Planning for biodiversity: Issues and examples.* Washington, D.C.: Island Press.

Pimm, S. L., K. Balent, T. Brooks, J. L. Curnutt, J. L. Lockwood, L. Manne, A. Mayer, M. P. Nott, and G. Russell. 1995. *Cape Sable sparrow annual report.* National Biological Service/National Park Service, Everglades National Park, Homestead, Fla.

Schindler, D. W. 1977. Evolution of phosphorus limitation in lakes. *Science* 195:260–262.

Wolff, W. F. 1994. An individual-oriented model of a wading bird nesting colony. *Ecological Modelling* 72:75–114.

CHAPTER 7

Role of Knowledge-Based Systems in Analysis and Communication of Monitoring Data

Keith M. Reynolds and Gordon H. Reeves

Concepts of ecosystem management have become basic principles of modern philosophies underlying natural resource management in United States and other nations in the 1990s (Committee of Scientists 1999). Ecosystem management per se has been defined as "the use of skill and care in handling integrated units of organisms and their environments . . . to produce desired resource values, uses, or services in ways that also sustain the diversity and productivity of ecosystems" (Overbay 1993); its primary goal is ecosystem sustainability (Daly and Cobb 1989; Dixon and Fallon 1989; Gale and Cordray 1991; Greber and Johnson 1992; Maser 1994). In its broadest sense, sustainability requires achieving an operational balance among concerns for ecological states and processes, economic feasibility of management actions, and social acceptability of expected management consequences (Bormann et al. 1994; Salwasser 1993).

Monitoring in Adaptive Management

Conceptual models for implementing an ecosystem management process have been proposed (FEMAT 1993; Maser et al. 1994; Noon et al. 1999). The process is conceived to be both continuous and adaptive, including basic stages of monitoring, evaluation, planning, and implementation (chaps. 1 and 3). A process that actively supports adaptation is necessary because ecosystems are extremely complex entities and our knowledge about them is far from perfect, and because both the social and biophysical components of ecosystems are highly dynamic and unpredictable. Evaluation of monitoring data is fundamental to an adaptive ecosystem management process because it simultaneously concludes the iteration on the cycle and generates revisions to current knowledge that become the basis for adaptation in the next iteration (Committee of Scientists 1999).

A dominant theme of this chapter is *integrated* evaluation because ecosystem management is fundamentally concerned with integrated management, which in turn presupposes integrated analysis. An extensive literature on ecosystem management, including most citations above, has clearly articulated the need for integration across disciplines and spatial scales. However, convincing demonstrations of integrated analysis to support ecosystem management have been lacking. The objective of this chapter is to discuss practical approaches to integrated evaluation across disciplines and spatial scales, and landscape-level application of the latter.

Aquatic Conservation Strategy

The Forest Ecosystem Management Assessment Team (FEMAT) developed long-term management alternatives for maintaining and restoring habitat conditions to support well-distributed and viable populations of species associated with late-successional and old-growth forests in northwest California and in western Oregon and Washington. Analysis of alternatives developed by FEMAT in an environmental impact statement (USDA and USDI 1994a) led the U.S. Forest Service (FS) and USDI Bureau of Land Management (BLM) to adopt the ecosystem management strategy now known as the Northwest Forest Plan, which is contained in the Record of Decision (ROD; USDA and USDI 1994b).

Major goals of the aquatic component of FEMAT and the Record of Decision were to halt habitat degradation, maintain habitat and ecosystems that currently are in good condition, and aid in the recovery of freshwater habitats of at-risk fish populations. The Aquatic Conservation Strategy is a regionwide strategy that seeks to retain, restore, and protect the processes and landforms that contribute habitat elements to streams and to promote good habitat conditions for fish and other aquatic and riparian-dependent organisms (FEMAT 1993). The Aquatic Conservation Strategy has four primary components that, along with late-successional reserves, are designed to maintain and restore the productivity and resiliency of riparian and aquatic ecosystems. The four components are riparian reserves, key watersheds, watershed analysis, and watershed restoration (FEMAT 1993).

There are nine objectives of the Aquatic Conservation Strategy. These objectives describe general characteristics of functional aquatic ecosystems and landscapes that are intended to restore and maintain good habitat in the context of geomorphic and ecologic disturbance. The Aquatic Conservation Strategy objectives are designed to provide an ecological context for implementing management activities at landscape, watershed and site scales.

The Record of Decision identified the need to develop an effectiveness monitoring strategy and program for the Northwest Forest Plan. Initial resources designated by federal agencies for monitoring were the northern spotted owl, marbled murrelet, late-successional and old-growth forests, and aquatic and riparian ecosystems. Detailed discussion of the overarching strategy for these effectiveness monitoring programs is found in Mulder et al. (1999), Noon et al. (1999), and chapter 2.

In response to direction in the Record of Decision, and in accordance with other effectiveness monitoring programs in the Northwest Forest Plan region, an effectiveness monitoring plan for aquatic and riparian ecosystems was developed by an interagency team of biologists, ecologists, and statisticians from the cooperating federal agencies (Reeves et al. in press).

The move to ecosystem management as described in FEMAT (1993) and the Record of Decision (USDA and USDI 1994a,b) requires examining and considering aquatic ecosystems at larger spatial and temporal scales than was previously done. Thus, primary questions arising from the implementation of the Northwest Forest Plan concern the status and trends of watershed condition and associated aquatic ecosystems. The interagency technical team determined that the basic watershed scale should be sixth-field hydrologic units, which range from 10,000 to 20,000 hectares in size. This unit size is one at which many watershed analyses are conducted and for which previous monitoring has been done by the Forest Service and BLM. Based on the technical team's conceptual model (chapter 1), three sets of indicators, representing natural or human-induced stressors (Noon et al. 1999) have been identified (Reeves et al. in press). These are upslope (e.g., road density and vegetation types and ages), riparian (e.g., riparian vegetation types, ages, and density), and in-channel (e.g., amount of large wood, and number of pools) indicators. Water quality and biological indicators are less well defined but are being incorporated into the monitoring program.

Conventional Approaches to Monitoring and Evaluation

Several broad-scale ecoregional assessments have been conducted in the United States in recent years (FEMAT 1993; Everett et al. 1994; Anonymous 1996, 1997), and several more are in progress. Assessment teams developed an approach (now standardized) to define the analytical problem and used a scoping process to identify and evaluate critical issues deserving consideration. A needs assessment was performed to identify data requirements and analytical methods needed to respond to the issues, and a wide array of statistical, simulation, and optimization procedures was used to address various components of the overall assessment problem. To assert that the suite of analyses used in these assessments was conducted ad hoc would be a disservice. However, it is fair to say that, although assessment teams may have carefully coordinated the conduct of analyses with integration in mind, there is little evidence in the reports that they were successful in achieving an effectively integrated analysis.

Knowledge-based Approaches to Monitoring and Evaluation

Expert systems (Jackson 1990; Waterman 1986), known more generically as knowledge base systems (Waterman 1986), began to be applied to natural resource management in 1983 (Davis and Clark 1989). More recently, Durkin (1993) catalogued over one hundred knowledge base applications in the environmental sciences. O'Keefe (1985) envisioned an important role for knowledge base systems as components of larger decision support systems in the future.

A knowledge base is a formal logical specification for interpreting information and is therefore a form of meta database in the strict sense. A knowledge base is a logical representation expressed in terms of the relevant entities and logical relations among them. Interpretation of data by a knowledge base engine (a logic processor) provides an assessment of system states and processes represented in the knowledge base. Use of logical representation for assessing the state of systems frequently is desirable or necessary. Often, the current state of knowledge about a problem is too imprecise for statistical or simulation models or optimization, each of which presume precise knowledge about relevant mathematical relations. In contrast, knowledge-based reasoning provides solutions for evaluating more imprecise information, and some knowledge base systems can provide useful analyses even in circumstances in which data are incomplete.

Knowledge-based solutions are particularly relevant to ecosystem management because the topic is conceptually broad and complex, involving numerous, often abstract, concepts (e.g., health, sustainability, ecosystem resilience, ecosystem stability) whose assessment depends on numerous interdependent states and processes. Logical constructs are useful in this context because the problem can be evaluated as long as the entities and their logical relations are understood in a general way and can be expressed by subject matter authorities.

Logic-based analysis is not in direct competition with other, more traditional forms of analysis. Instead, knowledge-based representations can be used as logical frameworks within which results from many specific mathematical models are integrated. Traditional knowledge-based systems, dating from the early 1970s, have employed rule-based reasoning. However, such systems only are suitable for narrow, well-defined problems (Waterman 1986; Jackson 1990). As discussed in the next and later sections, newer forms of knowledge-based representation, based on object models and fuzzy logic, substantially improve the ability to model large, general problem domains such as ecosystem assessment.

Zadeh (1965, 1968) first presented basic concepts of approximate reasoning with fuzzy logic. Subsequent concept papers (Zadeh 1975a,b, 1976) elaborated on the syntax and semantics of linguistic variables, laying the foundation for what has now become a significant new branch of applied mathematics. Fuzzy logic is concerned with quantification of set membership and associated set operations. Because fuzzy set theory is a generalization of Boolean set theory, most Boolean set operations have equivalent operations in fuzzy subsets (Kaufmann 1975).

Fuzzy logic representations are more intuitively satisfying than classical Boolean (bivalent) logic and are more precise and compact compared to classical rule-based representations. As a simple example of fuzzy set membership functions, consider our definition of the linguistic variable "warm" (fig. 7-1). In this particular definition, a temperature of 60 degrees F is definitely not warm (it's cool). A temperature of 90 degrees likewise is definitely not warm (it's hot). Temperatures in the open intervals (60, 70) and (80, 90) have partial membership in the concept "warm," and temperatures in the closed interval [70, 80] have full membership in the set. Terms such as warm, cold, hot, wide, tall, and so forth, convey a clear meaning even though they are imprecise (e.g., fuzzy). An important point that deserves highlighting here is that fuzzy logic is not about sloppy or "loose" logic. Rather, fuzzy logic is a precise and formal branch of mathematics concerned with *quantification of imprecise information.*

Application of fuzzy logic to natural resource science and management is still relatively new. General areas of application include classification in

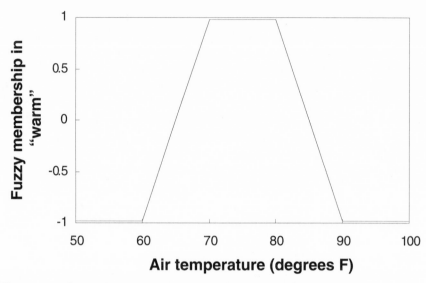

Air temperature (degrees F)

Figure 7-1. Quantifying membership in a fuzzy subset for air temperature where temperatures less than 60 degrees F are considered "cold," temperatures between 65 degrees F and 80 degrees F are considered optimal, and temperatures higher than 90 degrees F are considered "hot."

remote sensing (Blonda et al. 1996), environmental risk assessment (Holland 1994), phytosociology (Moraczewski 1993a,b), geography (Openshaw 1996), ecosystem research (Salski and Sperlbaum 1991), and environmental assessment (Smith 1995, 1997). More specific applications include catchment modeling (Anonymous 1994), cloud classification (Baum et al. 1997), evaluation of plant nutrient supply (Hahn et al. 1995), soil interpretation (Mays et al. 1997; McBratney and Odeh 1997), and land suitability for crop or forest production (Ranst et al. 1996). Our application to monitoring (Reeves et al. in press) is yet another new and promising application of fuzzy logic in natural resource science.

To appreciate the compactness of fuzzy set representation, consider a fuzzy logic implementation of landslide risk in which risk depends on three factors, A, B, and C. In the NetWeaver knowledge base system, which we describe in more detail later, each factor is represented by a data object, each having an associated fuzzy membership function. Risk is evaluated by a single network object that contains a graphically constructed logic expression, typically involving a limited number of combinations of the three data objects. In the simplest case, for example, the NetWeaver representation of the risk problem would require one network object with a single logical expression to evaluate the three data objects. A more complex formulation might require a single network object with a compound logical expression involving multiple references to the three data objects and with each elementary expression using varying combinations of fuzzy arguments on the data objects. Even in this more complex case, there would still only be one network object, three data objects (because they are reusable by multiple reference), and now perhaps six to nine fuzzy arguments. By comparison, a set of rules implementing a comparable model of landslide risk and evaluating a reasonable number of categorical outcomes (e.g., risk is low, moderate, or high) would need to consider about 256 rules.

Similarly, fuzzy logic has significant practical advantages over Bayesian belief networks (Ellison 1996; Howard and Matheson 1981) in some contexts. Bayesian belief networks may be preferable to fuzzy logic networks when conditional probabilities of outcomes are known. However, Bayesian belief networks, like rule sets, are difficult to apply to large, general problems because the number of conditional probabilities that must be specified can quickly become extremely large as the conceptual scope of a problem increases. In many natural resource contexts, model design not only becomes difficult to manage, but many probabilities will not be well characterized and will therefore need to supplied by expert judgment, thus negating much of the value to be gained by a more statistically based approach to knowledge representation.

The foregoing arguments are not meant to imply that fuzzy logic networks are inherently superior to Bayesian belief networks or other forms of knowledge representation. On the contrary, these alternative forms of representation may be highly complementary to one another in practice. In particular, fuzzy logic networks are well suited as logic frameworks for integrating model results from a variety of analytical systems.

Landscape Analysis with EMDS

The Forest Service Pacific Northwest Research Station released the first production version of the Ecosystem Management Decision Support (EMDS) system in February 1997. As of July 2001, about 500 sites worldwide have requested EMDS, including about seventy Forest Service sites. About 90 sites are national research institutes and another 125 are universities.

System Overview

EMDS integrates a knowledge base engine into the ArcView (Environmental Systems Research Institute, Redlands, Calif.) geographic information system (GIS) to provide knowledge-based reasoning for landscape-level ecological analyses. Major components of the EMDS system include the NetWeaver knowledge base system, the EMDS ArcView application extension, and the Assessment system. More detailed descriptions of the system are provided in Reynolds et al. (1996, 1997a, 1997b) and Reynolds (1999a, 1999b).

NetWeaver was initially developed in 1988 by Dr. Michael Saunders of Pennsylvania State University based on concepts originally proposed by Stone et al. (1986). It has steadily evolved since its introduction and now provides a form of knowledge representation that offers several critical advantages for landscape-level ecosystem analyses. Key features of the system include an intuitive graphical user interface, object-based logic networks of propositions, and fuzzy logic. Implementation of the user interface together with NetWeaver's object-based representation supports the design of highly modular knowledge bases. Modularity in turn enables effective, incremental evolution of knowledge base structures from simple to complex forms. It has been asserted as an axiom of modern systems theory that incremental evolution is a virtual requirement for design of complex systems (Gall 1986).

Reynolds (1999b) provides complete documentation on a public domain version of NetWeaver. The following three sections briefly summarize basic features of the system. Specific components of a prototype knowledge for evaluating salmon habitat suitability are used for illustrations. The complete knowledge base is described and illustrated in the section below entitled "The Nestucca Basin: An example."

Networks are the basic building blocks for designing logic-based representations of problem-solving knowledge in NetWeaver (fig. 7-2). Three key attributes of any network are (1) its mnemonic identifier, (2) the proposition that the network is designed to evaluate, and (3) the proposition's degree of truth. A *proposition* is defined as the smallest unit of thought to which a measure of truth can be assigned (Stillings et al. 1991). Each network object has an associated documentation object that specifies the name of the network (mnemonic), the proposition being evaluated, additional explanatory information, the source of knowledge that is the basis for the logic structure, relevant literature citations, and assumptions implied by the logic structure.

In the example (fig. 7-2), the network-named biophysical condition evaluates the proposition that the biophysical properties of a sixth-field watershed provide suitable habitat for salmon. The logic specification in the diagram makes the meaning of the proposition explicit and unambiguous. Other ovals in the figure (i.e., riparian, upland, and in-channel) also are networks. Because the evaluation of biophysical condition depends on these other networks, the other networks are said to be logically

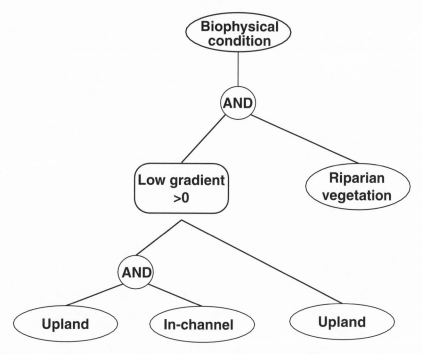

Figure 7-2. Logic specification for the biophysical condition network. Each oval represents a logic network. The upland, in-channel, and riparian vegetation networks are premises of biophysical condition.

antecedent to biophysical condition. In the sense of classical argument (Halpern 1989), the propositions of networks logically antecedent to biophysical condition are premises of biophysical condition.

Fuzzy Logic Operators

The basic logic operators in NetWeaver are AND, OR, XOR, SOR, and NOT. These are fuzzy logic, rather than Boolean, operators. That is, they perform fuzzy math operations (Kaufmann 1975) on their sets of antecedents, propagating truth values upward through the logic structure of a knowledge base. Full descriptions of these and other operators are given in Reynolds (1999b).

Data Links

Data links are a form of elementary network that read and typically evaluate data. The example uses two simple data links that each read a single datum and compare the value to a fuzzy membership function (similar to that shown in fig. 7-1) to compute a truth value for the proposition being evaluated by the data link. Arguments to data links may be simple Boolean expressions such as "less than 20," in which case the data link evaluates to simply true or false. A data link may have no argument, in which case the value it returns is simply used directly by the network object that is requesting the data. Finally, calculated data links read data from any number of other network objects, compute a result given a mathematical specification, and then evaluate the truth value of the result. The fuzzy argument defined earlier will most likely be replaced at a later stage of knowledge base development with a function whose shape parameters are themselves functions of other context information. For example, shape parameters for this function might vary by biophysical province.

Assessment System

The Assessment interface provides direct access to each object's documentation attributes for reference during an analysis. Alternative scenarios can be run with modifications of knowledge base structure (i.e., choosing to ignore certain knowledge base components) or with modifications to input data in copies of the original input database. The logic engine computes the influence of missing information, if any, and the data acquisition manager subsystem provides tools for synthesizing information about influence with information supplied by the analyst about the logistics of acquiring missing information to prioritize missing data for collection. Finally, the knowledge base browser subsystem provides a graphic interface for interactively navigating the evaluated state of an instantiated knowledge base.

The Nestucca Basin: An Example

The Nestucca Basin is a fifth-field hydrologic unit located in the northern portion of the Oregon Coast Range province and contains forty-five complete and composite sixth-field watersheds.

Complete watersheds are topographically defined, with no stream inlet and a single outlet. Composite watersheds have one or more stream inlets and can have more than one outlet. Streams in similar sized true watersheds are the same size. It is expected that biological and physical attributes will be similar. However, stream sizes in composite watersheds can vary widely, from small headwater streams to large rivers and may even be a combination of sizes. Biological and physical attributes and relations among them will vary accordingly.

In 1998, a watershed assessment team from the Siuslaw National Forest (USDA Forest Service) rated each true sixth-field watershed in the basin with respect to its suitability for providing salmon habitat, following the rating system of the National Marine Fisheries Service (NMFS) matrix (Reeves et al. 2001).

The NMFS matrix approach considered each parameter individually and did not consider interactions among parameters. Evaluation of parameters was either acceptable or not acceptable, which did not allow for intermediate considerations that the fuzzy arguments provide. Final interpretation of the effect of a proposed action or set of actions on fish habitat within the watershed under consideration was subjective. There were no clear rules for making the interpretation.

Knowledge Base Design

To improve on the basic approach to watershed evaluation for salmon habitat suitability as required to implement the Aquatic Conservation Strategy of the Northwest Forest Plan, the authors designed a prototype knowledge base suitable for application to true sixth-field watersheds in the Oregon Coast Range province. Starting with basic requirements identified in the NMFS matrix, logical requirements for the knowledge base were summarized from a synthesis of literature review and professional knowledge. We then reviewed the initial design with the assessment team from the Siuslaw National Forest to make refinements, corrections, and additions to the initial version.

The calculated data link, upland calc, is evaluated by a fuzzy node that is yet another method for implementing a fuzzy membership function. The xy nodes under a fuzzy node specify the x and y coordinates of the function. This form of function specification is more general than the fuzzy argument specification of a data link because a fuzzy node may have any number of xy nodes, and the x and y terms may be dynamically defined. That is, x and y coordinates may be supplied by a data link or computed in a calculated data link from one or more data links. The truth value returned by the fuzzy node is computed by linear interpolation between adjacent xy nodes, based on the position of the source datum on the abscissa.

EMDS Analysis Products

The most basic outputs from an EMDS analysis are maps of truth values for networks included in an analysis. In our example (fig. 7-3), the primary map of interest displays the strength of evidence for the proposition that sixth-field watersheds of the Nestucca Basin have a suitable bio-physical condition for salmon habitat. Composite watersheds in the basin, for which the knowledge base was not designed, are displayed without a fill pattern (fig. 7-3). All information that is logically antecedent to the selected networks must also be evaluated by the logic engine and so is dis-playable in maps of truth values, graphs, or tabular output generated by the logic engine.

The logic structure of the network for biophysical condition was de-signed to yield truth values representing a broad range, from definitely not suitable to completely suitable for salmon habitat. At times, it may be useful to translate ("defuzz-ify") truth value results into discrete or over-lapping categorical outcomes. For example, we used fuzzy arguments in each of three additional networks to define overlapping categories of truth value ranges on the full range of biophysical condition to indicate conditions that were considered degraded, compromised, or good for salmon habitat.

Maps characterizing landscape condition are powerful communication tools, but it is at least as important to be able to explain the basis for ob-served results in a clear, intuitive manner. Perhaps one of the most signif-icant features of knowledge base systems, in the latter regard, is their abil-ity to display a logic trace that explains the derivation of conclusions. The

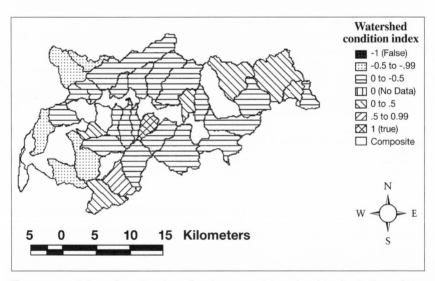

Figure 7-3. Map of truth values for the proposition that biophysical condition of sixth-field watersheds in the Nestucca Basin provides suitable salmon habitat. Features with a vertical hatch pattern are composite watersheds for which the knowledge base is not appropriate.

EMDS ArcView extension implements a graphic knowledge base browser via the Assessment system that allows an analyst to interactively navigate through the evaluated state of the knowledge base in order to trace the lines of evidence leading to conclusions. The top portion of the browser window displays an expandable outline of knowledge base structure, and the bottom portion displays the logic structure of the element currently selected in the outline.

The NetWeaver logic engine reasons with incomplete information, when needed, by using fuzzy math, and the engine evaluates the influence of any missing information with respect to its contribution to completeness of an analysis. Input data records for the Nestucca Basin analysis contained a few missing observations on data fields. However, the engine implements what might be described as evidence-based reasoning, producing partial evaluations of networks with incomplete information (fig. 7-3). The knowledge base is, in effect, a cognitive map of the problem. Its structure contains information that can be used to compute the influence of missing information (fig. 7-4). The computation of influence is based on the level at which a data link (or a network object in general) enters the logic structure, how many references there are to a data link in the knowledge base, and how frequently the associated database field contains missing observations.

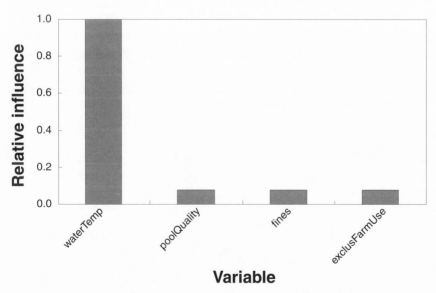

Figure 7-4. Influence of missing information in the Nestucca Basin analysis calculated by the NetWeaver logic engine.

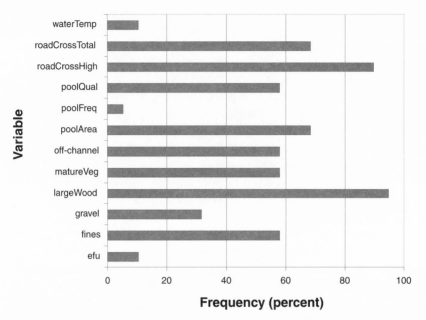

Figure 7-5. Frequency with which data in low-gradient drainages contributed to the conclusion of a compromised or degraded biophysical condition. Drainages with an average reach gradient less than or equal to 4 percent were classified as low-gradient and had data available on in-channel conditions.

Finally, the logic engine also generates tabular output from which an analyst can summarize additional information such as the frequency distribution of truth values or the frequency with which data links substantially contribute to a particular conclusion (fig. 7-5). Frequency distributions of truth values provide a simple synoptic view of conditions over an assessment area at a point in time, and useful statistical inferences about the efficacy of restoration programs, for example, could be drawn from comparisons of such frequency distributions over consecutive assessment times. Similarly, more detailed summaries of conditions in a monitoring area (fig. 7-5) provide useful background information for design of restoration programs as well as a basis for more detailed statistical inferences about the efficacy of these programs.

Discussion

Problem specification for ecological assessment may well deserve to be classified as a wicked problem (Allen and Gould 1986). As discussed in the introduction, ecosystem monitoring must consider potentially numerous states and processes of biophysical, social, and economic components of an ecosystem. Many entities may have both deep and broad networks of logical dependencies as well as complex interconnections. Although constructing such complex representations in NetWeaver is not trivial, it is at least rendered feasible by the precision and compactness of fuzzy logic relations, and by the graphic, object-based representation of logic networks in the system interface. No subject-matter authority, nor for that matter any group of authorities, is capable of holding a comprehensive cognitive map of such a complex problem domain in their consciousness. However, the graphic, object-based form of knowledge representation in NetWeaver is highly conducive to the incremental evolution of knowledge base design from simple to complex forms.

Two essential aspects of the knowledge-based approach to monitoring that we have described and illustrated are its goal orientation and use of logical formalism in problem specification. The implications for realizing something that approaches truly integrated analysis are significant. Knowledge base design begins with identification of the questions (formulated as propositions) that the analysis ultimately is designed to address. Problem specification proceeds by identifying lines of reasoning that decompose the original propositions into progressively more concrete ones until links to data are identified. The result, effectively, is a cognitive map of the problem, including not only identification of all entities pertinent to the problem, but also their logical interdependencies. The final specification thus not only provides logical links between data requirements and the original motivating questions, but also provides a specification for how the data are to be interpreted, taking into account interrelations among pieces of information.

Performing Evaluations with Incomplete Information

Almost all significant monitoring and assessment programs start out with substantial gaps in the data available to fulfill monitoring requirements. The NetWeaver logic engine is particularly well suited for use in applications such as effectiveness monitoring in particular and ecosystem analysis in general because it is able to reason with incomplete information and still provide meaningful results, and because it can evaluate the influence of missing information.

The number of logical entities in any reasonably realistic problem and the number of interdependencies among them can easily overwhelm the capacity of the human brain to reason effectively about such questions as "What is the most useful information that I could acquire at this point in time to improve the completeness of my analysis?" In this regard, it is important to appreciate that the influence of information is highly dynamic:

the influence of what is missing depends both on the observations that are currently available and on the values in those observations. Consequently, use of iterative analysis of data influence in EMDS following successive, incremental additions of new data could conceivably reduce resources devoted to data collection by 50 percent or more in typical applications.

Explanation of Monitoring Results

The ability to explain monitoring results to colleagues, managers, or the public is at least as important as the ability to perform an analysis in the first place. If the logic of underlying models cannot be presented in clear, unambiguous terms to such audiences, results will simply not be trusted. Very powerful and sophisticated linear programming systems such as FORPLAN largely failed on this account (Gustafson et al. 2002).

Our experience to date with EMDS, and with its knowledge base browser interface in particular, has been that the presentation of results from logic models is easy to grasp at an intuitive level so the results are accessible to broad audiences with only limited explanation. An analyst can easily run the browser interactively in front of an audience, navigating the structure of the evaluated knowledge base and explaining the derivation of the logic as he or she goes.

Evaluation within Larger Contexts

Almost all landscape analyses are performed within some broader context. Our knowledge base for evaluation of salmon habitat suitability is a prototype constructed for the specific context of the Oregon Coast Range. It is unlikely that the current version is sufficiently general that it would be suitable for application throughout the entire Northwest ecoregion.

Fortunately, it is not difficult to generalize prototypes such as ours for broader applicability. In some cases, it may be sufficient to replace a hard-wired fuzzy argument in a data link with a calculated data link that contains a more general fuzzy membership function whose parameters vary with additional context information. For more dramatic variations in logic structure, it may be necessary to add in new logic switches that similarly select alternative logic pathways based on additional context information. In a technical sense, knowledge base generalization as just discussed is a relatively minor issue. On the other hand, it leads naturally to discussion of the more interesting question of multiscale implementations.

Extending to Multiple Spatial Scales

It is relatively easy to extend integrated analysis via knowledge-based reasoning for monitoring applications to multiple spatial scales. Data from fine-scale landscape features such as sixth-field watersheds are first processed by a knowledge base designed for that scale. Knowledge base outputs then go through an intermediate filter (typically implemented in a spreadsheet or database application) to synthesize information for input to the next-coarser scale. A second knowledge base processes the synthesized information to provide an assessment of landscape-level attributes. Finally, knowledge base outputs at the broader landscape scale may feed back to the fine scale as context information that influences evaluations at the fine scale. This simple conceptual model provides the basis for a formal logical specification of analyses that are consistent across scales. Reynolds (2001b) provides a simple example of this approach based on evaluation of stream reaches and watersheds. Hierarchies—or, more generally, networks—of knowledge-based analyses for multiscale monitoring as suggested would be highly consistent with theories of the hierarchical organization of ecosystems.

Knowledge Bases as Frameworks for Collaboration

Most examples of regional ecosystem monitoring projects are broad in conceptual scope. Many variables are potentially of interest, and these variables influence a variety of ecosystem states and processes that frequently are conceptually interrelated. Knowledge from a collection of disciplines is typically brought to bear to design and implement the monitoring program. Not surprisingly, scientists and resource managers involved in monitoring programs have found conceptual models useful as a guide to deciding what variables to monitor and how to evaluate them to address the higher-order, more abstract questions that have motivated the monitoring program (Gustafson et al. 2002). We have already described knowledge bases as a form of meta-database. More broadly, knowledge bases can be thought of as conceptual frameworks, but with the added advantage of possessing a well-defined syntax that avoids the semantic ambiguities of less-formal representations. To the extent that knowledge bases can improve the precision of the semantics of evaluation monitoring, they facilitate sharing knowledge and therefore collaboration among parties to the project. Use of knowledge bases as integrating conceptual frameworks becomes perhaps even more valuable when multiple agencies or landowners are collaborating in regional assessments as in the case of the Aquatic/Riparian Effectiveness Monitoring Program

(AREMP) of the Northwest Forest Plan (Reeves et al. in press). Reynolds (2001a) presents an example of knowledge base design for evaluating forest ecosystem sustainability based on data requirements outlined in the Montreal Process, which included seven broad criteria and sixty-seven indicators, evidence that such approaches can be useful in integrating ecoregional analyses at continental and global scales.

We have seen some of the potential for sharing knowledge realized in adapting knowledge bases developed for AREMP to the North Coast Watershed Assessment Project (NCWAP) being conducted by the Department of Fish and Game and the Department of Forestry in northern California. In this situation, the authors worked with a team of twenty resource specialists from various state and federal agencies in the course of a three-day workshop to customize and adapt the AREMP knowledge bases to the geographic context of NCWAP. We also borrowed components related to evaluation of stream-flow properties from an earlier knowledge base designed for assessing hydrologic integrity of watersheds (Reynolds et al. 2000). Most adaptations were accomplished within the time frame of the workshop. However, NCWAP participants spent a few additional days completing adaptations to the knowledge base specifications on their own, so this example also illustrates the relative ease with which it is possible to transfer knowledge of how to build relatively complex knowledge bases. Lastly, the AREMP-NCWAP example suggests the possibility of evolving shared knowledge on an interregional level.

Repeatability and Adaptability

Monitoring programs are designed to repeat measurements to develop information about trends in ecosystem conditions over space and time. An advantage of the more formal knowledge-based approach in trend analysis, compared to informal conceptual models, is that knowledge bases permit consistent interpretations across time periods. By taking some care in knowledge base design, it also is possible to design models with broad geographic applicability by using dynamically defined functions and switches as discussed earlier. In addition to geographic adaptability, there is the larger issue of adapting knowledge base design as knowledge itself evolves about how to evaluate ecosystems. The modular architecture of NetWeaver knowledge bases, combined with the system's underlying object-based implementation, make editing the logic specification of a knowledge base relatively easy. Our earlier example of adapting the AREMP knowledge bases to the NCWAP context is a good case in point.

Conclusion

EMDS has been available as a production system for effectiveness monitoring of landscapes since 1997, but the system continues to evolve. The first demonstration of an integrated, multiscale assessment with EMDS (Reynolds 2001b) also demonstrated how decision models for planning, based on the analytic hierarchy process (AHP, Schmoldt et al. 2001), could be integrated with evaluation results at each scale of landscape evaluation. EMDS version 3.0, released in late fall of 2002, will implement an integrated AHP planning component, thus extending the analytical scope of EMDS to include specifications for (1) monitoring requirements, (2) evaluation of monitoring data, and (3) planning priorities for landscape elements within an assessment area. Somewhat farther into the future, we hope to provide a final EMDS component that effectively closes the loop on the adaptive management process by developing a specification for management of implementation data.

NOTE

The use of trade or firm names in this publication is for reader information and does not imply endorsement by the U.S. Department of Agriculture of any product or service.

LITERATURE CITED

Allen, G. M., and E. M. Gould Jr. 1986. Complexity, wickedness, and public forests. *Journal of Forestry* 84:20–23.

Anonymous. 1994. Fuzzy logic applied to catchment modelling. *Water and Wastewater International* 9:40–46.

———. 1996. *The southern Appalachian assessment summary report.* Technical Paper R8-TP-25. U.S. Department of Agriculture, Forest Service Region 8, Atlanta, Ga.

———. 1997. *Sierra Nevada Ecosystem Project, Final Report to Congress.* Addendum. University of California, Centers for Water and Wildland Resources, Davis, Calif.

Baum, B. A., V. Tovinkere, J. Titlow, and R. M. Welch. 1997. Automated cloud classification of global AVHRR data using a fuzzy logic approach. *Journal of Applied Meteorology* 6:1519–1526.

Blonda, P., A. Bennardo, G. Satalino, and G. Pasquariello. 1996. Fuzzy logic and neural techniques integration: An application to remotely sensed data. *Pattern Recognition Letters* 17:1343–1347.

Bormann, B. T., M. H. Brookes, E. D. Ford, A. R. Kiester, C. D. Oliver, and J. F. Weigand. 1994. *Eastside forest health assessment.* Vol. 5. *A framework for sustainable-ecosystem management.* General Technical Report PNW-GTR-331. U.S. Department of Agriculture, Forest Service, Pacific Northwest Research Station, Portland, Ore.

Committee of Scientists. 1999. *Sustaining the people's lands: Recommendations for stewardship of the national forests and grasslands into the next century.* U.S. Department of Agriculture, Washington, D.C.

Daly, H. E., and J. B. Cobb Jr. 1989. *For the common good: Redirecting the economy toward community, the environment, and a sustainable future.* Boston: Beacon Press.

Davis, J. R., and J. L. Clark. 1989. A selective bibliography of expert systems in natural resource management. *AI Applications* 3:1–18.

Dixon, J. A., and L. A. Fallon. 1989. The concept of sustainability: Origins, extensions, and usefulness for policy. *Society and Natural Resources* 2:73–84.

Durkin, J. 1993. *Expert systems: Catalog of applications.* Akron, Ohio: Intelligent Computer Systems.

Ellison, A. M. 1996. An introduction to Bayesian inference for ecological research and environmental decision-making. *Ecological Applications* 6:1036–1046.

Everett, R., P. Hessburg, M. Jensen, and B. Bormann. 1994. *Eastside forest ecosystem health assessment.* Vol. 1. *Executive summary.* General Technical Report PNW-GTR-317. U.S. Department of Agriculture, Forest Service, Pacific Northwest Research Station, Portland, Ore.

FEMAT [Forest Ecosystem Management Assessment Team]. 1993. *Forest ecosystem management: An ecological, economic, and social assessment.* U.S. Department of Agriculture, U.S. Department of the Interior [and others], Portland, Ore.

Gale, R. P., and S. M. Cordray. 1991. What should forests sustain? Eight answers. *Journal of Forestry* 89:31–36.

Gall, J. 1986. *Systematics: How systems really work and how they fail.* Ann Arbor, Mich.: General Systematics Press.

Greber, B. J., and K. N. Johnson. 1991. What's all the debate about overcutting? *Journal of Forestry* 89:25–30.

Gustafson, E., J. Nestler, L. Gross, K. Reynolds, D. Yaussy, T. Maxwell, and V. Dale. 2002. Evolving approaches and technologies to enhance the role of ecological modeling in decision-making. Chap. 10 in *Advances in ecological modeling,* edited by V. Dale. New York: Springer-Verlag.

Hahn, A., P. Pfeiffenberger, B. Wirsam, and C. Leitzmann. 1995. Evaluation and optimization of nutrient supply by fuzzy logic. *Ernahrungs-Umschau* 42:367–375.

Halpern, D. F. 1989. *Thought and knowledge: An introduction to critical thinking.* Hillsdale, N.J.: Lawrence Erlbaum Associates.

Holland, J. M. 1994. Using fuzzy logic to evaluate environmental threats. *Sensors* 11:57–62.

Howard, R., and J. Matheson. 1981. Influence diagrams. Pp. 721–762 in *Readings on the principles and applications of decision analysis.* Vol. 2. Edited by R. Howard and J. Matheson. Menlo Park, Calif.: Strategic Decisions Group.

Jackson, P. 1990. *Introduction to expert systems.* Reading, Mass.: Addison-Wesley.

Kaufmann, A. 1975. *Introduction to the theory of fuzzy subsets.* Vol. 1. *Fundamental theoretical elements.* New York: Academic Press.

Maser, C. 1994. *Sustainable forestry: Philosophy, science, and economics.* Delray Beach, Fla.: St. Lucie Press.

Maser, C., B. T. Bormann, M. H. Brookes, A. R. Kiester, J. F. Weigand. 1994. Sustainable forestry through adaptive ecosystem management is an open-ended experiment. Pp. 303–340 in *Sustainable forestry: Philosophy, science, and economics,* edited by C. Maser. Delray Beach, Fla.: St. Lucie Press.

Mays, M. D., I. Bogardi, and A. Bardossy. 1997. Fuzzy logic and risk-based soil interpretations. *Geoderma* 77:299–309.

McBratney, A. B., and I. O. A. Odeh. 1997. Application of fuzzy sets in soil science: Fuzzy logic, fuzzy measurements and fuzzy decisions. *Geoderma* 77:85–91.

Moraczewski, I. R. 1993a. Fuzzy logic for phytosociology 1. Syntaxa as vague concepts. *Vegetatio* 106:1–12.

———. 1993b. Fuzzy logic for phytosociology 2. Generalizations and prediction. *Vegetatio* 106:13–27.

Mulder, B. S., B. R. Noon, T. A. Spies, M. G. Raphael, C. J. Palmer, A. R. Olson, G. H. Reeves, and H. H. Welsh. 1999. *The strategy and design of the effectiveness monitoring program for the Northwest Forest Plan.* General Technical Report PNW-GTR-437. U.S. Department of Agriculture, Forest Service, Pacific Northwest Research Station, Portland, Ore.

Noon, B. R., M. G. Raphael, and T. A. Spies. 1999. Conceptual basis for designing an effectiveness monitoring program. Pp. 17–36. in *The strategy and design of the Effectiveness Monitoring Program for the Northwest Forest Plan,* edited by B. S. Mulder, B. R. Noon, T. A. Spies, M. G. Raphael, C. J. Palmer, A. R. Olson, G. H. Reeves, and H. H. Welsh. General Technical Report PNW-GTR-437. U.S. Department of Agriculture, Forest Service, Pacific Northwest Research Station, Portland, Ore.

O'Keefe, R. M. 1985. Expert systems and operational research mutual benefits. *Operations Research* 36:125–129.

Openshaw, S. 1996. Fuzzy logic as a new scientific paradigm for doing geography. *Environment and Planning* 28:761–767.

Overbay, J. C. 1993. Ecosystem management. Pp. 3–9 in *Talking an ecological approach to management.* 27–30 April 1992, Salt Lake City, Utah. Watershed and Air Management WO-WSA-3. U.S. Department of Agriculture, Forest Service, Washington, D.C.

Ranst, E. Van, H. Tang, R. Groenemans, and S. Sinthurahat. 1996. Application of fuzzy logic to land suitability for rubber production in peninsular Thailand. *Geoderma* 70:1–12.

Reeves, G. H., D. B. Hohler, D. P. Larsen, D. E. Busch, K. Kratz, K. Reynolds, K. F. Stein, T. Atzet, P. Hays, and M. Tehan. In press. *Aquatic and riparian effectiveness monitoring program for the Northwest Forest Plan.* General Technical Report. U.S. Department of Agriculture, Forest Service, Pacific Northwest Research Station, Portland, Ore.

Reynolds, K., P. Cunningham, L. Bednar, M. Saunders, M. Foster, R. Olson, D. Schmoldt, D. Latham, B. Miller, and J. Steffenson. 1996. A design framework for a knowledge-based information management system for watershed analysis in the Pacific Northwest U.S. *AI Applications* 10:9–22.

Reynolds, K., M. Jensen, J. Andreasen, and I. Goodman. 2000. Knowledge-based assessment of watershed condition. *Computers and Electronics in Agriculture* 27:315–333.

Reynolds, K., M. Saunders, B. Miller, S. Murray, and J. Slade. 1997a. An application framework for decision support in environmental assessment. Pages 333–337 in: GIS 97 Conference Proceedings of the Eleventh Annual Symposium on Geographic Information Systems. Vancouver, B.C. 17–20 February 1997. GIS World, Washington, D.C.

Reynolds, K., M. Saunders, B. Miller, J. Slade, and S. Murray. 1997b. Knowledge-based decision support in environmental assessment. Pp. 344–352 in *Resource Technology 97 Conference Proceedings.* Vol. 4. ACSM 57th Annual Convention. ASPRS 63rd Annual Convention. Seattle, Wash. 7–10 April 1997. American Society for Photogrammetry and Remote Sensing and American Congress on Surveying and Mapping, Bethesda, Md.

Reynolds, K. M. 1999a. *EMDS users guide (version 2.0): Knowledge-based decision support for ecological assessment.* General Technical Report PNW-GTR-470. U.S. Department of Agriculture, Forest Service, Pacific Northwest Research Station, Portland, Ore.

———. 1999b. *NetWeaver for EMDS version 2.0 user guide: A knowledge base development system*. General Technical Report PNW-GTR-471. U.S. Department of Agriculture, Forest Service, Pacific Northwest Research Station, Portland, Ore.

———. 2001a. Using a logic framework to assess forest ecosystem sustainability. *Journal of Forestry* 99:26–30.

———. 2001b. Overview of an approach to integrated assessment and priority-setting for protection and restoration of watersheds. Proceedings of the IUFRO 4.11 conference on forest biometry, modeling and information science. 26–29 June 2001, Greenwich, U.K. Published on CD-ROM. Greenwich University Press; Greenwich, U.K.

Salski, A., and C. Sperlbaum. 1991. Fuzzy logic approach to modelling in ecosystem research. *Lecture Notes in Computer Science* 20:520–527.

Salwasser, H. 1993. Perspectives in modeling sustainable forest ecosystems. Pp. 176–181 in *Modeling Sustainable forest ecosystems*, edited by D. C. Le Master and R. A. Sedjo. Washington, D.C.: American Forests.

Schmoldt, D. L, J. Kangas, G. A. Mendoza, and M. Pesonen, eds. 2001. *The analytic hierarchy process in natural resource and environmental decision making*. London: Kluwer Academic Publishers.

Smith, P. N. 1995. A fuzzy logic evaluation method for environmental assessment. *Journal of Environmental Systems* 24:275–279.

———. 1997. Environmental project evaluation: A fuzzy logic based method. *International Journal of Systems Science* 28:467–471.

Stillings, N. A., M. H. Feinstein, J. L. Garfield, E. L. Rissland, D. A. Rosenbaum, S. E. Weisler, and L. Baker-Ward. 1991. *Cognitive science: An introduction*. Cambridge, Mass: MIT Press.

Stone, N. D., R. N. Coulson, R. E. Frisbie, and D. K. Loh. 1986. Expert systems in entomology: Three approaches to problem solving. *Bulletin of the Entomological Society of America* 32:161–166.

USDA and USDI (U.S. Department of Agriculture and Department of the Interior). 1994a. *Record of decision for amendments to Forest Service and Bureau of Land Management planning documents within the range of the northern spotted owl and standards and guidelines for management of habitat for late-successional and old-growth related species within the range of the northern spotted owl*. U.S. Department of Agriculture and Department of the Interior, Washington, D.C.

———. 1994b. *Final supplemental environmental impact statement on management of habitat for late-successional and old-growth forest related species within the range of the northern spotted owl*. U.S. Department of Agriculture and Department of the Interior, Washington, D.C.

Waterman, D. A. 1986. *A guide to expert systems*. Reading, Mass.: Addison-Wesley.

Zadeh, L. A. 1965. Fuzzy sets. *Information and Control* 8:338–353.

———. 1968. Probability measures of fuzzy events. *Journal of Mathematical Analysis and Applications* 23:421–427.

———. 1975a. The concept of a linguistic variable and its application to approximate reasoning. Part 1. *Information Science* 8:199–249.

———. 1975b. The concept of a linguistic variable and its application to approximate reasoning. Part 2. *Information Science* 8:301–357.

———. 1976. The concept of a linguistic variable and its application to approximate reasoning. Part 3. *Information Science* 9:43–80.

CHAPTER 8

Approaches to Quality Assurance and Information Management for Regional Ecological Monitoring Programs

Craig J. Palmer

When evaluating the effectiveness of regional ecosystem initiatives, ecological monitoring data from many different sources must be combined. Several different organizations may collect and evaluate information essential to the overall success of the monitoring program. These organizations may also have archived data of potential value to regional assessments. The manner in which ecological monitoring data are collected, summarized, interpreted, and archived has a significant effect on the data's long-term utility. This can best be explained by providing two recent examples.

Example 1: An All Too Common Past

Scientists involved in the development of a regional ecological monitoring program interviewed agency staff regarding the availability of historical forest inventory data. They were excited to learn that three extensive regional timber inventories had been conducted in 1968, 1978, and 1988 with the specific measurements they needed to address their monitoring questions. These data had been collected at a significant expense to address the needs of a ten-year planning cycle for timber management.

Upon further examination, however, they learned that none of these data were readily accessible. The first two inventory results had at one time been available on a mainframe computer but were now available only by reviewing the original reports. The most recent survey was available in electronic format, but the format was that of a program no longer supported by the agency. All data were stored on outmoded media and would be difficult to extract due to changes in computer hardware. A few manuals were available to describe how the data were collected, but no information was available on the quality of the data nor the quality-control procedures used to ensure comparability in the data between field crews. The individual most knowledgeable about the surveys had recently retired, making the use of the data even more difficult. The end result was that these scientists determined that the considerable effort it would take to make these data available rendered them useless.

Example 2: A Possible Future

Two soil scientists involved in the development of a national forest monitoring program were asked to determine if the monitoring program could detect a 2 percent change per year in soil carbon at a regional scale. To answer this question, they decided to remeasure some forest plots that had been sampled eight years previously during a regional pilot study. As the original study had been funded by a regulatory agency that required quality assurance plans and documentation, the scientists were able to obtain detailed information about how soil samples had been collected, processed, and analyzed and the precision associated with each of these activities. The original database was still available as it had been maintained and updated periodically to new hardware and software systems. As a result, a successful remeasurement study could be undertaken.

Objectives of This Chapter

The problem faced by many regional ecosystem-monitoring programs is how to get from the all-too-common reality described in the first example to the possible future described in the second example. Unless steps are taken at the initiation of regional monitoring programs to address quality assurance (QA) and information management (IM), data that are collected today may have very little utility in the future.

The first objective of this chapter is to provide a conceptual basis for quality assurance and information management in ecological monitoring programs. The role of monitoring data in the adaptive management cycle will be described as well as the process for changing data into information and knowledge that might assist with natural resource management decisions. The goals of QA and IM systems will be reviewed and basic concepts presented. The second objective is to describe alternative approaches for structuring QA and IM systems in monitoring programs, and the final objective is to provide some suggestions on how to initiate QA and IM programs for regional ecosystem initiatives.

Conceptual Basis

When planning a regional ecosystem-monitoring program, it is helpful to consider alternative approaches one might take to the collection and management of data and information. The purpose of this section is to lay the groundwork for a discussion by reviewing how information flows in an adaptive ecosystem management cycle and what resource managers hope to achieve with the data collected in monitoring programs. Basic concepts regarding the management of ecological data will be discussed along with approaches to ensure that data collected are of the highest quality.

Information Flow in the Adaptive Ecosystem Management Cycle

In an adaptive ecosystem management cycle, a decision is implemented to achieve goals related to the ecosystem being managed. Monitoring of the ecosystem provides feedback to management regarding the effectiveness of management actions in achieving those goals. Based upon these results, new management decisions can then be implemented creating an adaptive cycle that, hopefully, improves itself over time.

For this cycle to work effectively, information must readily flow from the monitoring program to the decision makers. This process is depicted in figure 8-1. Field measurements must first be collated into a database. These data can then be summarized into information. Information, by itself, may not be of value to decision makers unless it is synthesized and interpreted along with other monitoring and research information into a better understanding and knowledge of how the ecosystem is responding to management actions. Based upon this knowledge, new decisions can then be made with the hope of improving management of ecosystem condition.

This process can be illustrated by the steps involved in developing a monitoring strategy for the Northwest Forest Plan (Palmer and Mulder 1999). In developing that plan, the authors considered this cycle and the steps that would need to be taken to have field measurements that are ultimately useful in the decision-making process. The danger, as the authors saw it, was that significant resources might be expended in the collection of data that ultimately were not used in the adaptive management process

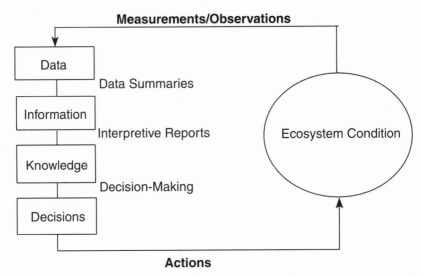

Figure 8-1. Flow of monitoring information in the adaptive management cycle.

due to information overload. As T. S. Eliot (1980) wrote, "Where is the wisdom we have lost in knowledge? Where is the knowledge we have lost in information?" One might add another line to this poem: Where is the information we have lost in data?

A primary recommendation made to management was that a reporting process be developed to summarize data into information through the preparation of annual data summaries (Palmer and Mulder 1999). These reports would be "preplanned" in terms of their format, content, and approach to data synthesis. The annual frequency of these reports would have the added benefit of encouraging the development of databases through the collation and checking of data in a timely manner. As a second step in the reporting process, the preparation of periodic interpretive reports was recommended to encourage the synthesis of information from the annual data summaries into a better understanding of ecosystem condition. In addition to evaluating monitoring data, the preparation of synthesis reports required an evaluation of other ongoing research activities, data from other monitoring programs, and the use of the latest modeling and statistical tools.

Data Management Goals of Natural Resource Managers

In addition to fostering the adaptive management cycle, land managers have other goals related to data management that should be considered when establishing a monitoring program. During a review of data management goals for a land management unit in the lower Colorado River basin (Landis and Palmer 2000), we learned that several objectives must be taken into consideration when developing a data management system. These objectives include permanence, accessibility, credibility, and accountability.

Ecological data can be lost, or lose significant value over time, unless care is taken to adequately document and maintain these data. This is a process called data entropy (Michener et al. 1997) and was described earlier in this chapter with the example of "an all-too-common past." The study of ecosystems often requires the analysis of long-term trends and relationships. These can only be undertaken if data management systems provide permanence to the data for at least several decades. Preventing data entropy must take into consideration both the gradual processes and the discrete events that can cause data loss. Gradual processes include the degradation of storage media, the obsolescence of storage technology, the destruction of data forms and notes, and the loss of specific facts from the memories of the investigators. Discrete events leading to data loss include changes in personnel (e.g., retirement, career changes, death), loss of computer records, loss of conceptual or computer models used to interpret data, or the loss of storage media through a catastrophic event.

A second requirement is that ecological data be readily accessible, preferably through the Internet. The synthesis of data into information and knowledge at regional scales frequently requires the participation of many individuals, often from different organizations. Unless the data are easily available, this participation is not likely to occur.

Users must also have access to information about data-collection methods and quality of data. Once they have ascertained that the data are credible, they can then proceed with the summarization and synthesis into useful information and knowledge. The information system must therefore provide documentation about the data through the development of appropriate metadata. Metadata may be defined as a higher level of information or instructions that describe the content, context, quality, structure, and accessibility of a data set (Michener et al. 1997). This information should also be available in a readily accessible manner and should include all the information needed to enable the long-term reuse of data sets by the original investigators as well as new uses by other scientists not involved in the original data-collection efforts.

An important trend from a resource manager's perspective is increasing accountability for the natural resources they are responsible for managing. To address this accountability, resource managers require ecological data and information that is legally and scientifically defensible. The quality assurance and information management systems supporting the acquisition, maintenance, and interpretation of these data must help foster such defensibility.

Quality-Assurance Concepts

An important goal of monitoring programs is to obtain data that are meaningful, representative, reliable, and comparable over space and time (Burton 1995; Mohnen 1996). Quality assurance refers to those activities undertaken during the planning, implementation, and assessment phases of a project to ensure that monitoring results meet these expectations (Taylor 1987). Regulatory agencies such as the Environmental Protection Agency and the Department of Energy require that a structured quality assurance program, as well as quality-assurance plans, be in place prior to data collection in monitoring programs. These requirements are detailed by the American National Standards Institute (ANSI 1995) and the basic elements are summarized in table 8-1. Definitions for some common quality-assurance terms are provided in table 8-2. A quality-assurance program is often documented in a quality-management plan that describes a system (Kulkarni and Bertoni 1996) of policies and procedures used to ensure that the monitoring data will be of the type and quality needed (Storey et al. 2000; EPA 2001b). The quality-management plan is used as a framework for quality-assurance activities during planning, data collection, and assessment phases of monitoring programs. These activities can include the preparation of detailed quality-assurance project plans

Table 8-1. Quality-assurance elements required by the American National Standard Institute (ANSI 1995) for environmental monitoring programs.

Quality-management elements	*Specific project elements*
Management and organization	Planning and scoping
Quality system and description	Design of data collection operations
Personnel qualification and training	Implementation of planned operations
Procurement of items and services	Assessment and response
Documents and records	Assessment and verification of data usability
Computer hardware and software	
Planning	
Implementation of work processes	
Assessment and response	
Quality improvement	

Table 8-2. Quality assurance terms and definitions (from EPA 2001a).

Term	Definition
Quality	The totality of features and characteristics of a product or service that bear on its ability to meet the stated or implied needs and expectations of the user.
Quality assurance (QA)	An integrated system of management activities involving planning, implementation, documentation, assessment, reporting, and quality improvement to ensure that a process, item, or service is of the type and quality needed and expected by the customer.
Quality assurance project plan	A document describing in comprehensive detail the necessary quality assurance, quality control, and other technical activities that must be implemented to ensure that the results of the work performed will satisfy the stated performance criteria.
Quality control (QC)	The overall system of technical activities that measures the attributes and performance of a process, item, or service against defined standards to verify that they meet the stated requirements established by the customer; operational techniques and activities that are used to fulfill requirements for quality.
Quality management	That aspect of the overall management system of the organization that determines and implements the quality policy. Quality management includes strategic planning, allocation of resources, and other systematic activities (e.g., planning, implementation, documentation, and assessment) pertaining to the quality system.
Quality management plan	A document that describes a quality system in terms of the organizational structure, policy and procedures, functional responsibilities of management and staff, lines of authority, and required interfaces for those planning, implementing, documenting, and assessing all activities conducted.
Quality system	A structured and documented management system describing the policies, objectives, principles, organizational authority, responsibilities, accountability, and implementation plan of an organization for ensuring quality in its work processes, products (items), and services. The quality system provides the framework for planning, implementing, documenting, and assessing work performed by the organization and for carrying out QA and QC.

or scientific-investigation plans (EPA 2001a), establishment of data-quality objectives (Barnard 1996; Wilson 1995), training and certification, quality-control activities, review of data (Edwards 2000), and assessment of data quality (Taylor 1996).

Although structured quality-assurance programs are a requirement for data collection in regulatory agencies, they are not currently a requirement for natural resource management agencies. Consequently, many regional ecosystem initiatives are still considering whether or not to establish structured quality-assurance systems based upon the expected benefits versus the potential costs. The author's recommendation has been to encourage the development of ecological quality-assurance programs (Lawrence and Palmer 1996) as the benefits outweigh the costs when properly conceived and implemented (Palmer and Mulder 1999).

Ecological Information Management Concepts

Ecological databases are often developed in regional ecosystem initiatives for the purpose of answering specific monitoring and assessment questions (Briggs and Su 1994). These databases need to have several attributes in order to achieve this goal. The data need to be of high quality. The correctness or integrity of these data needs to be maintained (Ingersoll et al. 1997). The data must be accessible, well documented, and easy to use (Brunt 2000) for the original intended objectives and must also be available to scientists with other planned objectives (Michener et al. 1997). The data must be secure for long periods of time (decades) even though computer hardware and software are likely to change frequently (Strebel et al. 1994).

The purpose of an information management system is to achieve the goals of fostering data defensibility, access, use, and permanence. As with a quality-assurance program, the information management system should be integrated with all phases of a project including planning, data collection, and data assessment. In fact, a very important role of information management in ecological monitoring programs is to facilitate quality-assurance activities. For example, electronic-data collection programs are often developed for monitoring programs to allow for quality-assurance checks to be undertaken at the time of data entry and thereby reduce the number field-data entry errors.

It is helpful to consider data management as a stepwise process (fig. 8-2). The first step is to plan the overall data management system, including data collection programs and the overall database design. Data is then acquired through field notes, portable data recorders, field data loggers, or laboratory analyses. Data are then entered into the database for processing.

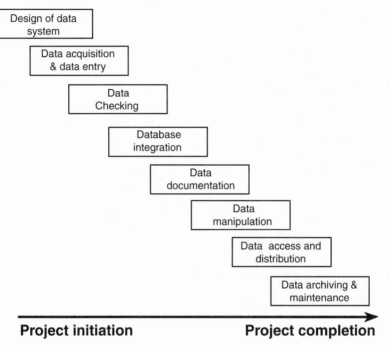

Figure 8-2. Steps in the data management process from project initiation through project completion.

An important first step is to identify data entry errors through quality-assurance checks (Edwards 2000) such as the evaluation of outlier values. The data are then integrated into the database with other data collected in the monitoring program at the same site or similar sites. Documentation of how the data were collected is developed and integrated with the original data as metadata. Data manipulation to summarize or collate the data can then take place, which improves the utility of the data. To encourage use of the data by others, data publishing can be undertaken. Centers that have been established to assist with the sharing of ecological data or data clearinghouses are notified of database content and accessibility. The data are then maintained through frequent backups to prevent loss or corruption. A process called data archiving is implemented to provide for the long-term preservation of the data, even after completion of the project. The resources required to undertake these activities are significant (Strebel et al. 1994), often requiring from 10 to 20 percent of the monitoring budget (Hale 1999).

The development of monitoring programs for regional ecosystem initiatives requires that some thought be given to the nature of the information management system that will be needed. In addition to maintaining the "internal" data collected within the program itself, it is important to note that data collected in other ongoing monitoring programs might be needed to effectively answer questions about the ecoregion and its management. For example, the national Forest Inventory and Analysis (FIA) program collects detailed information about all forests in the United States. The late-succession/old-growth forest module of the Northwest Forest Plan monitoring effort is planning to use portions of these data to evaluate status and trends in forests for the Pacific Northwest ecoregion. This is an example of a concept that is depicted in figure 8-3; required data to answer monitoring questions is shaded in gray. Some external programs such as FIA may provide data that is critical to the success of the regional ecosystem initiative; close ties will therefore need to be developed with these programs to ensure that these data are of high quality and are accessible. Other research or monitoring programs may also have data that are useful, and efforts will need to be taken to develop a data interface with these programs as well.

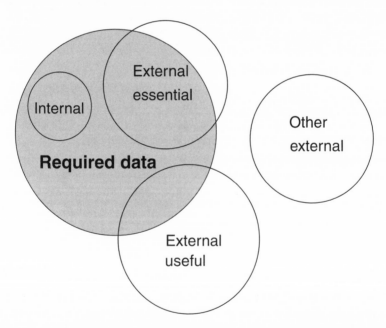

Figure 8-3. Sources of data for regional ecosystem monitoring programs. Data required to answer monitoring questions are shaded in gray. These data may come from internal or external programs. Close ties are needed to external programs providing essential or useful information to ensure high quality and accessible data.

Alternative Approaches to Quality Assurance and Information Management

There are a wide variety of approaches that can be taken to quality assurance and information management in regional ecosystem monitoring programs. The two extremes are the *craftsman-artisan approach* and the *formal programmatic approach* (Taylor 1987). The craftsman-artisan approach can be defined as one that depends on the use of highly competent scientists who can be trusted to provide high-quality data sets upon the completion of their project activities. Using this approach, overall quality is controlled through the peer review process and not through the implementation of a structured quality-assurance program.

A formal programmatic approach defines all the operational aspects that must be undertaken by participants in a monitoring program. These are defined in quality-assurance program plans and data management plans. The effectiveness of this approach depends on strict compliance to defined protocols implemented by competent staff. A support system of quality assurance and information management personnel provides assistance to ensure that data quality and data management goals are met.

When a monitoring program consists of many different complex investigations conducted by individual scientists in an independent manner, a craftsman-artisan approach might provide the most effective atmosphere to encourage excellence. When a large number of participants conduct similar measurements, a more formal approach is warranted. This is often the situation with regional ecosystem initiatives where many organizations and individuals are working together to achieve overall program goals. A common structured approach to quality assurance and information management serves to encourage the collection of comparable data that is of high quality and the documentation and sharing of these data in an efficient manner.

Recommendations for Regional Ecosystem Monitoring Programs

Without the establishment of some type of formal approach to quality assurance and information management, regional ecosystem monitoring programs by default select the craftsman-artisan approach. The selection of this approach puts these programs at the risk of many pitfalls associated with a lack of formalized planning of quality assurance and information management systems as detailed in table 8-3. Formal quality-assurance and information-management program development is needed to reduce this risk and improve the likelihood of being able to achieve high-quality data and information that are useful in adaptive management activities and have the characteristics of being long lived, accessible, and credible. These goals can be achieved by investing in the development and implementation of formal programmatic quality assurance and information management systems.

To start this process, a quality assurance program plan will need to be developed that addresses the factors identified in Table 8-1. An individual will need to be assigned to coordinate the development and implementation of this plan. Quality-assurance activities such as the development of quality-assurance project plans, training and certification of field crews, field audits, data checking, and data-quality assessments will need to be developed for each regional activity. Program participants will need to be trained in quality-assurance concepts to assist with the implementation of these plans. Adequate resources will also need to be allocated to allow for the successful implementation of the quality-assurance program.

In a similar manner, a structured data management system will need to be developed. This can be accomplished through a structured approach of information systems development that includes an analysis of the current information management situation, the definition of new requirements, the development of an information management system design to address these requirements, the programming and testing of the proposed system, the implementation of the system, and the evaluation and maintenance of the system over the long term. A data management plan is a document

Table 8-3. Some pitfalls associated with the lack of planning and execution of formal quality assurance and information management systems for regional ecosystem initiatives.

Pitfall	Explanation
Defensibility	Often there is no provision for the quality assurance of information gathered.
Access	Data access is often limited. Coordination among neighboring resource management units is restricted.
Use	Data are everywhere, but little of it is useful due to poor planning.
Permanence	Reporting and use of metadata will likely be limited leading to data entropy of newly collected data and rapid loss of legacy data.
Budgeting	Quality assurance and data management costs are often not included in project budget processes, making it difficult to implement or improve these systems later.

that incorporates the first three components of this approach. The data management plan should also answer specific questions related to the coordination and management of the data management system. A number of these questions are detailed in box 8-1. The data management plan must also consider and support quality assurance activities in addition to the other data management goals. These two documents can then serve as the framework for the development of monitoring activities within a regional ecosystem initiative.

Box 8-1. Questions to consider during the development of a data management plan for a regional ecosystem initiative.

1. Which interagency body oversees resolution of issues related to monitoring information management?
2. Should the information management system support only the monitoring program, or should related programs be integrated or linked?
3. Should such a system serve as a regional clearinghouse for monitoring information?
4. Should the system house and steward the data in a central location, or should these functions be distributed throughout the region?
5. What is the relationship of monitoring information management to existing agency data management systems?
6. Would the monitoring information network operate best as a metadata database, or should the data and metadata both be included in the regional system?
7. Can a basic level of support for monitoring information management (estimated to be 15–20 percent of the monitoring budget) be generated?
8. How can security and stability over time be assured in the face of institutional change?
9. What is the best way to incorporate valid legacy data into a monitoring information network?
10. How can a monitoring information system best deal with both the reality and notion of proprietary data?

Conclusion

Monitoring is a key component to the success of an adaptive process for natural resource management. Current approaches to quality assurance and information management in regional ecosystem monitoring programs are often informal and unstructured. It is recommended that a formalized structure be implemented for these programs to encourage data quality, access, utility, and permanence.

ACKNOWLEDGEMENTS

This chapter was funded in part by the U.S. Geological Survey through the Western Ecological Research Center under Cooperative Agreement No. 01WRAG0004 with the University of Nevada, Las Vegas.

LITERATURE CITED

ANSI (American National Standards Institute). 1995. *American National Standard: Specifications and guidelines for quality systems for environmental data collection and environmental technology programs.* Energy and Environmental Quality Division, Environmental Issues Group, Milwaukee, Wis.

Barnard, T. E. 1996. Extending the concept of data quality objectives to account for total sample variance. Pp. 169–184 in *Principles in environmental sampling*, edited by L. H. Keith. Washington, D.C.: American Chemical Society.

Briggs, J. M., and H. Su. 1994. Development and refinement of the Konza Prairie LTER research information management program. Pp. 87–100 in *Environmental information management and analysis: Ecosystem to global scales*, edited by W. K. Michener et al. Bristol, Penn.: Taylor and Francis.

Brunt, J. W. 2000. Data management principles, implementation and administration. Pp. 25–47 in *Ecological data: Design, management and processing*, edited by W. K. Michener and J. W. Brunt. Malden, Mass.: Blackwell Science.

Burton, G. A. 1995. Quality assurance issues in assessing receiving water. Pp. 275–284 in *Stormwater runoff and receiving systems: Impact monitoring and assessment*, edited by E. E. Herricks, Boca Raton, Fla.: CRC Press.

Edwards, D. 2000. Data quality assurance. Pp. 70–91 in *Ecological data: Design, management and processing*, edited by W. K. Michener and J. W. Brunt, Malden, Mass.: Blackwell Science.

Eliot, T. S. 1980. *The complete poems and plays, 1909–1950.* San Diego: Harcourt Brace Jovanovich.

EPA (Environmental Protection Agency). 2001a. *EPA requirements for quality assurance project plans.* EPA QA/R-5. EPA/240/B-01/003. Environmental Protection Agency, Washington, D.C.

———. 2001b. *EPA requirements for quality management plans.* EPA QA/R-2. EPA/240/B-01/002. Environmental Protection Agency, Washington, D.C.

Ingersoll, R. C., T. R. Seastedt, and M. Hartman. 1997. A model information management system for ecological research. *BioScience* 47:310–316.

Hale, S. S. 1999. How to manage data badly. Part 1. *Bulletin of the Ecological Society of America* (Oct.):265–268.

Kulkarni, S. V., and M. J. Bertoni. 1996. Environmental sampling quality assurance. Pp. 111–137 in *Principles in environmental sampling*, edited by L. H. Keith, Washington, D.C.: American Chemical Society.

Landis, E., and C. J. Palmer. 2000. Lake Mead National Recreation Area resource management data and information: Current status and future options. Harry Reid Center for Environmental Studies, University of Nevada, Las Vegas. Unpublished report.

Lawrence, J., and C. Palmer. 1996. Ecological quality assurance: A Canadian perspective. Pp. 172–181 in *North American workshop on monitoring for ecological assessment of terrestrial and aquatic ecosystems*, edited by C. Aguirre-Bravo, General Technical Report RM-GTR-284. U.S. Department of Agriculture, Forest Service, Rocky Mountain Forest and Range Experiment Station, Fort Collins, Colo.

Michener, W. K., J. W. Brunt, J. J. Helly, T. B. Kirchner, and S. G. Stafford. 1997. Nongeospatial metadata for the ecological sciences. *Ecological Applications* 7:330–342.

Mohnen, V. A. 1996. Quality assurance/quality control as prerequisite for harmonization of data collection and interpretation within the WMO-Global Atmosphere Watch (GAW) Programme. Pp. 77–84 in *Global monitoring of terrestrial ecosystems*, edited by W. Schroder et al., Berlin, Germany: Ernst and Sohn.

Palmer, C. J., and B. S. Mulder. 1999. Components of the effectiveness monitoring program. Pp. 69–97 in *The strategy and design of the effectiveness monitoring program for the Northwest Forest Plan*, edited by B. S. Mulder et al., General Technical Report PNW-GTR-437. U.S. Department of Agriculture, Forest Service, Pacific Northwest Research Station, Portland, Ore.

Storey, A., R. Briggs, H. Jones, and R. Russell. 2000. Quality assurance. Pp. 49–67 in *Monitoring bathing waters: A practical guide to the design and implementation of assessments and monitoring programmes*, edited by J. Bartram and G. Rees. New York: E and FN Spon.

Strebel, D. E., B. W. Meeson, and A. K. Nelson. 1994. Scientific information systems: A conceptual framework. Pp. 59–85 in *Environmental information management and analysis: Ecosystem to global scales*, edited by W. K. Michener et al. Bristol, Penn.: Taylor and Francis.

Taylor, J. K. 1987. *Quality assurance of chemical measurements*. Chelsea, Mich.: Lewis Publishers.

———. 1996. Defining the accuracy, precision, and confidence limits of sample data. Pp. 77–83 in *Principles in environmental sampling*, edited by L. H. Keith. Washington, D.C.: American Chemical Society.

Wilson, N. 1995. *Soil water and ground water sampling*. Boca Raton, Fla.: Lewis Publishers.

CHAPTER 9

Estimation of Change in Populations and Communities from Monitoring Survey Data

John R. Sauer, William A. Link, and James D. Nichols

Monitoring surveys provide fundamental information for use in environmental decision making by permitting assessment of both current population (or community) status and change in status, by providing a historical context for the present status, and by documenting response to ongoing management (Williams and Johnson 1995). Conservation of species and communities has historically been based upon monitoring information (Caughley 1994), and prioritization of species and habitats for conservation action often requires reliable, quantitative results (e.g., Carter et al. 2000). Although many monitoring programs exist for populations, species, and communities, as well as for biotic and abiotic features of the environment, estimation of population and community change from surveys can sometimes be controversial (Link and Sauer 1998a), and demands on monitoring information have increased greatly in recent years. Information is often required at multiple spatial scales for use in geographic information systems, and information needs exist for description of regional patterns of change in populations, communities, and ecosystems. Often, attempts are made to meet these needs using information collected for other purposes or at inappropriate geographic scales, leading to information that is difficult to analyze and interpret.

In this chapter, we address some of the constraints and issues associated with estimating change in wildlife species and species groups from monitoring surveys, and use bird surveys as our primary examples. Birds are a useful group for discussion, as many geographically extensive, multispecies surveys have been developed, and emphasis is often placed on description at the scale of ecoregions (Commission for Environmental Cooperation 1997). Bird surveys exemplify both the values and flaws inherent in monitoring programs, and many methods of estimating change of both species and communities have been developed in the context of bird surveys.

Understanding of survey design has greatly advanced in recent years as a consequence of statistical research and the development of techniques of adaptive management (Williams and Johnson 1995). Unfortunately, many surveys used for monitoring are based on flawed designs, due to logistical limitations or to limitations of available survey designs when the surveys were developed (Thompson et al. 1998). Many of the surveys were designed and implemented by biologists to provide general population information, and specific quantitative goals for monitoring have only recently been superimposed on these surveys. For surveys such as the North American Breeding Bird Survey (BBS) and the Audubon Christmas Bird Count (CBC) the designs impose important constraints on analysis. In particular, issues such as site selection for sampling and flaws in the counting methods both limit the scope of inference from the surveys and greatly complicate their analysis. Simple statistics, such as means of counts, that are commonly used for summary of regional patterns and for modeling of population change over time often contain large bias as estimators of the true population means. We examine modern goals for monitoring programs and discuss essential design elements needed to meet these goals. We describe the design of several sample monitoring programs and discuss analytical approaches for estimation of population and community change from the surveys.

Estimation of population change at the species level is often accomplished using generalized linear models and generalized additive models. Deficiencies of count data are accommodated through covariates that adjust for differences in detectability and through modeling of overdispersion of counts. Composite estimates of change for groups of species are often of interest; we describe empirical Bayes and hierarchical models for estimations of attributes associated with collections of species estimates. Hierarchical models fit by Markov Chain Monte Carlo methods provide a general approach to analysis of change from these data. Changes in community attributes from count data can be estimated directly using capture recapture estimation procedures, and we describe the application of these estimation procedures to document species richness and community dynamics attributes.

Fundamental Design Constraints of Monitoring Programs

Goals

The most useful monitoring programs are designed to provide estimates of population or community attributes that have direct relevance to management. For example, waterfowl surveys in North America are designed to provide yearly estimates of the population size of the major harvested species for use in setting harvest regulations (Smith 1995). A consequence of this well-defined goal has been a clear definition of the parameter to be estimated, a well-defined scope of inference, and a concern for the details of estimation. Yearly analyses are conducted from survey data, and the performance of the survey is evaluated both through critical evaluation of yearly results (e.g., U.S. Fish and Wildlife Service 2001) and periodic reviews (e.g., Smith 1995).

Adaptive management (Williams and Johnson 1995) also provides a specific context for the use of monitoring data. In adaptive management, the consequences of alternative management actions are assessed by making model-based predictions of the consequences of the proposed actions on the populations. Optimal management actions are taken based on the prediction of the consequences of the management for a defined set of goals specified in an objective function. The essential role of monitoring in adaptive management is in assessing system status and change in response to management, allowing us to update our models of the response of the system to management. Adaptive harvest management uses survey data to assess response of waterfowl populations to harvest regulations (Williams and Johnson 1995).

Bird activities associated with the North American Bird Conservation Initiative have focused on population management at the scale of ecoregions (Commission for Environmental Cooperation 1997). Estimates of population change from surveys are used in defining priority species for conservation activities (Carter et al. 2000), and management activities frequently use bird survey data in development of models of bird-habitat associations at the ecoregional scale. These uses of survey data for model development at the landscape scale are only beginning to be explored by modelers and statisticians, and although bird conservation activities generally rely on bird survey data, it is unclear whether these data are appropriate for use in estimation and modeling.

Unfortunately, many bird surveys did not have well-defined monitoring goals when they were developed but instead relied upon vague goals of tracking bird populations. Consequently, there was little concern for estimation of specific parameters associated with population change. This lack of specific quantitative goals has led to many complications and uncertainties in estimation of population attributes from monitoring surveys. Appropriate estimation of population change (or any attribute) from these surveys requires many assumptions to compensate for the absence of survey design components such as detectability estimation and a sample that allows for extrapolation to a defined target population (Link and Sauer 1998a). Although the focus of this chapter is on estimation, we must first provide a brief description of some of the essential design components for surveys.

Detectability

Analysis of change from bird surveys is complicated by the nature of count data. Bird counts at sample sites are generally not censuses, but instead the counts only detect some portion of the population. If population size at a survey point for a species is N, and the expected value of a count is denoted by $\mathbf{E}(Y)$, detectability can be represented as

$$\mathbf{E}(Y) = pN,$$

where p is the detection probability. In many bird counting methods commonly used in surveys (e.g., point counts; Ralph et al. 1995), only count data are collected. Obviously, it is inappropriate to use Y as a surrogate of N in any analysis unless (1) p is known to be constant; (2) variation in p is independent of variation in parameters of interest or can be accommodated through use of covariates; or (3) p is estimated and tests are conducted to see whether p differs over space or time. Because it is unlikely that (1) is correct, enormous amounts of effort have been devoted to developing methods for estimating p from wildlife surveys (e.g., Thompson et al. 1998; Skalski and Robson 1992). However, many ongoing surveys have not yet been modified to incorporate these procedures, and historical analyses of the data often rely on (2) for most analysis. We note that these covariate analyses are model based in that we generally assume a relationship between some observable covariate and detection, and adjust for variation in detection as estimated from that model. Other sources of variation in detection rate are not accommodated by this approach, and many statisticians suggest that p must be routinely estimated during surveys to allow unbiased estimation (e.g., Thompson et al. 1998; Lancia et al. 1994; Thompson 1992). Extensive surveys that estimate population attributes over many habitats and regions are particularly likely to encounter situations in which detection rates vary among samples, making estimation of p an important priority for these surveys (Sauer et al. 1995).

Sample Frame Considerations

To implement a probabilistic sample of a wildlife population, it is necessary to ensure that all components of the population have some chance of being sampled (e.g., Cochran 1977). To ensure this, a sample frame must be defined; that is, the target population must be divided into sample units from which a sample can be drawn. There are many ways of efficiently selecting samples in such a way that inference can be made for the entire population. Simple random samples, stratified random samples, cluster samples, and adaptive sampling methods are a few of the many methods used to ensure efficient allocation of samples, and we refer readers to the literature on this topic for details of these approaches (e.g., Cochran 1977; Thompson 1992; Thompson and Seber 1996). However, many bird surveys are based on sample designs that exclude certain components of the population (i.e., the sampled population is different from the target population). As noted below, for many continent-scale bird surveys, sample frames are either incomplete or poorly defined, and any inference from these surveys relies on assumptions that the sample is representative of the target population.

Survey designs are permanent attributes of the survey, and design components such as strata can limit and complicate future uses of surveys. For example, providing information on bird populations at the scale of ecoregions requires additional analyses to estimate attributes at the ecoregional scale in the context of the original strata used in development of the surveys. For some sample designs, it is difficult to alter the original design and still retain the probabilistic sample. Consequently, care is needed in both choosing the appropriate sample design that allows for efficient sampling and in modifying the original design to accommodate new uses of the survey.

Examples of Bird Monitoring Surveys

Because birds have traditionally been a subject of public interest, a large number of bird monitoring programs has been developed throughout the world. In North America, several major bird monitoring programs were developed specifically to provide information for harvest management (Martin et al. 1979). However, there are many other bird monitoring programs developed for a variety of recreational, educational, and scientific reasons. Consequently, the quality of information from these surveys tends to vary greatly. Martin et al. (1979), Ralph and Scott (1981), and Sauer and Droege (1990) provide summaries of many of the ongoing programs. The North American Bird Conservation Initiative developed in recent years has provided the impetus for the development of additional monitoring surveys, leading to a great deal of research into statistical design and analysis of monitoring data (e.g., Link and Sauer 1998a). We describe a few surveys of particular interest, to motivate the discussion of statistical analysis of population change.

North American Breeding Bird Survey

The North American Breeding Bird Survey (BBS) is a primary source of information on population change for more than four hundred bird species, and is used to set research and management priorities by many management agencies and bird conservation initiatives. It is a roadside survey conducted by volunteer observers along 39.2-kilometer survey routes randomly located on secondary roads. Routes consist of fifty stops, evenly spaced along the roadsides, with a three-minute point count used to sample bird populations at each stop. All birds heard and seen at the stop during the count are recorded, and the sum by species of the fifty counts is used to characterize bird abundance on the route. Each route is surveyed once per year. The survey was started in 1966 on about six hundred routes in the eastern United States and has gradually increased in extent until the present, when it has over four thousand routes over the continental United States, Alaska, and most of Canada (Robbins et al. 1986).

Primary criticisms of the BBS involve its counting method and sample frame. Because point counts are used, no censuses occur, and the detection probability has been shown to vary due to a variety of factors such as observer competence and experience (Sauer et al. 1994; Kendall et al. 1996) and habitat (Sauer et al. 1995). Partial accommodation of variation in p associated with observers can be accomplished with covariates, but many potential sources of variation in p can not be accommodated through observable covariates. The sample frame concerns relate to the possibility that bird abundance and population change differ between the sampled population (roadside birds) and the target population (all birds).

The Audubon Christmas Bird Count

The Audubon Christmas Bird Count (CBC) was initiated in 1900 as a recreation and conservation activity for birdwatchers and is now carried out worldwide, although its most consistent information is from North America. It is an early-winter survey, conducted within two weeks of Christmas (25 December). Counts are conducted within 15-mile-diameter circles, in a single day, by many observers who vary greatly in experience. Counts are made in several different ways, by groups who count using a variety of transportation means such as on foot, from automobiles, or from boats. Other participants conduct feeder watches, or count specifically for nocturnal species such as owls. The locations of the count circles are a consequence of where people want to count rather than a statistical design. See Butcher (1990) for details of CBC design and history.

Because the original intent of the CBC was to increase interest in birds through public participation, it has many features that limit its application as a survey. In particular, the circle sample units are selected to be accessible to people and tend to be clustered near cities and coasts. Effort varies greatly over time and has consistently increased over time. Effort is recorded by the survey coordinators, and demonstrable increases in counts are associated with the increased effort in counting (Butcher 1990). Most analyses of CBC data have attempted to accommodate for variation in p associated with changes in effort by using effort as a covariate in analysis and use strata to accommodate spatial issues associated with nonrandom site selection. Recent scientific uses of CBC data include mapping of bird distributions and estimating population change (e.g., Dunn and Sauer 1997).

Spring Breeding Ground Waterfowl Survey

The U.S. Fish and Wildlife Service and Canadian Wildlife Service conduct an extensive survey of waterfowl populations in Canada, Alaska, and northern portions of the United States. This survey is composed of aerial transects located in a variety of strata, with subsampling of ground counts of segments of the transects to allow for estimation of detection (air-ground) rates. In some remote regions, ground crews cannot access the survey area, and helicopters are used to survey the more intensive sample. Combined ratio estimation procedures are used to estimate population totals for the sampled areas (Smith 1995). These estimation procedures are well established (e.g., U.S. Fish and Wildlife Service 2001) and lead to yearly population total estimates that can be analyzed using a variety of standard time-series and regression procedures (e.g., Smith 1995; Box et al. 1994; Martin et al. 1979). We refer readers to Smith (1995) or the yearly population trend reports (e.g., U.S. Fish and Wildlife Service 2001) for details of the analysis and results of this survey.

Estimation of Population Change from Count Surveys

There is an enormous statistical literature on estimation of "trend" (e.g., Dagum and Dagum 1988) and cyclical components (e.g., Box et al. 1994) from time series data. Time series analyses using autoregressive integrated moving averages (Box et al. 1994), and a large variety of alternative statistical modeling procedures such as locally weighted least squares (LOESS, James et al. 1996), and generalized additive models (Fewster et al. 2000) have been applied to biological time series of counts.

In our discussion, we emphasize a general formulation for estimation of population change based on log-linear models that has proven useful for the analysis of bird survey data from count-based surveys (e.g., Link and Sauer 1994, 1998a, 1999). We also briefly discuss the use of hierarchical models for estimation of population change from survey data. Count data are often assumed to have Poisson distributions, and overdispersed Poisson distributions can be used to relax the restrictive Poisson assumption of equal variance and mean. Generalized linear models (GLMs) based on overdispersed Poisson distributions are widely used in analyses of count data (McCullagh and Nelder 1989; Diggle et al. 1994) and can be fitted using software such as GLIM (Frances et al. 1994).

Surveys such as the BBS and the CBC contain the common feature of counts at a collection of fixed sites. These replicated, fixed sites provide time series of counts and allow us to define local populations (denoted as N_{it}) at site i in year t from which we model population change. Note that this local population is often vaguely defined, and (as is the case in the BBS) is not necessarily associated with a known area. Population change can be modeled as a log-linear model, with the expected value of local populations:

$$\log(\mathbf{E}(N_{it})) = \theta_i + f_i(t).$$

In this formulation, e^{θ_i} is a baseline abundance level for site i, and $e^{f_i(t)}$ is the "population trajectory," which can be defined in a variety of ways depending on the form of population change, with constraint $f_i(t_1) = 0$. For example, $f_i(t)$ could be defined as a single slope parameter if population change can be described as a "trend," or it could be defined with parameters for each year if "year effects" are of interest.

Count Proportions

Of course, the local population size (N_{it}) is never observed in count surveys such as the BBS or the CBC. Instead, counts are observed, and any analysis of these data requires assumptions to relate counts (Y_{it}) to population sizes (N_{it}). To do this, we define the count-proportion as the logarithm of the ratio of expected counts and expected population sizes, or

$$\lambda_{it} = \log\{E(Y_{it}) \, / \, E(N_{it})\}$$

The log-linear model based on expected value of counts is

$$\log(E(Y_{it})) = \theta_i + \lambda_{it} + f_i(t). \tag{9-1}$$

If the count proportion is constant (i.e., $\lambda_{it} \equiv \lambda$), then the log-linear model can be written as

$$\log(E(Y_{it})) = \theta_i + \lambda + f_i(t).$$

This model reflects the naive analysis of count data, and although it is clearly unreasonable, it is often applied when no distinction is made between counts and actual population sizes. It is worth noting that even with the strong assumption, actual abundances are not estimable unless $\lambda \equiv 1$. Without additional information, we can only estimate $q_i = \theta_i + \lambda$. This relative abundance is the only quantity that can be estimated from counts unless additional information exists that can be used to estimate λ.

If the count proportion is not constant, then the goal of the analysis must be to model λ_{it} to define indices of relative abundance. This means that we model λ_{it} using observable covariates to develop model-based adjustments of counts for differences in detection probabilities. A general class of models for λ_{it} is

$$\lambda_{it} = \delta_{j(i)} + h(\xi_{it}),$$

in which the covariates associated with count proportions can be either site-specific classification variables $(\delta_{j(i)})$ or count adjustments $(h(\xi_{it}))$. Site-specific classification variables are factors nested within sites, such as survey methods or observers, that influence count proportions. For example, different observers on BBS routes may count at varying levels of efficiency, and fixed-wing aircraft surveys may have detection probabilities different than those of helicopters surveys. To estimate change over time in these surveys, it is reasonable to assume that the count proportion will be constant *within* methods, but differ *among* methods. Count adjustments $(h(\xi_{it}))$ represent continuous covariates that could influence count proportion. Water levels or effort expended in counting are examples of covariates that have been shown to influence count proportions (e.g., Bennetts et al. 1999). In our examples, we will discuss two cases of models for λ_{it}. In the BBS, it is well known that observers are reasonable site-specific classification variables, as they are nested within routes and tend to differ in count proportion (Sauer et al. 1994). In Christmas Bird Counts, effort in counting is a well-documented count adjustment (Butcher 1990).

The structure of the log-linear model (eq. [9-1]) is useful for the estimation of population change from time series of counts because it separates factors influencing detectability of counts from factors influencing change in population. Unless features that influence count proportion are identified and appropriately modeled, they can act as lurking variables in the estimation of population change. For example, increasing effort in counting over time causes the counts to increase even if the population is stable.

Modeling of Trajectories

Estimation of components of the population trajectory is the primary focus of the statistical time series literature. The simplest trajectory is a trend, which is generally defined as a consistent change in the level of a time series (Dagum and Dagum 1988). For analysis of population change, it is useful to redefine trend as an interval-specific estimate of population change. For the log-linear model, trend can be modeled as a single parameter $f_i(t) = \beta_i(t - t_1)$. A variety of more complicated population trajectories can be fit, from polynomials of time (e.g., the quadratic, $f_i(t) = \beta_{i1}(t - t_1) + \beta_{i2}(t - t_1)^2$) to year effects trajectories (denoted by $f_i(t) = \varphi_j$) in which $J-1$ parameters are estimated (e.g., Bennetts et al. 1999). Note that, even if year effects are estimated, it is often of interest to present an estimate of "trend," which is often summarized from the more complicated trajectories (e.g., Link and Sauer 1998a).

Estimation

Link and Sauer (1997, 1998a,b) discuss estimation under the log-linear model. GLMs (generalized linear models) are a convenient tool for estimation, allowing for treatment of counts as overdispersed Poisson random variables. For the BBS, in which observers are nested within routes, we condition our analysis on total counts for each observer, allowing us to consider the vector of counts as following a multinomial distribution. Often, however, counts are overdispersed, having larger variance than would be expected from a Poisson distribution, a situation that can be modeled by adding an additional distributional assumption. For example, if counts are distributed as a Poisson distribution mixed with a gamma distribution, the resulting composite distribution is a negative binomial distribution. Conditional on their totals, such overdispersed counts follow an overdispersed multinomial distribution and can be approximated using a quasilikelihood. Fitting this model is accomplished by (1) fitting the model treating the counts as Poisson random variables; (2) estimating overdispersion by calculating chi-square goodness of fit statistics within each observer on a route; (3) refitting the model, weighting the observations by their overdispersion parameters; and (4) repeating the procedure until the estimates are similar among iterations. This procedure can be implemented in standard statistical software packages (e.g., SAS or GLIM).

Link and Sauer (in press) developed a hierarchical log-linear model that is similar to the GLM discussed above. The hierarchical model has the attractive feature that observers and other covariates can be treated as random variables, permitting a very flexible modeling approach. Link and Sauer (in press) fit the model using Markov Chain Monte Carlo methods (Spiegelhalter et al. 1995). We discuss hierarchical models in more detail in a later section of this chapter.

For CBC analyses, some function of effort ($h(\xi)$) must be incorporated as a covariate. Link and Sauer (1999a,b) suggest that a Box/Cox-style transformation $h_p(\xi) = B (\xi^p - 1) / p$ be used, as it allows an asymptote in proportion counted, and becomes a simple scaling by effort (log-transform) in the limit as $p \to 0$. This leads to the log-linear model

$$\ln(\mu_{ij}) = \theta_i + B \frac{\xi_{it}^p - 1}{p} + f_i (t)$$

for fixed p. The model is also fit as an overdispersed GLM, in which overdispersion varies among sites. Fitting p is a complication in the analysis and is accomplished by selecting a variety of candidate values for p, fitting the model for each value and selecting the p that minimizes scaled deviance from the analysis.

Complications Associated with Geographic Scale

Both the BBS and the CBC are continent wide, and the scales of the surveys lead to a large variety of complications for analysts. In particular, geographic strata are needed in both surveys, as the number of sample sites tends to differ regionally. Aggregation of results among strata is complicated by the need to accommodate (1) estimates with widely differing levels of precision, (2) areas of the regions in each stratum, and (3) abundances of birds associated with each stratum. Geissler and Sauer (1990) suggested a weighting procedure using these three factors that is still used in a variety of simple analyses of BBS data (e.g., Peterjohn and Sauer 1999), but it has been criticized for the ad hoc nature of the weightings. Empirical Bayes procedures (e.g., Link and Sauer 1998a) have been suggested as an alternative approach for accommodating differences in precision among regional estimates of population change. In the empirical Bayes approach, information from the year effects in all regions is used to estimate prior mean year effects for each year. Then, for each region and year an empirical Bayes estimate is calculated as a weighted average of the original year effect and the prior mean year effect, with weights determined by the relative precision of the year effect. Note that these Bayes estimates fall between the original estimates and the prior mean, with the amount of "shrinkage" toward the prior mean reflecting the amount of information in the original estimate.

We note that the hierarchical model developed by Link and Sauer (in press) provides an alternative means of aggregating among strata. Because hierarchical models allow for specification of strata information as random effects, the entire aggregation to provide overall regional estimates can be directly accomplished in the hierarchical model. This avoids the need to conduct separate analyses for stratum-level estimation and aggregation, as required by the GLM-empirical Bayes approaches described earlier. Link and Sauer (in press) provide an example analysis using BBS data for Cerulean Warblers (*Dendrioca cerulea*).

Alternatives to GLM Approaches

Modeling of population trajectories using covariates to model count proportions is a very general approach to the summary of count data. A variety of estimation procedures have been used to model count data, including (1) Poisson regression (Link and Sauer 1994; Underhill and Prys-Jones 1994), (2) mixed models (VanLeeuwen et al. 1996), (3) LOESS smoothing of population change (James et al. 1996), and (4) generalized additive models (Fewster et al. 2000).

Breeding Bird Survey Example

Mourning dove (*Zenaida macroura*) data were available from twelve BBS routes in the ridge and valley physiographic region (Bystrak 1981) in Maryland. Data consisted of 190 observations based on data collected in years 1966–1997. We fit the model

$$\log(E\ (Y_{it})) = \gamma + \delta_{j\,(i)} + \varphi_t$$

using program GLIM4 (Francis et al. 1994). Here, $\delta_{j(i)}$ is the observer effect for observer $j(i)$, γ is an additional effect that allows observers to have lower counts the first year they conduct a survey (first year observer effect, Kendall et al. 1996), and φ_t are the year effects.

Fitting the model led to an estimate of $\gamma = 0.04147$ (se $= 0.1082$), hence the first year effect was not significant for this species. Exponentiated year effects indicated a generally increasing population. Note that year effects are scaled to the final year, and in figure 9-1, we scaled the year effects to an estimated abundance in the final year (29.265) of the time series, representing the trajectory-adjusted count for the average observer in that year, and then multiplied each year effect by the value. To estimate "trend" for the interval, we conducted a least-squares regression through the year effects and estimated yearly proportional change to be 0.045 (0.00597), corresponding to a yearly percentage change of 4.6 percent per year.

Christmas Bird Count Example

For the Christmas Bird Count (CBC) example, we summarized population change for the American kestrel (*Falco sparverius*) regionally and at the continental level for the interval 1955–1999. It is generally convenient to use physiographic strata to segregate CBC circles into homogeneous regions (e.g., Link and Sauer 1998a). In this analysis, we used Bird

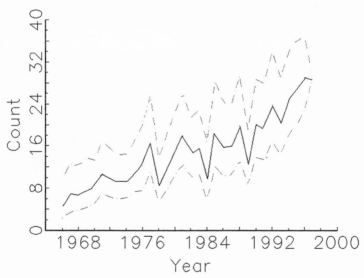

Figure 9-1. Graph of year effects for mourning doves in Maryland as estimated from North American Breeding Bird Survey data over the interval 1966–1997, with associated 95 percent confidence intervals.

Conservation Regions (Commission for Environmental Cooperation 1997) to divide the continent into physiographic strata, estimated change for each of these strata, and then estimated a combined change for the entire survey area. To do this, we first had to establish the appropriate effort adjustment for the species. This was accomplished by fitting a generalized linear model for each strata with year effects and the effort adjustment for a set of candidate values for p ranging from -2 to 2 in increments of 0.5. For each effort adjustment, we summed the deviances among the results for each stratum to calculate a composite deviance for the effort adjustment. We then chose the effort adjustment that minimized the summed deviance and used the GLIM results calculated using that effort adjustment for additional analyses.

The effort adjustment that minimized deviance was $p = -2.0$, leading to a combined deviance of 36287. Other deviances ranged from 39771 ($p = 0.5$) to 39814 ($p = -1.5$). Sample results of year effects for the southeastern coastal plain region are presented in figure 9-2. These results are scaled to an effort-adjusted mean abundance for the final year. We estimated a trend for this interval of -0.088 percent per year (weighted least squares fit through the year effects slope = -0.0008812 with se = 0.001776). Finally, we used empirical Bayes procedures to "shrink" stratum-specific estimates toward the prior mean, and then calculated the overall trend for the species as the weighted average of the empirical

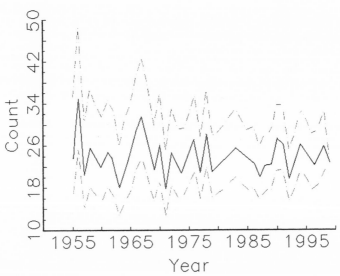

Figure 9-2. Graph of year effects with associated 95 percent confidence intervals, of American kestrels from the southeastern coastal plain region from 1955 to 1999 as estimated from CBC data.

Bayes estimates of regional trends weighted by the mean effort-adjusted regional abundances and areas of each region. The empirical Bayes estimate (-0.085 percent per year, 0.1254) for the Southeastern Coastal Plain Region was little changed from the original, and the overall estimate of change for the survey area was 0.233 percent per year (se = 0.08605, Test of null hypothesis of no trend: $P = 0.007$).

Analysis of Composite Change in Species Groups

Collections of estimates of population change from a group of species are often used to monitor community-level population change. In North America, population change and abundance for groups of bird species that share a common characteristic such as migration status (e.g., Neotropical migrants) or breeding habitats (e.g., grassland birds) have been the subject of public, management, and scientific interest due to perceived declines (e.g., Robbins et al. 1989). Managers require summary information for these taxa, and frequently use collective responses of all species to measure response of the taxa for management. Unfortunately, any grouping of species can be controversial, both due to concerns about the validity of the group (Mannan et al. 1984) and due to variation in quality of information among species in the group. As a consequence of the differences in estimated precision of trend estimates among species in the BBS, many common summaries of group population change, such as average population trend, number of "declining" species, number of species with significant trend estimates, ranks of species by population trend, and lists of species with extreme changes, are flawed measures of group response (Link and Sauer 1995).

Estimates of population trend tend to differ among species in both estimated magnitude and precision of individual estimates. Because of this, imprecise trend estimates may be quite large while still having a confidence interval large enough to include zero, indicating that the trend is not significantly different from zero; the magnitude of trend does not always reflect significance. On the other hand, statistical significance often reflects the quality of the data, and a very small rate of change may be precisely estimated but of no significance to the population. Separating notions of statistical significance from magnitude of trend has proven to be quite difficult when summarizing trends within groups of species.

For single-species analyses, statistics are based on distribution of data Y_s, conditional on an unknown parameter, $f(Y_s | \theta_s)$. For multiple species, we are actually interested in attributes of the parameters θ_s. One approach to this is to view both data and parameters as random variables in a hierarchical (multi-level) model. To do this, distributional assumptions about the parameters must be included in the analysis.

We use Bayesian methods to make model-based probability statements about θ_s (Gilks et al. 1996). To do this, we must define the sampling distribution of the data given the unknown parameter, or $f(Y_s|\theta_s)$, and the prior distribution of the parameters, or $\pi(\theta_s|\Psi)$. Ψ is a hyperparameter that describes the distribution of the parameters θ_s. Bayesian inference is used to make statements about the θ_s based on a posterior distribution $f(\theta_s|Y_s)$. Although Bayesian analyses have been often been limited by technical complexity of derivation of posterior distributions, computer-intensive statistical approaches, in particular Markov Chain Monte Carlo methods, now provide an extremely powerful tool for estimating the posterior distribution.

Approaches to Fitting Hierarchical Models for Species Groups

Empirical Bayes methods have been used to implement hierarchical models for species group attributes (Link and Sauer 1995). In empirical Bayes, hyperparameters are estimated using information from the data, and we applied a simple model to the case of estimation of a prior mean trend for the group, and then estimated posterior means of the species trend parameters. These "shrunken" estimates are a weighted average of the prior mean (the estimate of the group mean trend parameter) and the estimated trend. The resulting estimate for each species is intermediate between the original estimated trend and the prior mean, with the actual value dependent on the relative precision of the original trend. We used this model to rank trends (Link and Sauer 1996), and estimated the number of increasing species using a bootstrapping procedure (Link and Sauer 1995). However this approach is limited to quite simple models. Markov Chain Monte Carlo (MCMC) methods can be used to fit a much more useful set of hierarchical models.

In MCMC, simulation is used in estimation of hyperparameters. A model is defined in terms of distributions of parameters and hyperparameters, and the distribution information for each variable is described in terms of "full conditionals," distributions with all other parameters fixed, and an iterative sampling is conducted using these full conditionals. An iterative procedure (e.g., Gibbs Sampling) produces results that converge after many iterations on posterior distributions for the parameters. After this convergence occurs, subsequent iteration results provide replicate estimates based on sampling from the distributions. Means and variances from the simulation results after convergence can be used as estimates of parameters and hyperparameters and their variances. See Spiegelhalter et al. (1995), Sauer and Link (2002), or Sauer and Link (in press) for more details of the estimation procedure.

The Hierarchical Model for Composite Change in a Collection of Estimates

We assume that trend estimates $\hat{\beta}_s$, $s = 1, 2, \ldots, n$, are normal random variables with parameters β_s and σ_s^2, and the variance estimated from the data $\hat{\sigma}_s^2$ is chi-square distributed. Parameter β_s has a normal distribution with hyperparameters μ and τ^2. These hyperparameters also follow distributions, with μ distributed normally (mean 0, variance = 100000), and τ^2 and σ_s^2 are assumed to follow inverse gamma distributions. The model was fit using the Gibbs sampling approach described above, in program BUGS (Spiegelhalter et al. 1995).

From the MCMC analysis, posterior mean trend estimates for individual species are those associated with the parameter β_s. These are precision-adjusted estimates that are "shrunken" toward the overall prior mean estimate (μ). Although similar to the ranked trend estimates described by Link and Sauer (1995), the posterior mean estimates from the MCMC approach better accommodate imprecision in the estimates of precision than did the empirical Bayes approach. The number of species with positive trend estimates (N_{Inc}) are estimated directly from the MCMC results by simply counting the number of positive posterior mean trend estimates from each MCMC replicate and using these as replicates to get a mean and variance.

Hierarchical Model Analysis Example

Sauer and Link (in press) use the MCMC approach to estimate group attributes for twenty-eight species of grassland-breeding birds from the BBS for the time period 1966–2000. Estimated trends (percent change per year, Link and Sauer 1994) over the survey interval provide estimates that differ greatly in precision and magnitude among species, although many species are declining. They calculated the posterior mean estimates β_s and estimated the number of species with positive trend estimates. Sixty-one percent of species have significant negative trends, and only 18 percent of species have trend estimates greater than zero.

Species rankings are summarized in figure 9-3, in which the trends are ranked by size of posterior mean and the estimated trends are displayed for each posterior mean estimate. The posterior mean estimates show less variation, especially for the species with extreme estimates of increases and declines. This is evident from observation of individual species estimates. For example, the estimated trend for Henslow's sparrow (*Ammodramus henslowii*) was -7.46 percent per year, $n = 155$, but the posterior mean trend was -3.91 percent per year. For Baird's sparrow (*Ammodramus bairdii*) the estimated trend was -2.88 percent per year, $n = 124$, but the posterior mean was -2.35 percent per year. The number of species with positive trend estimates was 5.16 species (SE 1.216), while the naive estimate (i.e., based on the estimated trends) was 5. The estimated number of positive trends is no different than the count of positive trend estimates. This is not always the case, the proportion of positive trend estimates being typically biased toward 50 percent. Use of the hierarchical modeling approach has the additional benefit of allowing the calculation of confidence intervals for the number of increasing trends. In this case, it indicated that most species in the group were quite precisely estimated.

Benefits of Hierarchical Models

Thinking about summaries of group attributes is often complicated by the need to think in terms of estimated individual species trends that are observed, while the actual summary of interest should be based on the collection of parameters. Hierarchical models explicitly recognize this structure and provide an appropriate conceptual framework for estimating attributes associated with the parameters. Also, derived attributes such as

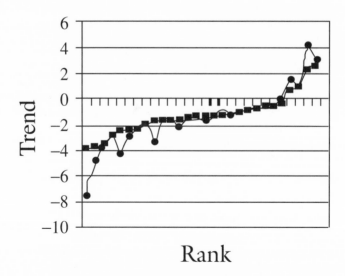

Figure 9-3. Posterior means of trends in grassland birds (boxes) plotted with the original trend estimates for the species (circles). Data are from twenty-eight species of grassland-breeding birds from the BBS for the time period 1966–2000 (Sauer and Link in press).

number of increasing species can be conveniently estimated as part of the MCMC simulations. The hierarchical model approach is superior to empirical Bayes in that it provides much greater flexibility in defining models and implementing the estimation. In the case described here, the model had an additional component that accommodated uncertainty in estimation of the variance of the estimated trends, and earlier empirical Bayes procedures could not accommodate this source of variation.

Direct Analysis of Change in Species Richness

Analysis of species data from surveys such as the BBS and CBC are fundamentally compromised by the failure of the surveys to provide for differences in detectability of individuals over space and time. However, it is possible to directly estimate species richness from count data using estimators that accommodate heterogeneity in detection rates among individual species (e.g., Nichols and Conroy 1996; Burnham and Overton 1979), and these estimators can be used to both estimate species richness and address hypotheses about change in species richness over time and space (e.g., Nichols et al. 1998a,b). As change in species richness is often a reasonable measure of community change (Weins et al. 1996), the development of these estimators provides an important new tool for evaluation of community dynamics from count survey data (Boulinier et al. 1998). Nichols et al. (1998a,b) have developed procedures for estimation of change in species richness along with metrics for estimation of colonization and extinction rates, and Cam et al. (2000) have used these estimators to develop an approach to estimation of completeness of a local site relative to regions. Finally, these methods have been applied to BBS data through the use of program COMDYN (Hines et al. 1999).

Estimation of Species Richness

Nichols and Conroy (1996) and Boulinier et al. (1998) have suggested the use of the closed-population capture-recapture models (Otis et al. 1978; Burnham and Overton 1979) for species richness estimation from count surveys. Although detection rates of species can often differ among sample sites and during the time of sampling, it appears that heterogeneity among species is an important source of variation in detection rates in the BBS (Boulinier et al. 1998). Estimation of detection rates when heterogeneity occurs among species is accomplished through use of Burnham and Overton's (1979) jackknife estimator (denoted as M_h by Otis et al. [1978]), or modifications of the jackknife or other estimators (e.g., Chao 1989; Chao et al. 1992). In their standard formulations, closed-population capture-recapture models require "capture histories" of individual animals for a series of capture occasions. The capture histories are a series of "1" or "0" outcomes that reflect whether an animal was captured (or not captured) during each trapping session. The capture histories are then used in estimation of detection rates. Otis et al. (1978) described the general approaches to closed-population estimation procedures. In the case of species richness, each capture history corresponds to a species (rather than an individual), and each "capture occasion" refers to a sample site. So, for a BBS route, a capture history for a species would have up to fifty capture occasions (corresponding to the fifty stops), and for each stop the species would be coded as either occurring, "1," or not detected, "0." Technical issues associated with application of these procedures to species richness are discussed in Boulinier et al. (1998).

Robust Designs and Estimation of Change

Closed-population capture recapture procedures provide a great deal of flexibility in estimation of temporal and spatial variation in detection rates but do not permit the number of species to change. Hence, development of estimators for change in species richness requires strategic application of the closed-population procedures at different times (or locations), and change has to be estimated between the separately estimated species richness estimates. This procedure, in which there are secondary sampling periods (e.g., stops on BBS routes) used for estimation of N, then primary sampling periods over space and time through which change in N is computed, is called a *robust design* (Pollock et al. 1993; Kendall et al. 1997). Estimation using the robust design is complicated by the need to estimate variances from several sources and occasionally requires the use of bootstrapping procedures (Nichols et al. 1998a). Use of the robust design allows for estimation of rate of change of species richness, local extinction probabilities, local species turnover, and number of colonizing species over time (where primary periods are measured over time), and space (where primary periods are defined in terms of space). We also describe a measure of local species completeness, a measure developed by Cam et al. (2000) to summarize the proportion of a regional species pool present at a site.

Rate of Change of Species Richness

Change in species richness is the ratio of estimated species richness (N) at two times i and j, or

$$\hat{\lambda}_{i,j} = \frac{\hat{N}_j}{\hat{N}_i}.$$

Note that, if average species detection rates are the same between the two times, the ratio of counts of species at each site can be used to estimate richness (c.f., Skalski and Robson 1992). Nichols et al. (1998a) describe a test for the null hypothesis of equal detection rates.

Local Extinction Probabilities

The probability that a species present at one time (i) is not present at a subsequent time (j) can be estimated as 1 minus the proportion of observed species at time i that still exist at time j, or

$$1-\hat{\phi}_{ij} = 1-\frac{\hat{M}_{j}^{R_i}}{R_i}.$$

where R_i is the number of species observed in period i, and the quantity in the numerator is the estimated number of those R_i species that are estimated to be present in time j. This quantity is estimated using the sample data for time j, but only using data from those R_i species seen at time i.

Local Species Turnover

Species turnover is defined as the probability that a species randomly selected at time j was not a member of the community at previous time i, and is hence a "new" species. See Nichols et al. (1998a) for a description of the motivation of this estimator. In practice, this estimator is similar to the extinction estimator, but the time periods are ordered differently:

$$1-\hat{\phi}_{ji} = 1-\frac{\hat{M}_{i}^{R_j}}{R_j}.$$

Estimation is accomplished similarly to the extinction probability estimate, but by conditioning on the species observed at the jth time.

Number of Local Colonizing Species

The number of species that colonize between two times (i and j) is the estimated number of species at time j minus the number of species surviving from time i, or

$$\hat{B}_{ij} = \hat{N}_j - \hat{\phi}_{ij}\hat{N}_i.$$

Estimation over Space

Nichols et al. (1998b) discuss methods for spatial analysis of change. Generally, these estimates are similar in form to the temporal estimates of change, but the primary periods are defined spatially, not temporally.

They define *relative species richness* as spatial rate of change of species richness, *species co-occurrence* as the estimated spatial φ_{ij} (i.e., the probability that a species present at location i is also present at location j), and *number of species at only one of two locations* as a spatial analogue of number of local colonizing species.

Local Species Completeness

It is often of interest to evaluate species richness at sites relative to some standard list of potential species, and Cam et al. (2000) developed a metric for relative species richness at a site for use with count data. This estimator

$$\hat{\phi}_j = 1 - \frac{\hat{M}_j^R}{R},$$

is the estimated species richness at site j relative to some predefined species pool of R species. They applied the estimator to assess the consequences of urbanization on bird communities in the eastern United States.

Example of Species Richness Estimation

We demonstrate the estimation of some of the change in community attributes using data from the BBS from the Floyd, Maryland Route. We used program COMDYN (Hines et al. 1999) to estimate species richness and related information for 1966 and 1996 surveys. In 1966, fifty-seven species were seen on the route, and sixty-five (se = 4.806) were estimated to occur, while in 1996, sixty-six species were seen and 109 (se = 14.21157) estimated. The test of the null hypothesis of no difference in average p provided some evidence of difference ($\chi^2 = 9.37$, $df = 5$, $P = 0.095$). Estimated change in species richness was 1.6458 (0.2514). The estimated number of species present in 1996 that were observed in 1966 was 53.71 (11.10), hence $\varphi_{1966,1996}$ was 0.8828 (0.1278) and the estimated extinction probability was 0.1172. The estimated number of species present in 1966 that were observed in 1996 was 56.49 (8.78), hence $\varphi_{1996,1966}$ was 0.8353 (0.1072) and the estimated species turnover probability was 0.1647.

Conclusion

Often, our perceptions of status and trends of natural resources are based on very poor information. Regional estimates are often based on aggregation of information from a subset of sites that do not form an adequate sample for the region, and counts collected from those sites are often biased estimates of the parameters of interest. Understanding the limitations of existing data, developing new surveys that are adequate for monitoring regional population and community attributes, and implementing analytical tools that allow appropriate summaries of survey results are all essential components of using survey data for science and management. Because survey information is important to any management activity, survey development requires a clear statement of goals that ensure that survey data will meet information needs. Surveys also require constant evaluation and updating to ensure that information needs continue to be met, and that technological innovations are adequately incorporated into the conduct of the survey and its analysis. Often, the concept of standardization is put forward as a constraint on changes in survey methods. Unfortunately, emphasis on standardization often perpetuates flawed designs, and comparability should be viewed as less important than improved design and estimation.

Count surveys often require complicated analyses to accommodate the constraints imposed by factors that influence counts but not the actual population size. Evaluation of the consequences of regional and temporal differences in detectability is a major component of any count-based survey of biological resources. Often, major factors influencing detectability of birds can be accommodated by use of covariates, and generalized linear models can be used to accommodate the distributional constraints of count data by permitting overdispersed Poisson distributions. Hierarchical models provide a reasonable approach to accommodating the difficulties in evaluating pattern among collections of estimates for species groups, and capture-recapture analyses provide direct estimation of detectability of species and species richness.

LITERATURE CITED

Bennetts, R. E., W. A. Link, J. R. Sauer, and P. W. Sikes Jr. 1999. Factors influencing counts in an annual survey of snail kites in Florida. *Auk* 116:316–323.

Boulinier, T, J. D. Nichols, J. R. Sauer, J. E. Hines, and K. H. Pollock. 1998. Estimating species richness to make inference in community ecology: The importance of heterogeneity in species detectability as shown from capture-recapture analyses of North American Breeding Bird Survey data. *Ecology* 79:1018–1028.

Box, G. E. P., G. Jenkins, and G. Reinsel. 1994. *Time series analysis: Forecasting and control.* New York: Prentice Hall.

Burnham, K. P., and W. S. Overton. 1979. Robust estimation of population size when capture probabilities vary among animals. *Ecology* 60:579–604.

Butcher, G. S. 1990. Audubon Christmas bird counts. Pp. 5–13 in *Survey designs and statistical methods for the estimation of avian population trends,* edited by J. R. Sauer and S. Droege. Biological Report 90(1). U.S. Fish and Wildlife Service.

Bystrak, D. 1981. The North American breeding bird survey. Pp. 34–41 in *Estimating numbers of terrestrial birds,* edited by C. J. Ralph and J. M. Scott. Studies in Avian Biology No. 6. Cooper Ornithological Society. Lawrence, Kansas: Allen Press.

Cam, E., J. D. Nichols, J. R. Sauer, J. E. Hines, and C. H. Flather. 2000a. Estimation of relative species richness for assessing the degree of completeness of ecological communities: Avian communities and urbanization in the mid-Atlantic states. *Ecological Applications* 10:1196–1210.

———. 2000b. Geographic analysis of species richness and community attributes of forest birds from survey data in the mid-Atlantic integrated assessment region. *Environmental Monitoring and Assessment* 63:81–94.

Carter, M. F., W. C. Hunter, D. N. Pashley, and K. V. Rosenberg. 2000. Setting conservation priorities for landbirds in the United States: The Partners in Flight Approach. *Auk* 117:541–548.

Caughley, G. 1994. Directions in conservation biology. *Journal of Animal Ecology* 63:215–244.

Chao, A. 1989. Estimating population size for sparse data in capture-recapture experiments. *Biometrics* 45:427–438.

Chao, A., S. M. Lee, and S. L. Jeng. 1992. Estimation of population size for capture-recapture data when capture probabilities vary by time and individual animals. *Biometrics* 48:201–216.

Cochran, W. G. 1977. *Sampling techniques.* 3d ed. New York: John Wiley and Sons.

Commission for Environmental Cooperation. 1997. *Ecological regions of North America: Toward a common perspective.* Commission for Environmental Cooperation, Montreal, Canada.

Dagum, C., and E. B. Dagum. 1988. Trend. Pp. 321–324 in *Encyclopedia of statistical sciences.* Vol. 9. Edited by S. Kotz and N. L. Johnson. New York: John Wiley and Sons.

Diggle, P. J., K.- Y. Liang, and S. L. Zeger. 1994. Analysis of longitudinal data. New York: Oxford University Press.

Dunn, E. H., and J. R. Sauer. 1997. Monitoring Canadian bird populations with winter counts. Pp. 49–55 in *Monitoring bird populations: The Canadian experience,* edited by E. H. Dunn, M. D. Cadman, and J. Bruce Falls. Occasional Paper 95. Canadian Wildlife Service.

Fewster, R. M., S. T. Buckland, G. M. Siriwardena, S. R. Baillie, and J. D. Wilson. 2000. Analysis of population trends for farmland birds using generalized additive models. *Ecology* 81:1970–1984.

Frances, B., M. Green, C. Payne, and T. Swan. 1994. *The GLIM System: Release 4 manual.* Oxford: Clarendon Press.

Geissler, P. H., and J. R. Sauer. 1990. Topics in route regression analysis. Pp. 54–57 in *Survey designs and statistical methods for the estimation of avian population trends,* edited by J. R. Sauer and S. Droege. Biological Report 90(1). U.S. Fish and Wildlife Service.

Gelman, A., J. B. Carlin, H. S. Stern, and D. R. Rubin. 1995. *Bayesian data analysis.* New York: Chapman and Hall.

Gilks, W. R., S. Richardson, and D. J. Spiegelhalter. 1996. *Markov chain Monte Carlo in Practice.* New York: Chapman and Hall.

Hines, J. E., T. Boulinier, J. D. Nichols, J. R. Sauer, and K. H. Pollock. 1999. COMDYN: Software to study the dynamics of animal communities using a capture-recapture approach. *Bird Study* 46 (supplement):S209–217.

James, F. C., C. E. McCulloch, and D. A. Wiedenfeld. 1996. New approaches to the analysis of population trends in land birds. *Ecology* 77:13–27.

Kendall, W. L., J. D. Nichols, and J. E. Hines. 1997. Estimating temporary emigration and breeding proportions using capture-recapture data with Pollock's robust design. *Ecology* 78:563–578.

Kendall, W. L., B. G. Peterjohn, and J. R. Sauer. 1996. First-time observer effects in the North American breeding bird survey. *Auk* 113:823–829.

Lancia, R. A., J. D. Nichols, and K. H. Pollock. 1994. Estimating the number of animals in wildlife populations. Pp. 215–253 in *Research and management techniques for wildlife and habitats*, edited by T. A. Bookhout. Bethesda, Md.: The Wildlife Society.

Link, W. A., and J. R. Sauer. 1994. Estimating equations estimates of trends. *Bird Populations* 2:23–32.

———. 1995. Estimation of empirical mixing distributions in summary analyses. *Biometrics* 51:810–821.

———. 1996. Extremes in ecology: Avoiding the misleading effects of sampling variation in summary analyses. *Ecology* 77:1633–1640.

———. 1997. Estimation of population trajectories from count data. *Biometrics* 53:63–72.

———. 1998a. Estimation of population change from count data with applications to the North American breeding bird survey. *Ecological Applications* 8:258–268.

———. 1998b. Estimating relative abundance from count data. *Austrian Journal of Statistics* 27(1): 83–97.

———. 1999a. On the importance of controlling for effort in analysis of count survey data: Modeling population change from Christmas Bird Count data. *Vogelwelt* 120, Supplement 1:15–20.

———. 1999b. Controlling for varying effort in count surveys: An analysis of Christmas Bird Count Data. *Journal of Agricultural, Biological and Environmental Statistics* 4(2):116–125.

———. In press. A hierarchical model of population change with application to Cerulean Warblers. *Ecology* (2002).

Louis, T. A. 1984. Estimating a population of parameter values using Bayes and empirical Bayes Methods. *Journal of the American Statistical Association* 79:393–398.

Mannan, R. W., M. L. Morrison, and E. C. Meslow. 1984. Comment: The use of guilds in forest bird management. *Wildlife Society Bulletin* 12:426–430.

Martin, F. W., R. S. Pospahala, and J. D. Nichols. 1979. Assessment and population management of North American migratory birds. Pp. 187–239 in *Environmental biomonitoring, assessment, prediction, and management: Certain case studies and related quantitative issues*, edited by J. Cairns Jr., G. P. Patil, and W. E. Waters. Fairland, Md.: International Cooperative Publishing House.

McCullagh, P., and J. A. Nelder. 1989. *Generalized linear models*, 2d ed. London: Chapman and Hall.

Nichols, J. D., T. Boulinier, J. E. Hines, K. H. Pollock, and J. R. Sauer. 1998a. Estimating rates of local species extinction, colonization, and turnover in animal communities. *Ecological Applications* 8:1213–1225.

———. 1998b. Inference methods for spatial variation in species richness and community composition when not all species are detected. *Conservation Biology* 12:1390–1398.

Nichols, J. D., and M. J. Conroy. 1996. Estimation of species richness. Pp. 226–234 in *Measuring and monitoring biological diversity: Standard methods for mammals*, edited by D. E. Wilson, F. R. Cole, J. D. Nichols, R. Rudran, and M. S. Foster. Washington, D.C.: Smithsonian Institution Press.

Otis, D. L., K. P. Burnham, G. C. White, and D. R. Anderson. 1978. Statistical inference from capture data on closed populations. *Wildlife Monographs* 62.

Peterjohn, B. G., and J. R. Sauer. 1999. Population status of North American grassland birds from the North American Breeding Bird Survey, 1966–1996. *Studies in Avian Biology* 19:27–44.

Pollock, K. H., W. L. Kendall, and J. D. Nichols. 1993. The "robust" capture-recapture design allows components of recruitment to be estimated. Pp. 245–252 in *Marked individuals in the study of bird populations*, edited by J.-D. Lebreton and P. M. North. Basel, Switzerland: Birkhauser Verlag.

Ralph, C. J., J. R. Sauer, and S. Droege, eds. 1995. *Monitoring bird populations by point counts*. General Technical Report PSW-GTR-149. Berkeley, Calif.: U.S. Department of Agriculture, Forest Service, Pacific Southwest Research Station, Berkeley, Calif.

Ralph, C. J., and M. J. Scott, eds. 1981. Estimating numbers of terrestrial birds. *Studies in Avian Biology 6*, Lawrence, Kansas: Cooper Ornithological Society.

Robbins, C. S., D. Bystrak, and P. H. Geissler. 1986. *The breeding bird survey: Its first fifteen years, 1965–1979*. Resource Publication 157. U.S. Fish and Wildlife Service.

Robbins, C. S., J. R. Sauer, R. S. Greenberg, and S. Droege. 1989. Population declines in North American birds that migrate to the Neotropics. *Proceedings of the National Academy of Sciences USA* 86:7658–7662.

Rubin, D. B. 1984. Bayesianly justifiable and relevant frequency calculations for the applied statistician. *Annals of Statistics* 12:1151–1172.

Sauer, J. R., and S. Droege, eds. 1990. *Survey designs and statistical methods for the estimation of avian population trends*. Biological Report 90(1). U.S. Fish and Wildlife Service.

Sauer, J. R., and W. A. Link. 2002. Hierarchical modeling of population stability and species group attributes using Markov Chain Monte Carlo methods. *Ecology* 86:1743–1751.

Sauer, J. R., and W. A. Link. In press. Hierarchical models and the analysis of bird survey information. *Proceedings Bird Numbers 2001 Conference*. Ornis, Hungaria.

Sauer, J. R., G. W. Pendleton, and S. Orsillo. 1995. Mapping of bird distributions from point count surveys. Pp. 151–160 in *Monitoring bird populations by point counts*, edited by C. J. Ralph, J. R. Sauer, and S. Droege. General Technical Report PSW-GTR-149. U.S. Department of Agriculture, Forest Service, Pacific Southwest Research Station, Berkeley, Calif.

Sauer, J. R., G. W. Pendleton, and B. G. Peterjohn. 1996. Evaluating causes of population change in North American insectivorous songbirds. *Conservation Biology* 10:465–478.

Sauer, J. R., B. G. Peterjohn, and W. A. Link. 1994. Observer differences in the North American breeding bird survey. *Auk* 111:50–62.

Skalski, J. R., and D. S. Robson. 1992. *Techniques for wildlife investigation: Design and analysis of capture data*. New York: Academic Press.

Smith, G. W. 1995. *A critical review of the aerial and ground surveys of breeding waterfowl in North America*. Biological Science Report 5. Washington, D.C.: U.S. Department of the Interior, National Biological Service, Washington, D.C.

Spiegelhalter, D. J., A. Thomas, N. G. Best, and W. R. Gilks. 1995. *BUGS: Bayesian inference using Gibbs sampling, Version 0.50*. Cambridge, U.K.: MRC Biostatistics Unit.

Temple, S. A., and J. A. Wiens. 1989. Bird populations and environmental changes: Can birds be biological indicators? *American Birds* 43:260–270.

Thompson, S. K. 1992. *Sampling*. New York: John Wiley and Sons.

Thompson, S. K., and G. A. F. Seber. 1996. *Adaptive sampling*. New York: John Wiley and Sons.

Thompson, W. L., G. C. White, and C. Gowan. 1998. *Monitoring vertebrate populations*. New York: Academic Press.

Underhill, L. G., and R. P. Prys-Jones. 1994. Index numbers for waterbird populations. I. Review and methodology. *Journal of Applied Ecology* 31:463–480.

U.S. Fish and Wildlife Service. 2001. *Waterfowl population status 2001*. U.S. Department of the Interior, Washington, D.C.

Van Leeuwen, D. W., L. W. Murray, and N. S. Urquhart. 1996. A mixed model for both fixed and random trend components over time. *Journal of Agricultural, Biological, and Environmental Statistics* 1:435–453.

Weins, J. A., T. O. Crist, R. H. Day, S. M. Murphy, and G. D. Hayward. 1996. Effects of the Exxon Valdez oil spill on marine bird communities in Prince William Sound. *Ecological Applications* 6:828–841.

Williams, B. K., and F. A. Johnson. 1995. Adaptive management and the regulation of waterfowl harvests. *Wildlife Society Bulletin* 23:430–436.

PART IV

Monitoring Habitats, Populations, and Communities

CHAPTER 10

Competing Goals of Spatial and Temporal Resolution: Monitoring Seagrass Communities on a Regional Scale

James W. Fourqurean and Leanne M. Rutten

Situated on the southern tip of the Florida peninsula, the Florida Keys are the downstream terminus of the wetland-dominated south Florida ecosystem that starts at the Kissimmee River and includes Lake Okeechobee and the Florida Everglades (see chap. 13 for a description of a regional monitoring program in these wetland ecosystems). Water flows and levels have been extensively engineered in this ecosystem to allow for agriculture and urban development, but a monumental effort is now underway to reevaluate and redesign the management of freshwater flows in the south Florida ecosystem (chap. 5). This effort has as one of its major goals preventing irretrievable loss of components of the ecosystem as a consequence of altered supply and quality of freshwater as a result of satisfying competing human water-related needs in the region (USACE 1999; Redfield 2000; Kiker et al. 2001).

Although most of the engineering of natural water flows has taken place in the rivers, lakes, and marshes of the ecosystem (Light and Dineen 1994), it is widely recognized that water management practices have also had effects on the estuarine and marine water bodies that receive these flows. Alterations in quantity, timing, and/or quality of freshwater discharge have been implicated in changing the structure of benthic and planktonic communities in the Caloosahatchee estuary (Kraemer et al. 1999), the St. Lucie estuary (Chamberlain and Hayward 1996; Havens and Aumen 2000), and Florida Bay (reviewed in Fourqurean and Robblee 1999). It has also been suggested that these alterations may be causing general eutrophication of the marine environment in south Florida, possibly impacting seagrass beds surrounding the Florida Keys (Lapointe et al. 1994) and the Florida Keys barrier coral reef (Porter et al. 1999). The impacts, both proven and perceived, of human alteration of water flow in south Florida on the estuarine and marine environment underscore the high interdependency of the various components of the south Florida ecosystem. Partially in response to concerns about upstream water management effects on the marine end of the south Florida ecosystem, the U.S. Congress established the Florida Keys National Marine Sanctuary in 1990 (fig. 5-3) with the aim to "preserve and protect the physical and biological components on the south Florida estuarine and marine ecosystem to ensure its viability for the use and enjoyment of present and future generations" (NOAA 1996).

The impacts of regional water management on the marine ecological communities of south Florida are compounded by local anthropogenic influences. The net annual budget of rainfall and evapotranspiration is near zero in Florida Bay and the Florida Keys (Nuttle et al. 2000); hence any human transport of water can significantly influence local water budgets. Almost all of the potable water used by the approximately one hundred thousand residents of the Florida Keys is piped in from groundwater wells on mainland Florida. Wastewater generated by the population of the Keys is, for the most part, disposed of in on-site sewage disposal systems (OSDS: septic tanks and cesspits) that release freshwater and nutrients into the shallow groundwater and eventually to the marine surface waters (Lapointe et al. 1990; Paul et al. 1997; Dillon et al. 1999).

Despite the widely held opinion that human alterations of water and nutrient budgets on both regional and local scales are having deleterious effects on the estuarine and marine ecosystems of south Florida, direct causal links have mostly eluded detection. One major reason for this is the high degree of ecosystem variability in both time and space (Fourqurean and Robblee 1999). Water quality varies not only across the ecosystem, but also seasonally as a consequence of natural processes (Boyer et al. 1999, Boyer and Jones 2002). Interannual variability in climate can override the effects of human water management on the water budget of coastal bays (Nuttle et al. 2000). Standing stock and productivity of primary producers is highly seasonal, even in tropical-subtropical south Florida (Fourqurean et al. 2001). Extreme meteorological events like hurricanes and winter freezes can have impacts that last for years (Tilmant et al. 1994). Poorly understood outbreaks (e.g., Rose et al. 1999) or die-offs (e.g., Lessios et al. 1984; Robblee et al. 1991; Butler et al. 1995) of organisms can have consequences that cascade through the ecosystem. Finally, long-term modification of predator and grazer populations by human activities also may be responsible for long-term gradual changes in the structure of marine ecosystems (Jackson 2001; Jackson et al. 2001). So, to determine the effects of water management practices on estuarine and marine resources of south Florida requires a monitoring program that is spatially expansive, temporally intense, long-lived, and multidisciplinary.

There are few places on earth where seagrass beds are as expansive as the near-shore marine ecosystem of south Florida, where there are at least 14,000 square kilometers of seagrass beds (Fourqurean et al. 2002). In the shallow water nearest shore, seagrasses are especially prevalent; over 90 percent of the area in water less than 10 meters deep supports seagrass. Seagrass beds are recognized as among the most productive (Zieman and Wetzel 1980) and economically valuable (Costanza et al. 1997) of ecosystems, and the economy of the Florida Keys is inextricably tied to seagrass beds. Fisheries landings in the Florida Keys total over 12 × 10⁶ kilograms annually of mostly seagrass-associated organisms (Bohnsack et al. 1994), and over half of all employment in the Florida Keys is dependent on outdoor recreation (NOAA 1996). But, the growing human population is placing visible strain on the marine communities surrounding south Florida, leading some to question the sustainability of the very ecological resources at the base of this portion of the regional economy.

Despite their recognized importance, worldwide loss of seagrass beds continues at an alarming rate (Short and Wyllie-Echeverria 1996). This loss largely has been attributed to anthropogenic inputs of sediment and nutrients. The difficulty of monitoring seagrass beds has led to obfuscation of the real extent of seagrass loss, as our best estimates of even the current global extent of this important habitat are at best within an order of magnitude (Duarte 2002). In Florida alone, anthropogenic seagrass losses have been reported in Pensacola Bay, St. Joseph Bay, Tampa Bay, Charlotte Harbor, the Florida Keys, Biscayne Bay, and the Indian River Lagoon (see Sargent et al. 1995; Short and Wyllie-Echeverria 1996 for reviews), but accurate estimates of the current areal extent of seagrasses even in a populated, first-world location like Florida are only recently available.

Design of the Monitoring Scheme for Seagrass Communities in South Florida

The monitoring plan for the seagrass-dominated marine communities of south Florida had to address two competing goals. First, the program had to define the spatial extent and present condition of the resource. Second, the monitoring program had to produce information that could be used to detect trends in the condition of the resource through time and be able to ascertain the causes of these trends. The reasons these goals were competing are quite simple: monitoring is an expensive, time-consuming process, the area to be monitored is vast (about 18,000 square kilometers) and heterogeneous, and the financial resources to accomplish the goals were limited.

Choosing What to Measure

The decision about what parameters to measure in a monitoring program is influenced by both conscious decision and subconscious bias. It is important that the measured parameters be:

- Unambiguously related to the conceptual model guiding the monitoring plan.
- Relatively easy to assess, for both economic reasons and because monitoring programs often outlast the investigators who originally designed the programs, making it a necessity that the techniques can be easily passed on to successive project personnel.
- Precise enough to allow for exact characterization and change detection.
- Robust enough so that slight differences among observers do not influence the interpretation of the data.

Research and technology advances constantly make possible new potential monitoring methods. These new methods should be incorporated and tested by comparison to more established methods as part of large monitoring programs, but using untried or unproven methods necessitates periodic assessment of the efficacy of such methods and a willingness to modify the monitoring program if the new methods do not prove effective.

Conceptual Models Used to Guide Monitoring

A successful monitoring program must be based on a solid understanding of the processes responsible for driving environmental change. Without such an understanding, it is not possible to allocate monitoring resources in the appropriate temporal and spatial scale. The temporal scale is important because sampling intensity must match the time course of ecological responses in the environment. Further, this understanding is important to managing the expectations of managers: ecosystem responses to changes in stressors can often play out over decades, which is much longer than the fiscal cycles normally used to fund monitoring programs. The spatial scale is important, because monitoring programs often are asked to be as spatially extensive and regionally representative as possible, requiring a broad spacing of monitoring sites. This regional representation can lead to the failure to detect important localized ecological responses if the response varies at a smaller scale than the average spacing between monitoring sites.

The pattern of anthropogenically driven loss of seagrass beds across the globe leads to a generalized model of the effects of eutrophication on seagrass beds (Duarte 1995). In general, eutrophication in aquatic environments shifts the competitive balance to faster-growing primary producers. The consequence of this generality in seagrass-dominated environments is that seagrasses are the dominant primary producers in oligotrophic conditions. As nutrient availability increases, there is an increase in the importance of macroalgae, both free-living and epiphytic, with a concomitant decrease in seagrasses because of competition for light. Macroalgae lose out to even faster-growing microalgae as nutrient availability continues to increase: first, epiphytic microalgae replace epiphytic macroalgae on seagrasses; then, planktonic microalgae bloom and deprive all benthic plants of light under the most eutrophic conditions.

Using knowledge of the life history characteristics of local species and experimental and distributional evidence, this general model can be adapted to seagrass beds of south Florida. The south Florida case is more complicated than the general case described above because there are six common seagrass species in south Florida, and these species have different nutrient and light requirements, and hence they have differing responses to eutrophication. Large expanses of the shallow marine environments in south Florida are so oligotrophic that biomass and growth of even the slowest-growing local seagrass species, *Thalassia testudinum*, are nutrient-limited (Fourqurean et al. 1992a,b); at this very oligotrophic end of the spectrum, increases in nutrient availability actually cause increases in seagrass biomass and growth rate (Powell et al. 1989). As nutrient availability increases beyond what is required by a dense stand of *T. testudinum*, other seagrass species will outcompete it. At locations with more constant marine conditions, there is evidence that *Syringodium filiforme* may be a superior competitor to *T. testudinum* in areas of enhanced nutrient availability (Williams 1987). In estuarine areas of south Florida, nutrient addition experiments show that *Halodule wrightii* will prevail over

T. testudinum under fertilized conditions (Fourqurean et al. 1995). Evidence from the distribution of primary producers around point sources of nutrient input show that in estuarine areas, there are zones of dominance of different species with respect to nutrient availability, from *T. testudinum* at lowest nutrient availability, to *H. wrightii* at higher availability, to *Ruppia maritima* at and even-higher availability, followed by a microalgae-dominated zone at highest nutrient availabilities (Powell et al. 1991). The abundance of macroalgal epiphytes also increases along the same gradient, up until the point that microalgae become dominant (Frankovich and Fourqurean 1997). The relative importance of the various primary producers, then, can be used to assess the trophic state of the community (fig. 10-1). Spatial pattern in the distribution of trophic states potentially can be used to assess nutrient sources in an ecosystem, and temporal changes in trophic state can be used as an indicator of trends in nutrient availability at a site.

Each species in the species dominance-eutrophication gradient model (fig. 10-1) can potentially dominate over a range of nutrient availability,

Eutrophication model

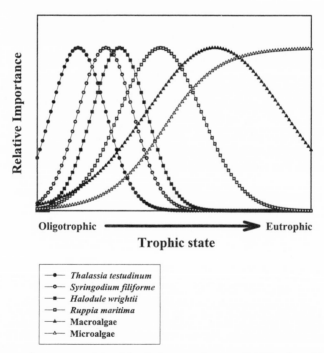

Figure 10-1. A conceptual model relating nutrient availability (trophic state) to relative abundances of primary producers in near-shore marine waters of south Florida.

and the model predicts a change in species dominance as nutrient availability changes. These changes are not instantaneous, however. Field evidence suggests that species replacements may take place on a time scale of a decade or more (Fourqurean et al. 1995). It is desirable that we be able to predict the tendency of the system to undergo these changes in species dominance before they occur, so that management actions can be taken. Tissue nutrient concentrations can be monitored to assess the relative availability of nutrients to the plants. For phytoplankton communities, this idea is captured in the interpretation of elemental ratios compared to the familiar "Redfield ratio" of 106 moles of carbon to 16 moles of nitrogen to 1 mole of phosphorus (i.e., C:N:P = 106:16:1) (Redfield 1958). Similar analyses can be made with data from seagrasses and macroalgae with the recognition that the taxon-specific Redfield ratio may be different from the phytoplankton ratio (Gerloff and Krombholtz 1966; Atkinson and Smith 1983; Duarte 1992). For the seagrass *T. testudinum*, the critical ratio of N:P in green leaves that indicates a balance in the availability of N and P is approximately 30:1, and monitoring deviations from this ratio can be used to infer whether N or P availabilities are limiting this species' growth (Fourqurean and Zieman 2002). Hence, *T. testudinum* is likely to be replaced by faster-growing competitors if nutrient availability is such that the N:P of its leaves is approximately 30:1. A change in the N:P in time to a value closer to 30:1 is indicative of eutrophication (fig. 10-2). Spatial pattern in the N:P can be used to infer

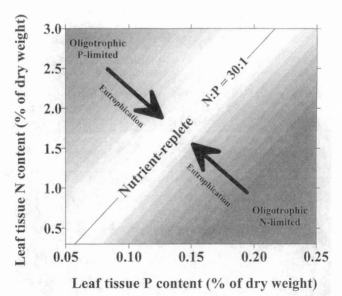

Figure 10-2. A conceptual model of the relationship between seagrass leaf nutrient content and nutrient availability in south Florida.

sources of nutrients for supporting primary production in the ecosystem (Fourqrean et al. 1992a, 1997; Fourqrean and Zieman 2002).

These models lead directly to a definition of trends likely to be encountered in the seagrass communities of south Florida if humans are causing regional changes in nutrient available because of alterations to quantity and quality of freshwater inputs to the marine ecosystem: (1) regional eutrophication will cause N:P ratios of seagrasses to approach 30:1 from higher or lower values indicative of oligotrophic conditions, and (2) regional eutrophication will cause a shift in species dominance in south Florida seagrass beds. The first responses to eutrophication will be evidenced by an increase in the relative abundance of fast-growing seagrass species (*H. wrightii* and *S. filiforme*) at the expense of the now-dominant, slow-growing *T. testudinum*. At later stages of eutrophication, macroalgae and microalgae will become the dominant primary producers.

Parameters Measured

Benthic plant community structure was measured using the rapid, visual assessment technique developed early in the twentieth century by the plant sociologist Braun-Blanquet (Braun-Blanquet 1972). This method is very quick, requiring only minutes at each sampling site; yet it is robust and highly repeatable, thereby minimizing among-observer differences. In this method, a series of quadrats are randomly placed on the bottom at a given location. All taxa occurring in the quadrat are listed, and a ranking based on abundance of the species in that quadrat is assigned for each species. We have adopted a modified Braun-Blanquet scale for our work in south Florida (table 10-1; Fourqrean et al. 2002). Ten randomly

Table 10-1. Braun-Blanquet abundance scale used to assess seagrass density.

Cover Class	Description
0	Absent
0.1	Solitary individual ramet, less than 5% cover
0.5	Few individual ramets, less than 5% cover
1	Many individual ramets, less than 5% cover
2	5–25% cover
3	25–50% cover
4	50–75% cov
5	75–100% cover

Cover is defined as the fraction of the bottom that is obscured by the species when viewed by a diver from directly above.

placed replicate 0.25-square-meter quadrat observations were made at each survey point.

Three metrics were computed for each taxon at a site: abundance, frequency, and density. Abundance was calculated as

$$A_i = \frac{\sum_{j=1}^{n} S_{ij}}{N_i} \qquad (10\text{-}1)$$

where S_{ij} is the Braun Blanquet score for taxon i in quadrat j; n is the total number of observed quadrats at a site; and N_i is the number of quadrats at a site at which taxon i was observed. For any taxon, A can range between 0 and 5, the maximum Braun-Blanquet score. Frequency was calculated as

$$F_i = \frac{N_i}{n} \qquad (10\text{-}2)$$

where n is the total number of quadrats observed at a site, such that $0 \leq F_i \leq 1$. Density (D_i) was calculated as the product $A_i \times F_i$. It should also be noted that a taxon may be observed at a site by the sample collector, but unless the taxon falls within one of the randomly placed observation quadrats, the taxon receives A_i, F_i, and D_i scores of 0. In addition to taxon-specific measures, seagrass taxon richness R was calculated for each site by summing the number of seagrass taxon for which $D > 0$.

Elemental content (carbon, nitrogen, and phosphorus) of green leaves was determined for all seagrasses (Fourqurean et al. 1992a). Elemental content was calculated on a dry weight basis; elemental ratios were calculated on a mole:mole basis. Net aboveground productivity and short-shoot morphology of *T. testudinum* were measured on a quarterly basis at each permanent site using a leaf-marking technique (Fourqurean et al. 2001).

One experimental monitoring approach was included in the monitoring program. Detecting change in ecological communities through repeated mapping takes many years, and at best can only illuminate changes that have occurred. It is desirable to monitor parameters that could predict the future trajectory of a resource before drastic changes occur. To this end, it has been proposed that estimates of future seagrass population size can be generated from the analysis of the age structure of seagrass shoots (Duarte et al. 1994). This method relies on the ability to age individual shoots of seagrasses by counting leaf scars and applying a plastochron interval (Erickson and Michelini 1957). The age-frequency distribution of shoots is a reflection of recruitment and mortality of individual shoots to the seagrass population. A potential advantage of this approach is that estimates of recruitment and mortality can be obtained from a single sampling event. Populations of shoots were collected from each site by excavating a sod, approximately one square meter that contained more than one hundred shoots. The number of leaves produced over the life span of each shoot in a sample was determined by counting leaf scars and extant green leaves. Shoot age was estimated by the number of leaves produced by a shoot, scaled by the site-specific annual leaf production rates. Instantaneous mortality and gross recruitment rates for each population were calculated from the age distribution at each site following the methods given in Peterson and Fourqurean 2001; projected net population growth was calculated as the difference between gross recruitment and mortality.

Choosing Where to Measure

In order to describe the spatial extent and pattern in the benthic communities, monitoring sites must be selected across the extent of the monitoring area. This monitoring program was designed to assess status and trends in seagrass communities across the entire extent of the Florida Keys National Marine Sanctuary, a 9,000-square-kilometer area of ocean surrounding the Florida Keys (fig. 5-3). As if monitoring such a vast extent were not task enough, it was recognized early in the monitoring program that the expansive shallow marine habitats immediately to the north of the sanctuary on the southwest Florida Shelf also were important for determining the status of seagrass communities within the sanctuary itself, so the monitoring program was extended to cover these additional 8,000-square-kilometers as well.

It was a goal of the program to describe spatial pattern in the indicators of interest, hence it was important to sample the entire region. However, without data on the underlying variance in the indicators to be measured, it was not possible to determine a priori how many sites would be needed to assess mean state of any indicator. The number of sites needed to generate a synoptic map of the condition of any measured indicator was therefore determined in a very ad hoc way: the maximum number of sites that could be sampled by the manpower provided for in the budget were selected. The locations for each of this number (n) of sites were chosen by laying a probability-based grid with n cells over the area of interest and then randomly choosing a location within each grid cell (fig. 10-3). This method allowed sampling locations to be spaced quasi-evenly across the landscape while still maintaining the assumptions required for a random sample, or in another words all locations had an equal probability of being sampled. In each of the first five years of the monitoring program (1996–2000), the same arrangement of grid cells was employed, but new random points were selected within each cell each year. This allowed for the development of synoptic maps of measured indicators during each monitoring year, as well as a combined data set of quasi-evenly-spaced random points collected over five years. The monitoring plan calls for revisiting the first year's sites during the sixth year, the second year's site during the seventh year, and so forth—so that trends in the resource over a five-year interval can be tested with *n* pair-wise comparisons for five years in a row.

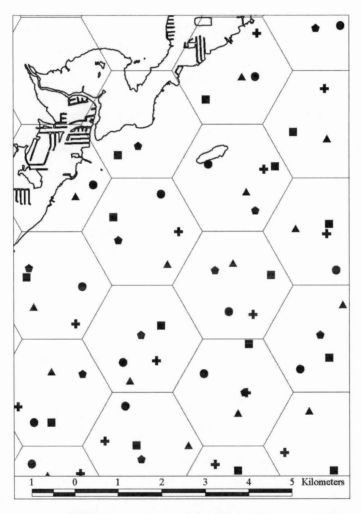

Figure 10-3. An illustration of the method used to distribute random samples quasi-evenly across the landscape. A number of hexagonal cells matching the desired number of sampling locations are packed into the sample area, and one random point is sampled from each cell for each year's sampling. Different symbols indicate a different year's sampling location within a hexagon.

More detailed time series of data describing the state of the benthic communities than could be generated by the synoptic mapping efforts were required to evaluate the intra- and interannual patterns in indicators. To this end, thirty permanent monitoring sites were established in regions that could be routinely visited given the budget of the project. To allow for analyses of the relationships of indicators and trends in the benthic communities with overlying environmental variables, these permanent sites were located at a subset of sites used for a related water-quality monitoring program (Boyer and Jones 2002); that program's sites were originally located using the same probability-based design as described above for the synoptic stations. Further constraints were placed on the selection of permanent sites. In each of the segments of the FKNMS (Klein and Orlando 1994), two inshore, two offshore, and two intermediate sites were chosen. And, because many of the indicators are specific to *T. testudinum* (see above), permanent monitoring sites were required to be located where this species occurred. Permanent monitoring sites were visited eight times per year, with two visits separated by seven to ten days every three months.

Visualizing Monitoring Data

Data by themselves are not generally useful to environmental managers, but require synthesis and interpretation before they yield useful management information. This synthesis is a daunting task that can take years after the data are collected: data reports need to be digested, hypotheses about relationships and driving variables need to be tested, and finally peer-reviewed analyses of these data must be published in scientific journals for scientists running the monitoring program to feel they have produced a defensible product. But, intermediate results are often desperately needed by resource managers. How, then, to present a mass of monitoring data to resource managers in a way that will allow them to make their own interpretations of the data while the scientists continue to refine their analyses? New computer tools, from spatial analysis and mapping programs to Internet access of hyperlinked data reports, provide effective means to allow managers to make their own data analyses by representing spatial pattern at an instant in time or by generating time series at fixed points from their own desks. In this seagrass monitoring program, we accomplish this by making the data freely available on the Web site http://serc.fiu.edu/seagrass/!CDreport/DataHome.htm as quickly as possible.

Synoptic views of the data are generated by interpolation between sample points using a kriging algorithm (Watson 1992) and then mapping these data using commercially available software. Such maps allow clear presentation of spatial pattern in a form that is most useful to resource managers. Although it would be tempting to use synoptic maps generated in successive years for a change analysis, there can be problems with this approach if (1) the spatial points are not identical in the two years so that small-scale spatial variability creates apparent changes, or (2) differences in timing in the collection of the data lead to apparent changes that are caused by seasonality.

Seasonality in the data from the permanent stations was assessed by fitting a sine model to the data:

$$\hat{Y} = Mean + amp \bullet sin(DOY + \Phi) \qquad (10\text{-}3)$$

where $\hat{Y}$ is the estimated value of the time series as a function of a yearly mean and a time-varying sine function with amplitude *amp*, day of year (*DOY*) in radians, and a phase angle (Φ) that determines the timing of the seasonal peak of the time series. This model assumes no long-term monotonic trend in the time series, as none have been yet detected in our monitoring program, but this model could also be easily applied to data with such a trend with the addition of a trend slope parameter. Once the seasonality was understood, it became possible to do change analyses on data from these permanent sites by using data collected in the same phase of the seasonal cycle in successive years.

Some Monitoring Examples

Data from the synoptic sampling sites allowed for the creation of maps of the densities of benthic plants in the region (fig. 10-4). While the general nature of the mapped distribution was similar from year to year, there were differences in maps produced in different years. We detected no consistent patterns in direction or rate of change in seagrass populations through time at our permanent monitoring sites; this made it possible to combine data from all years of the synoptic study to produce a more detailed view of the distribution of seagrasses. The species distribution maps produced by this monitoring program are the first detailed inventory of benthic plants in south Florida, despite the recognized importance of seagrasses for the ecological continuity and economic health of the region.

Analyses of the spatial extent of seagrasses in the study region indicate that seagrass species respond differently to environmental conditions so that densities of different species are not correlated (Fourqurean et al. 2002). These analyses support the conceptual model of the response of seagrass beds in south Florida to eutrophication (fig. 10-1). Each species has a separate ecological niche one axis of which is the relative nutrient

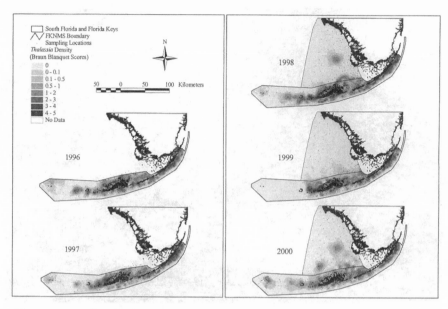

Figure 10-4. Synoptic maps of the density (D_i) of the seagrass *Thalassia testudinum* based on data collected in the years 1996–2000.

availability of the environment as outlined in figure 10-1. Of course, there are many other axes of these niches. For example, seagrass species are capable of tolerating different levels of salinity and salinity variability (McMillan and Moseley 1967). They also have different light requirements so that some species can grow in deeper water than others (Wiginton and McMillan 1979).

The Braun-Blanquet survey techniques provided three separate measures of the distribution of benthic plants at each survey site. Frequency (F_i) data provide information on the homogeneity of the distribution of the taxa; F_i values approaching 1.0 indicate that the taxa of interest is rather evenly distributed and that the patches in the landscape are larger than the spatial extent of the sampling at a site. Abundance (A_i) data provide an indication of the potential maximum cover attainable by a taxa given the environmental conditions. For example, in the area immediately behind the barrier coral reef, *Thalassia testudinum* often had quite high abundance $(A > 3)$ but relatively low frequency—because in this environment, *T. testudinum* meadows can be quite lush, but physical disturbance causes patchiness of this otherwise dense canopy, expressed by low values of F. The product $F \times A$ at a site yields the density (D_i). Hence, D can be low if A is consistently low at all quadrats at a site or if especially lush patches are isolated in otherwise barren habitat. The independence of the measures F

and A is underscored by the lack of any correlation between D and A for
T. testudinum from 1,500 sites visited over the five years of this study.

Analysis of the age structure of populations of *T. testudinum* short
shoots produced estimates of mortality and gross recruitment that varied
considerably across the spatial domain of our monitoring program. Both
mortality and gross recruitment estimates ranged from 0.01 to 1.30 y^{-1} at
118 sampling locations; mortality and gross recruitment at a site were
highly correlated (Peterson and Fourqurean 2001). Variability in these
population parameters led to a spatial pattern in predicted net population
growth rates, which ranged from -0.2 to 0.5 y^{-1}, with a mean of approximately 0.0 y^{-1} (fig. 10-5). Although these predictions of future population trajectory from the age structure of the population are as yet
untested, the promise of this type of predictive monitoring is that it may
be possible to detect change in a resource before the changes would be detected in a repetitive-mapping change analysis. There are still problems
with applying these techniques to management, as some of the assumptions necessary for applying demographic models to field data can not
necessarily be met (Jensen et al. 1996; Kaldy et al. 1999). The robustness
of the estimates to violations of these assumptions must be understood
before these tools can be routinely applied.

Spatial pattern in the elemental content of seagrass leaves indicated
strong patterns in nutrient availability, and therefore trophic state, in
Thalassia testudinum meadows in the study area. Elemental content did
not vary randomly; rather there was a strong spatial pattern in the relative
amounts of nitrogen and phosphorous in the samples (fig. 10-6). The spatial pattern suggested two main trends. The first trend was a decrease in
the availability of phosphorous from the Gulf of Mexico eastward into
Florida Bay, and the second pattern was an increase in relative phosphorous availability as distance from shore increases on the Atlantic Ocean
side of the Florida Keys (Fourqurean and Zieman 2002). Although a similar onshore-offshore pattern has been documented for sediment nutrients (Szmant and Forrester 1996), the seagrass leaf element content data
adds to this previously established trend by documenting that changes in
ratios of sediment nutrients are reflected in primary producers, and analysis of the seagrass leaf N:P data suggest that there is a qualitative shift
from phosphorous-limitation to nitrogen-limitation of the seagrasses in
different regions of our study area.

Time series data from the permanent monitoring stations underscore
the importance of understanding seasonal cycles for monitoring programs. We have summarized data from the period of record from an example monitoring station (fig. 10-7). Because this site is a relatively deep
(3.5 meters), nearshore location, it displays average seasonal amplitude

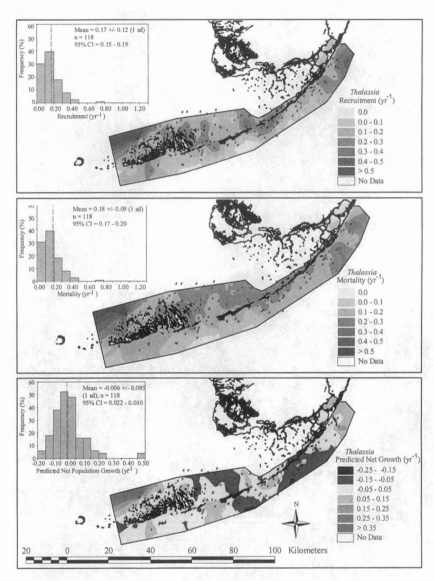

Figure 10-5. Net population change predictions from analysis of the age structure of populations of short shoots of *Thalassia testudinum* (redrawn from Peterson and Fourqurean 2001).

compared to the rest of the sites in the program. Water temperature at this site varies in a predictable seasonal pattern, with wintertime lows of around 22 degrees C and summertime highs of around 32 degrees C (fig. 10-7A). The mean temperature from the sine model (equation [10-3]) was 26.5 degrees C, with an amplitude of 4.9 degrees C. Seasonality can be ex-

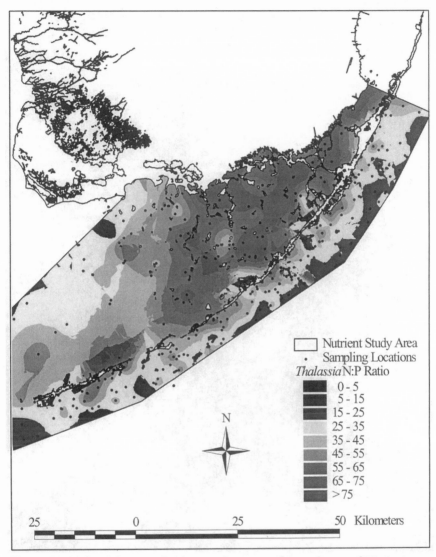

Figure 10-6. Spatial pattern in elemental content of green leaves of *Thalassia testudinum* in the eastern half of the study area (redrawn from Fourqurean and Zieman 2002).

pressed as the amplitude/mean: seasonal peaks were 18.5 percent higher than the mean, while seasonal minima were 18.5 percent lower. The maximum temperature occurs on 14 June for an average year. Many of the measured seagrass parameters exhibited seasonal patterns very similar to temperature. The abundance (fig. 10-7B), productivity (fig. 10-7C), and C:P (fig. 10-7D) of *Thalassia testudinum* all had peaks roughly coincident

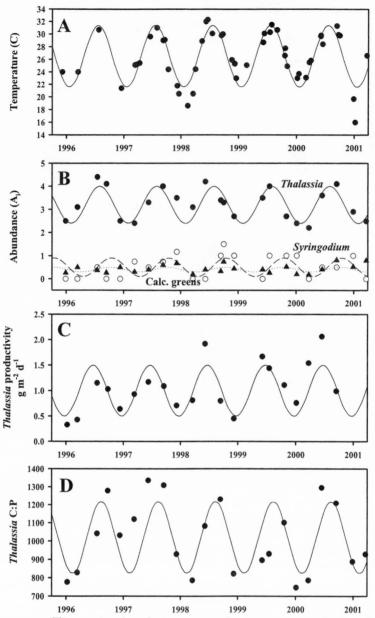

Figure 10-7. Time series data from an example permanent monitoring site. A. Temperature record. The line represents the fit of the sine model (Eq. 10-7) to the data; $Y = 26.5 + 4.9\sin(\text{date in radians} + 4.4)$, $r^2 = 0.82$. B. Abundance of three conspicuous taxa of benthic plants: *Thalassia testudinum* •, $Y = 3.2 + 0.8\sin(\text{date in radians} + 4.2)$, $r^2 = 0.71$); *Syringodium filiforme* ○, $Y = 0.5 + 0.4\sin(\text{date in radians} + 2.9)$, $r^2 = 0.31$); and calcareous green algae ▲, $Y = 0.4 + 0.1\sin(\text{date in radians} + 3.0)$, $r^2 = 0.77$). C. Productivity of green leaves of *Thalassia testudinum*, $Y = 1.0 + 0.5\sin(\text{date in radians} + 4.9)$, $r^2 = 0.63$. D. Elemental content (C:P) of green leaves of *Thalassia testudinum*, $Y = 1021 + 196\sin(\text{date in radians} + 4.0)$, $r^2 = 0.53$.

with the temperature peak. Note that the abundance of *T. testudinum* had a seasonality (amplitude/mean) of 25 percent; hence abundance was highly dependent on the timing of sampling. Seasonality in productivity was even greater (50 percent), while seasonality in nutrient content was somewhat less (19 percent). It is important to note that seasonality in these parameters is spatially variable across the monitoring region (Fourqurean et al. 2001).

Understanding how biological communities behave through time is one of the goals of any ecological monitoring program, but collecting and analyzing these biological data is an expensive and labor-intensive task. Because our understanding that the mechanisms leading to changes in biological communities in this environment are related to nutrient availability in the water column (fig. 10-1), it is imperative that we relate the biological responses we monitor to nutrient availability in the water column. Were these relationships fully understood, it may prove practical to only monitor the driving variables (water quality) and use mathematical models to predict change in the ecosystem. For this reason, linking the seagrass monitoring effort to a water-quality monitoring effort was vital. The analysis of data needed to relate all of the factors monitored in this program to water quality is now underway, with the hopes of parameterizing a predictive model. For example, because of the concurrent water-quality monitoring program, we know that seagrass leaf tissue elemental content is reflective of the relative concentration of nitrogen and phosphorous in the water column (median N:P of *Thalassia testudinum* leaves was related to the median N:P in the water column measured during six years of quarterly monitoring [1995–2000] of seagrasses and water quality [J. N. Boyer, Florida International University, unpublished data]; N:P of *T. testudinum* = 9.8 + 0.4[N:P of the water column], $r^2 = 0.38$, $p < 0.01$). Also, involvement of researchers in the monitoring effort to perform manipulative experiments can elucidate the mechanisms relating environmental change to biological change.

Seagrass bed responses to changing water quality often occur over long time frames, but acute events can also be important in structuring these communities. Hurricanes are easily the most intense acute events to influence shallow water marine communities of south Florida. Fortunately for the residents of south Florida, few hurricanes have passed over the study area during the monitoring period, and these hurricanes (Georges and Mitch in 1998, Irene in 1999) were relatively mild. Hurricane Georges was the most intense storm during this period; it was a category 2 storm when it passed over Key West on 25 September 1998. The hurricane force winds were restricted to a relatively narrow band in the lower and middle Florida Keys. Such relatively small hurricanes had very substantial, but localized, effects. Over the network of thirty stations, the seagrass beds were completely destroyed by Georges at three stations. During the storm, two stations on the Atlantic Ocean side of the Florida Keys just behind the barrier reef experienced large waves coming from offshore that eroded sediments, and the rhizome mat of *T. testudinum* was washed away. In contrast, one site on the Gulf of Mexico side of the Florida Keys experienced strong easterly winds that caused sediment deposition so that a *T. testudinum* community was buried by 50 centimeters of sediment (fig. 10-8). Following the storm, early successional species (calcareous green algae and *Halodule wrightii*) quickly became established at the buried site, and after two years there was evidence of recovery of the original *T. testudinum*–dominated community. Hurricane impacts were very patchy in distribution—stations close to the affected stations were often unaffected by the storm.

Moving from Determining Status to Detecting Trends

This monitoring program has been very successful in defining the spatial extent and present condition of the seagrass beds in south Florida. The synoptic monitoring effort has provided a description of the distribution and species composition of seagrass beds whose detail and spatial extent

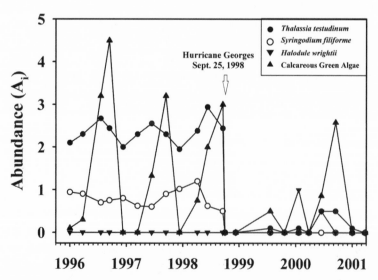

Figure 10-8. Time series of the abundance (A_i) of *Thalassia testudinum*, *Syringodium filiforme*, *Halodule wrightii* and calcareous green algae illustrating the impact of Hurricane Georges on the benthic plant communities at one site that was buried by 50 centimeters of sediment during the storm.

are without precedent in marine benthic community monitoring. Spatial patterns in elemental content of one of the dominant seagrasses, *T. testudinum*, has led to the better understanding of the pattern of relative nutrient availability across the monitoring area. Analysis of the population age-structure has allowed for predictions of population density trends for this species. Further, monitoring at fixed stations has led to a description of the intra-annual pattern of species densities, growth rates and nutrient contents of seagrasses, as well as allowed for the detection of longer term trends in these parameters.

The spatial extent and species composition data comprise an important benchmark against which future changes may be gauged. The program is now poised to begin revisiting the thousands of synoptic monitoring points (fig. 10-4) in order to assess change in these parameters. It is tempting to apply change analysis techniques to data collected in separate years—for example, compare the surfaces generated from 2000 data to the surfaces generated from 1996 data—but such analyses are not to be trusted because of spatial variability smaller in scale than the hexagons originally used to select the random sampling points. The consequences of small-scale spatial variability are that it is not possible to detect small changes in the status of indicators until after two rotations of sampling the exact synoptic mapping points—or in this case, until after ten years of data collection.

The spatial sampling intensity of the synoptic mapping stations was sufficient for describing the gross features of the distribution of benthic habitats across the study area, but it has proven to be too coarse to detect already extant anthropogenic influences to these communities. Casual observation and some published accounts (e.g., Tomasko and Lapointe 1991; Lapointe and Clark 1992) document increases in epiphytism and some loss of seagrasses in the Florida Keys in the immediate nearshore environment—in other words, the first 0–100 meters from shore. Further, natural point sources (like bird rookeries, Powell et al. 1991) of nutrients in the Florida Keys have effects that are confined to approximately 100 meters offshore. These observations indicate that the natural oligotrophic nature of the nearshore environment is strongly affected by nutrient releases and is an efficient attenuator of nutrient effects because of rapid and efficient uptake. It is now obvious that the spatial sampling intensity must be increased in areas adjacent to human activity if such acute, localized effects are to be characterized and monitored. Hence, sampling site density was increased within the first kilometer of shoreline in 2000, the fifth year of the monitoring program. This new intensity of sampling is allowing for definitions of fine-scale, nearshore features that were not detected in the original monitoring plan (fig. 10-9), but the effort required to monitor these smaller-scale phenomena necessarily diverts resources from the original, regional scope of the program.

The elemental composition of seagrass leaves was related to concentrations of nitrogen and phosphorus in the water column. This relationship is a useful one, because water quality is often variable in time at a single location, so that multiple determinations of concentrations of

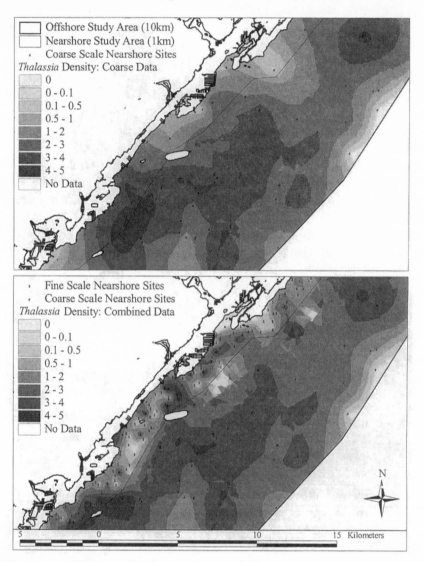

Figure 10-9. Comparison of contour maps of the density (D_i) of *Thalassia testudinum* generated using data from only the coarse-scale synoptic stations (top) and generated using data from the coarse-scale stations as well as an additional near-shore station on a finer spatial scale.

nutrients often are required to describe water quality at a site. Because seagrass nutrient content is a function of the average conditions at that site, a single sampling of seagrass leaves can be used as an integrative indicator of relative nutrient availability at that site. Some caution must be used with this approach, however, because elemental content of seagrass leaves is seasonal (Fourqurean et al. 1997 and fig. 10-7); but the use of tissue content as a proxy for water-quality data can potentially save much time and effort in a monitoring program. The spatial pattern of N:P in seagrass tissues shows a clear pattern in relative availability of nitrogen and phosphorous in the study area (fig. 10-6) and indicates that nearshore seagrasses may be P-limited while offshore seagrasses are potentially N-limited. The conceptual models that form the basis for the monitoring program predict that as nutrient-limited *T. testudinum*–dominated sites receive anthropogenic nutrients, the N:P of *T. testudinum* leaves will move toward 30:1 (fig. 10-2); then biomass of *T. testudinum* will increase, followed by replacement of this species by faster-growing species (fig. 10-1). A further prediction of the conceptual models is that *T. testudinum*–dominated sites with leaf N:P of around 30 are the most likely to experience a change in species dominance as a result of nutrient loading.

Changes in species dominance were caused by factors other than anthropogenic nutrient loading at one permanent monitoring station: a *T. testudinum*–dominated community was replaced by a *H. wrightii*–calcareous green algae community at station 309 as a consequence of the passage of Hurricane Georges (fig. 10-8). Were it not for the time series of quarterly observations, this change could have been attributed to the consequences of anthropogenic nutrient loading. This explanation would have been particularly attractive because of the proximity of station 309 to the city of Key West. But, the time series data point to a clear explanation of the change: the healthy *T. testudinum* seagrass bed was destroyed by the acute event of the storm, and the reestablishment of the seagrass community through the process of ecological succession produced two years after the storm a community dominated by the same fast-growing species that are favored by increases in nutrient loading rates. Without the fine-scale temporal resolution, it would not be possible to positively ascribe a cause to the dynamics observed at station 309.

Three of the thirty permanent stations experienced the catastrophic loss of their seagrass communities as a result of hurricane damage during the study period, but this does not mean that storms are causing a 10 percent loss of seagrass habitat over each five-year period. Physical disturbance creates a patchwork of disturbance in this community, and disturbed areas are slowly recolonized by early successional seagrass species (Patriquin 1975). Since the permanent stations were purposefully established at locations supporting seagrass beds, it was only possible at the beginning of this monitoring effort to observe loss of seagrasses caused by storms since there were initially no monitoring stations placed in disturbed areas with the potential for recolonization. We hope that monitoring these permanent stations will continue long enough that it becomes possible to record the time course of recovery from such disturbances at many sites. Our experience with hurricanes underscores the importance of designing ecosystem monitoring programs to accommodate the natural pattern of disturbance.

One of the benefits of operating a monitoring program out of an academic or research institution is that experimental approaches to monitoring can be evaluated during the course of the monitoring program. In this program, we have been evaluating the use of the age structure of seagrass population to make predictions about the trajectory of the populations (fig. 10-5). Although this is a common practice in fisheries management, it has only recently been reintroduced to seagrass ecologists (Duarte et al. 1994). It is a topic of ongoing debate whether the violations of the assumptions of the models necessary for making such population trajectory predictions will render these techniques invalid (Jensen et al. 1996; Durako and Duarte 1997; Kaldy et al. 1999), but this monitoring program is providing the structure for a long-term test of the applicability of the techniques by allowing a comparison of population trajectory predictions with more conventional long-term monitoring data. After the appropriate assessment, age structure analysis may prove to be a valuable, time-saving, predictive tool that can be used to assess the status of the community before changes in density or species composition are evident. Like age frequency analyses, new measures of the physiological status of seagrasses are now being proposed as potential new monitoring tools; we are currently evaluating the utility of in situ measures of the fluorescence characteristics of plants as a monitoring tool (Beer et al. 1998; Ralph et al. 1998; Beer and Björk 2000). Such new techniques have the potential to revolutionize monitoring programs, but properly assessing the utility of new techniques requires the involvement of research scientists in monitoring programs.

No monitoring program can, practically, encompass an entire ecosystem. This is because of the many factors that delineate responsibility for monitoring programs, including the motivation for the monitoring program, the spatial extent of ecosystems, and jurisdictional boundaries. Careful coordination between management agencies and research groups is necessary to ensure that data collected by different monitoring programs, for different goals funded by different agencies, can be pooled and analyzed as a whole for a more holistic view of regional ecosystem. Such careful coordination has been achieved in south Florida, where many different monitoring programs funded by federal, state, and local agencies have agreed to use comparable monitoring techniques, thereby providing a unique opportunity to paint a truly regional view of the status of seagrass ecosystems in south Florida (Fourqurean et al. 2002). From this coordination, it is possible to expand the spatial scale of our understanding of the patterns in variation of these communities beyond the considerable spatial extent of the monitoring program described in this chapter. For example, patterns in species diversity can be depicted from suburban Biscayne Bay, through Florida Bay and Everglades National Park, across the Florida Keys and out to the Dry Tortugas (fig. 10-10); this entire area is connected by water flows and is likely to respond to anthropogenic changes in water delivery. This type of cooperation should serve as a model to other groups embarking on the assessment of resources over large geographic ranges.

Remote sensing has been used successfully to monitor some aspects of seagrass beds, and remotely sensed data has been successfully used to map seagrass distributions in many environments (e.g., Orth and Moore 1983; Lusczkovich et al. 1993; Ferguson and Korfmacher 1997). Large changes

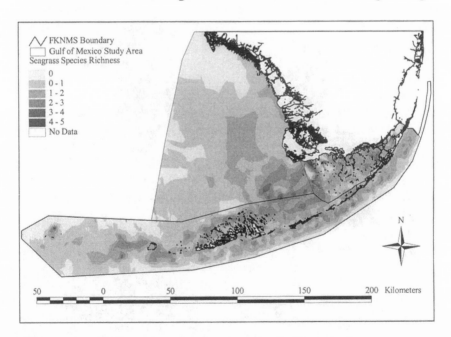

Figure 10-10. Species richness on the regional scale. If monitoring programs ensure that their methods are easily comparable to other monitoring groups working in similar systems, it is possible to elucidate patterns that occur on a larger scale than the original monitoring program.

in seagrass cover have been documented by analysis of remotely sensed data (e.g., Orth and Moore 1983; Robbins 1997; Ward et al. 1997). The large spatial scale, relatively small cost, and historic archive of aerial photographs and satellite images make the use of these data very appealing, but there are some serious limitations to its application to seagrass monitoring programs. In general, it is not possible to distinguish seagrass beds from other dark areas on the bottom—so macroalgal beds and seagrass beds can not reliably be differentiated. Seagrass species are also not generally distinguishable from one another in remotely sensed data (Moore et al. 2000). As a consequence, remotely sensed data are currently not adequate to evaluate the trophic status of the benthic habitats in south Florida, since species composition is a key indicator of this status (fig. 10-1). Further, both satellite images and aerial photographs require an optically clear water column and dense stands of macrophytes if they are to be useful for monitoring seagrasses, and it is common in south Florida for there to be very sparse seagrass communities overlain by an optically dense water column. With continuing advances in active remote sensing techniques and multispectral imaging, these limitations may be overcome in the future, but until then it is still necessary to send divers into the environment to collect the data needed for monitoring benthic communities in south Florida.

It is quite obvious that this monitoring program has led to a better understanding of the biology of seagrass beds in south Florida, but has the program been successful in terms of providing data useful to managers assessing impacts of human alteration of water and nutrient cycles in the region? At the present time, the program has yet to detect any common trends in seagrass nutrient content or community composition at the permanent monitoring locations. This can be interpreted to mean that there are no regional trends in the status of seagrass beds, but such an interpretation may be shortsighted. The lack of detectable trends poses a difficult challenge both to the monitoring team and to the management agencies responsible for funding monitoring programs. In such an instance, the monitoring team must reevaluate its monitoring scheme for proper spatial and temporal resolution, and the management agency must decide whether the potential for change in the monitored ecosystem is great enough to continue the program. In south Florida, the case can easily be made to continue the monitoring effort. Natural disturbances (e.g., hurricanes) have caused changes at the monitoring stations that are unrelated to anthropogenic factors. Additionally, manipulative experiments in seagrass beds demonstrate that the time course of the response of seagrass beds to eutrophication is on the order of decades, and our monitoring data have not been collected for a sufficient period to detect such decadal change. And finally, we do not understand completely the interaction man has with the natural dynamics of these systems.

Lessons Learned about Ecological Monitoring from This Program

There are several lessons learned from this monitoring program that should be heeded by others designing regional ecosystem monitoring plans:

- A strong conceptual model must underlie the monitoring strategy.
- No one monitoring design is adequate for all questions asked within a monitoring plan. In particular, trade-offs between allocating resources to broad spatial coverage or fine-scale temporal coverage need to be addressed at the outset of the program. The split design we employed was an attempt to allocate resources to both of these goals.
- Not every monitoring technique can adequately answer the questions addressed by the monitoring program. Often, inadequacies are not apparent until years of data have been collected and analyzed. Monitoring techniques should be reviewed periodically and changes made as appropriate.
- Research scientists involved with monitoring projects should be responsible for analyzing data and publishing it in peer-reviewed journals.
- Methods should be kept as simple and universal as possible. Often, seemingly small methodological differences between groups of investigators and monitoring programs make comparison of results difficult or impossible. Further, careful consideration should be given to any change in techniques during the monitoring program, since small technique changes may render trend detection impossible.
- Disturbances are a natural structuring force in most ecosystems, so their impact should be taken into account when designing the monitoring program. Redundancy of sampling locations will lessen the impact of a natural disaster (such as a hurricane or fire) on the monitoring program, and it will also provide interesting research opportunities.

ACKNOWLEDGMENTS

This monitoring program was funded by the Environmental Protection Agency (EPA) as part of the Water Quality Protection Program for the Florida Keys National Marine Sanctuary (X994620-94-5). Fred McManus at the EPA has managed the project admirably and is in large part responsible for the success of the larger Water Quality Protection Program. J. C. Zieman and M. D. Durako were both intimately involved in the planning of the monitoring program; L. M. Lagera coordinated the planning effort. Many technicians, post-docs and graduate students also participated in this program; especially notable for their contributions were A. Willsie, S. P. Escorcia, B. J. Peterson, C. D. Rose, B. Machovina, B. C. Davis, K. M. Cunniff, M. L. Ferdie, D. A. Byron, V. C. Cornett, C. Furst, and C. Barrosso. Captain Dave Ward of the R/V Magic and Captain M. O'Connor of the R/V Expedition II were both instrumental for logistical support. Close coordination with the water chemistry monitoring program was maintained by J. N. Boyer and R. D. Jones, who also supplied the water chemistry data used in this chapter. J. C. Trexler and D. E. Busch provided useful editorial suggestions and guidance in the preparation of this chapter which is based on contribution no. 172 of the Southeast Environmental Research Center at Florida International University.

LITERATURE CITED

Atkinson, M. J., and S. V. Smith. 1983. C:N:P ratios of benthic marine plants. *Limnology and Oceanography* 28:568–574.

Beer, S., and M. Björk. 2000. Measuring rates of photosynthesis of two tropical seagrasses by pulse amplitude modulated (PAM) fluorometry. *Aquatic Botany* 66:69–76.

Beer, S., B. Vilenkin, A. Weil, M. Veste, L. Susel, and A. Eshel. 1998. Measuring photosynthetic rates in seagrasses by pulse amplitude modulated (PAM) fluorometry. *Marine Ecology Progress Series* 174:293–300.

Bohnsack, J. A., D. E. Harper, and D. B. McClellan. 1994. Fisheries trends from Monroe County, Florida. *Bulletin of Marine Science* 54:982–1018.

Boyer, J. N., and R. D. Jones. 2002. View from the bridge: External and internal forces affecting the ambient water quality of the Florida Keys National Marine Sanctuary. Pp. 609–628 in *The Everglades, Florida Bay, and the coral reefs of the Florida Keys*, edited by J. W. Porter and K. G. Porter: Boca Raton, Fla.: CRC Press.

Boyer, J. N., P. Sterling, and R. D. Jones. 1999. Maximizing information from a water quality monitoring network through visualization techniques. *Estuarine, Coastal and Shelf Science* 50:39–48.

Braun-Blanquet, J. 1972. *Plant sociology: The study of plant communities*. New York: Hafner Publishing.

Butler, M. J., IV, J. H. Hunt, W. F. Herrnkind, M. J. Childress, R. Bertelsen, W. Sharp, T. Matthews, J. M. Field, and H. G. Marshall. 1995. Cascading disturbances in Florida Bay, USA: Cyanobacterial blooms, sponge mortality, and implications for juvenile spiny lobsters *Panulirus argus*. *Marine Ecology Progress Series* 129:119–125.

Chamberlain, R., and D. Hayward. 1996. Evaluation of water quality and monitoring in the St. Lucie estuary, Florida. *Water Resources Bulletin* 32:681–696.

Costanza, R., R. d'Arge, R. de Groot, S. Farber, M. Grasso, B. Hannon, K. Limburg, S. Naeem, R. V. O'Neill, J. Paruelo, R. G. Raskin, P. Sutton, and M. van den Belt. 1997. The value of the world's ecosystem services and natural capital. *Nature* 387:253–260.

Dillon, K. S., D. R. Corbett, J. P. Chanton, W. C. Burnett, and D. J. Furbish. 1999. The use of sulfur hexafluoride (SF6) as a tracer of septic tank effluent in the Florida Keys. *Journal of Hydrology* 220:129–140.

Duarte, C. M. 1992. Nutrient concentrations of aquatic plants: Patterns across species. *Limnology and Oceanography* 37:882–889.

———. 1995. Submerged aquatic vegetation in relation to different nutrient regimes. *Ophelia* 41:87–112.

———. 2002. The future of seagrass meadows. *Environmental Conservation* 29:192–206.

Duarte, C. M., N. Marba, N. Agawin, J. Cebrian, S. Enriquez, M. D. Fortes, M. E. Gallegos, M. Merino, B. Olesen, K. Sand-Jensen, J. Uri, and J. Vermaat. 1994. Reconstruction of seagrass dynamics: Age determinations and associated tools for the seagrass ecologist. *Marine Ecology Progress Series* 107:195–209.

Durako, M. J., and C. M. Duarte. 1997. On the use of reconstructive aging techniques for assessing seagrass demography: A critique of the model test of Jensen et al. (1996). *Marine Ecology Progress Series* 146:297–303.

Erickson, R. O., and F. J. Michelini. 1957. The plastochron index. *American Journal of Botany* 44:297–305.

Ferguson, R. L., and K. Korfmacher. 1997. Remote sensing and GIS analysis of seagrass meadows in North Carolina, USA. *Aquatic Botany* 58:241–258.

Fourqurean, J. W., M. J. Durako, M. O. Hall, and L. N. Hefty. 2002. Seagrass distribution in south Florida: A multi-agency coordinated monitoring program. Pp. 497–522 in *The Everglades, Florida Bay, and the coral reefs of the Florida Keys*, edited by J. W. Porter and K. G. Porter, Boca Raton, Fla.: CRC Press.

Fourqurean, J. W., T. O. Moore, B. Fry, and J. T. Hollibaugh. 1997. Spatial and temporal variation in C:N:P ratios, $\delta^{15}N$, and $\delta^{13}C$ of eelgrass *Zostera marina* as indicators of ecosystem processes, Tomales Bay, California, USA. *Marine Ecology Progress Series* 157:147–157.

Fourqurean, J. W., G. V. N. Powell, W. J. Kenworthy, and J. C. Zieman. 1995. The effects of long-term manipulation of nutrient supply on competition between the seagrasses *Thalassia testudinum* and *Halodule wrightii* in Florida Bay. *Oikos* 72:349–358.

Fourqurean, J. W., and M. B. Robblee. 1999. Florida Bay: A history of recent ecological changes. *Estuaries* 22:345–357.

Fourqurean, J. W., A. W. Willsie, C. D. Rose, and L. M. Rutten. 2001. Spatial and temporal pattern in seagrass community composition and productivity in south Florida. *Marine Biology* 138:341–354.

Fourqurean, J. W., and J. C. Zieman. 2002. Seagrass nutrient content reveals regional patterns of relative availability of nitrogen and phosphorus in the Florida Keys, USA. *Biogeochemistry* 61:221–245.

Fourqurean, J. W., J. C. Zieman, and G. V. N. Powell. 1992a. Phosphorus limitation of primary production in Florida Bay: Evidence from the C:N:P ratios of the dominant seagrass *Thalassia testudinum*. *Limnology and Oceanography* 37:162–171.

———. 1992b. Relationships between porewater nutrients and seagrasses in a subtropical carbonate environment. *Marine Biology* 114:57–65.

Frankovich, T. A., and J. W. Fourqurean. 1997. Seagrass epiphyte loads along a nutrient availability gradient, Florida Bay, USA. *Marine Ecology Progress Series* 159:37–50.

Gerloff, G. C., and P. H. Krombholtz. 1966. Tissue analysis as a measure of nutrient availability for the growth of angiosperm aquatic plants. *Limnology and Oceanography* 11:529–537.

Havens, K. E., and N. G. Aumen. 2000. Hypothesis-driven experimental research is necessary for natural resources management. *Environmental Management* 25:1–7.

Jackson, J. B. C. 2001. What was natural in the coastal ocean? *Proceedings of the National Academy of Sciences* USA 98:5411–5418.

Jackson, J. B. C., M. X. Kirby, W. H. Berger, K. A. Bjorndahl, L. W. Botsford, B. J. Bourque, R. H. Bradbury, R. Cooke, J. Erlandson, J. A. Estes, T. P. Hughes, S. Kidwell, C. B. Lange, H. S. Lenihan, J. M. Pandolfi, C. H. Peterson, R. S. Steneck, M. J. Tegner, and R. R. Warner. 2001. Historical overfishing and the recent collapse of coastal ecosystems. *Science* 293:629–638.

Jensen, S. L., B. D. Robbins, and S. S. Bell. 1996. Predicting population decline: Seagrass demographics and the reconstructive technique. *Marine Ecology Progress Series* 136:267–276.

Kaldy, J. E., N. Fowler, and K. H. Dunton. 1999. Critical assessment of *Thalassia testudinum* (turtle grass) aging techniques: Implications for demographic inferences. *Marine Ecology Progress Series* 181:279–288.

Kiker, C. F., J. W. Milon, and A. W. Hodges. 2001. Adaptive learning for science-based policy: The Everglades restoration. *Ecological Economics* 37:403–416.

Klein, C. J. I., and S. P. J. Orlando. 1994. A spatial framework for water-quality management in the Florida Keys National Marine Sanctuary. *Bulletin of Marine Science* 54:1036–1044.

Kraemer, G. P., R. H. Chamberlain, A. D. Steinman, and M. D. Hanisak. 1999. Physiological responses of transplants of the freshwater angiosperm *Valisneria americana* along a salinity gradient in the Caloosahatchee estuary (southwestern Florida). *Estuaries* 22:138–148.

Lapointe, B. E., and M. W. Clark. 1992. Nutrient inputs from the watershed and coastal eutrophication in the Florida Keys. *Estuaries* 15:465–476.

Lapointe, B. E., J. D. O'Connell, and G. S. Garrett. 1990. Nutrient couplings between on-site sewage disposal systems, groundwaters, and nearshore surface waters of the Florida Keys. *Biogeochemistry* 10:289–307.

Lapointe, B. E., D. A. Tomasko, and W. R. Matzie. 1994. Eutrophication and trophic state classification of seagrass communities in the Florida Keys. *Bulletin of Marine Science* 54:696–717.

Lessios, H. A., J. D. Cubit, D. R. Robertson, M. J. Shulman, M. R. Parker, S. D. Garrity, and S. C. Levings. 1984. Mass mortality of *Diadema antillarum* on the Caribbean coast of Panama. *Coral Reefs* 3:173–182.

Light, S. S., and J. W. Dineen. 1994. Water control in the Everglades: A historical perspective. Pp. 47–84 in *Everglades: The ecosystem and its restoration*, edited by S. M. Davis and J. C. Ogden. Delray Beach, Fla.: St. Lucie Press.

Lusczkovich, J. J., T. W. Wagner, J. L. Michelak, and R. W. Stoffle. 1993. Discrimination of coral reefs, seagrass meadows, and sand bottoms from space: A Dominican Republic case study. *Photogrammetric engineering and remote sensing* 59:385–389.

McMillan, C., and F. N. Moseley. 1967. Salinity tolerances of five marine spermatophytes of Redfish Bay, Texas. *Ecology* 48:503–506.

Moore, K. A., D. J. Wilcox, and R. J. Orth. 2000. Analysis of the abundance of submersed aquatic vegetation communities in the Chesapeake Bay. *Estuaries* 23:115–127.

NOAA. 1996 *Florida Keys National Marine Sanctuary final management plan/environmental impact statement*. Vol. 1. *Management Plan*. National Oceanographic and Atmospheric Administration, U.S. Department of Commerce, Washington, D.C.

Nuttle, W. K., J. W. Fourqurean, B. J. Cosby, J. C. Zieman, and M. B. Robblee. 2000. Influence of net freshwater supply on salinity in Florida Bay. *Water Resources Research* 36:1805–1822.

Orth, R. J., and K. A. Moore. 1983. Chesapeake Bay: An unprecedented decline in submerged aquatic vegetation. *Science* 222:51–53.

Patriquin, D. G. 1975. "Migration" of blowouts in seagrass beds at Barbados and Carricou, West Indies, and its ecological and geological implications. *Aquatic Botany* 1:163–189.

Paul, J. H., J. B. Rose, S. C. Jiang, X. T. Zhou, P. Cochran, C. Kellog, J. B. Kang, D. Griffin, S. Farrah, and J. Lukasik. 1997. Evidence of groundwater and surface marine water contamination by waste disposal wells in the Florida Keys. *Water Research* 31:1448–1454.

Peterson, B. J., and J. W. Fourqurean. 2001. Large-scale patterns in seagrass (*Thalassia testudinum*) demographics in south Florida. *Limnology and Oceanography* 46:1077–1090.

Porter, J. W., S. K. Lewis, and K. G. Porter. 1999. The effects of multiple stressors on the Florida Keys coral reef ecosystem: A landscape hypothesis and a physiological test. *Limnology and Oceanography* 44:941–949.

Powell, G. V. N., J. W. Fourqurean, W. J. Kenworthy, and J. C. Zieman. 1991. Bird colonies cause seagrass enrichment in a subtropical estuary: Observational and experimental evidence. *Estuarine, Coastal and Shelf Science* 32:567–579.

Powell, G. V. N., W. J. Kenworthy, and J. W. Fourqurean. 1989. Experimental evidence for nutrient limitation of seagrass growth in a tropical estuary with restricted circulation. *Bulletin of Marine Science* 44:324–340.

Ralph, P. J., R. Gademann, and W. C. Dennison. 1998. In situ seagrass photosynthesis measured using a submersible, pulse-amplitude modulated fluorometer. *Marine Biology* 132:367–373.

Redfield, A. C. 1958. The biological control of chemical factors in the environment. *American Scientist* 46:205–221.

Redfield, G. W. 2000. Ecological research for aquatic science and environmental restoration in south Florida. *Ecological Applications* 10:990–1005.

Robbins, B. D. 1997. Quantifying temporal change in seagrass areal coverage: The use of GIS and low-resolution aerial photography. *Aquatic Botany* 58:259–267.

Robblee, M. B., T. R. Barber, P. R. Carlson, M. J. Durako, J. W. Fourqurean, L. K. Muehlstein, D. Porter, L. A. Yarbro, R. T. Zieman, and J. C. Zieman. 1991. Mass mortality of the tropical seagrass *Thalassia testudinum* in Florida Bay (USA). *Marine Ecology Progress Series* 71:297–299.

Rose, C. D., W. C. Sharp, W. J. Kenworthy, J. H. Hunt, W. G. Lyons, E. J. Prager, J. F. Valentine, M. O. Hall, P. E. Whitfield, and J. W. Fourqurean. 1999. Overgrazing of a large seagrass bed by sea urchins in outer Florida Bay. *Marine Ecology Progress Series* 190:211–222.

Sargent, F. J., T. J. Leary, D. W. Crewz, and C. R. Kruer. 1995. *Scarring of Florida's seagrasses: Assessment and management options.* FMRI Technical Report TR-1. Florida Marine Research Institute, St. Petersburg, Fla.

Short, F. T., and S. Wyllie-Echeverria. 1996. Natural and human-induced disturbance of seagrasses. *Environmental Conservation* 23:17–27.

Szmant, A. M., and A. Forrester. 1996. Water column and sediment nitrogen and phosphorus distribution patterns in the Florida Keys, USA. *Coral Reefs* 15:21–41.

Tilmant, J. T., R. W. Curry, R. D. Jones, A. Szmant, J. C. Zieman, M. Flora, M. B. Robblee, D. Smith, R. W. Snow, and H. Wanless. 1994. Hurricane Andrew's effect on marine resources. *BioScience* 44:230–237.

Tomasko, D. A., and B. E. Lapointe. 1991. Productivity and biomass of *Thalassia testudinum* as related to water column nutrient availability and epiphyte levels: Field observations and experimental studies. *Marine Ecology Progress Series* 75:9–17.

USACE. 1999. *Central and southern Florida comprehensive review study: Final integrated feasibility report and programmatic environmental impact statement.* U.S. Army Corps of Engineers, Jacksonville District, Jacksonville, Fla.

Ward, D. H., C. H. Markon, and D. C. Douglas. 1997. Distribution and stability of eelgrass beds at Izembek Lagoon, Alaska. *Aquatic Botany* 58:229–240.

Watson, D. F. 1992. *Contouring: A guide to the analysis and display of spatial data*. Vol. 10. New York: Pergamon Press.

Wiginton, J. R., and C. McMillan. 1979. Chlorophyll composition under controlled light conditions as related to the distribution of seagrasses in Texas and the U.S. Virgin Islands. *Aquatic Botany* 6:171–184.

Williams, S. L. 1987. Competition between the seagrasses *Thalassia testudinum* and *Syringodium filiforme* in a Caribbean lagoon. *Marine Ecology Progress Series* 35:91–98.

Zieman, J. C., and R. G. Wetzel. 1980. Productivity in seagrasses: Methods and rates. Pp. 87–116 in *Handbook of seagrass biology, an ecosystem prospective*, edited by R. C. Phillips and C. P. McRoy. New York: Garland STPM Press.

CHAPTER 11

Late-Successional Forest Monitoring in the Pacific Northwest

Miles A. Hemstrom

Late-successional and old-growth forests in the Pacific Northwest have undergone substantial change and decline since Euro-American settlement in the middle nineteenth century (FEMAT 1993; NRC 2000). One consequence has been increasing concern over the viability of some native species and the maintenance of ecological processes and functions provided by late-successional and old-growth forests. Conflicts between those concerned about the loss of old-growth forests and those concerned with the high commercial value of the same forests increased to the point that litigation and subsequent legal decisions lead to the development of a comprehensive environmental impact statement and record of decision to guide the management of federal forests. This chapter describes an effort to monitor the effectiveness of those policies in managing and conserving late-successional and old-growth forests in the Pacific Northwest.

Old-growth forests, as addressed in this chapter, have accumulated specific structural and compositional attributes related to tree size, canopy structure, snags, woody debris, and plant composition (NRC 2000). These forests support assemblages of plants and animals, environmental conditions, and ecological processes not typically found in younger stands (younger than about 150–250 years) or in small patches of old trees (NRC 2000). Late-successional forests have variable definition. Spurr and Barnes (1973) defined late-successional forests as those in which shade-tolerant species, such as western hemlock (*Tsuga heterophylla* [Raf.] Sarg.) and Pacific silver fir (*Abies amabilis* [Dougl.] Forbes), begin to attain dominance. FEMAT (1993) defined late-successional forests as those during the period from first merchantability to the culmination of mean annual increment (NRC 2000). Because this chapter addresses monitoring the effectiveness of policies based on FEMAT (1993), the FEMAT definition of late-successional forests is used.

Monitoring for the Northwest Forest Plan

The *Final Supplemental Environmental Impact Statement on Management of Habitat for Late-Successional and Old-Growth Forest Related Species within the Range of the Northern Spotted Owl* (the "Northwest Forest Plan" or the "Forest Plan," FSEIS 1994) addresses conservation and management of late-successional and old-growth forests (LSOG) on federal lands in the range of the northern spotted owl (*Strix occidentalis caurina*) in Oregon, Washington, and portions of northern California (fig. 11-1). The Forest Plan was based on a scientific assessment of conditions within the range of the northern spotted owl (FEMAT 1993). The Forest Plan directs management on over 9 million hectares of mostly forested land managed by the USDA Forest Service (FS) and USDI Bureau of Land Management (BLM) in twelve biophysical provinces.

The Forest Plan requires monitoring the effectiveness of conservation and management to achieve late-successional and old-growth forest objectives and habitat for the northern spotted owl, marbled murrelet, and other species. An effectiveness monitoring plan provides the conceptual framework and general methods for monitoring late-successional and old-growth forests on federal lands under the Forest Plan (Hemstrom et al. 1998). The LSOG effectiveness monitoring plan is one part of a larger monitoring effort. Noon (chap. 2) describes seven steps that are included in the overall effectiveness monitoring plan: (1) specify goals, (2) identify stressors, (3) develop conceptual models, (4) select indicators, (5) establish sampling design, (6) define methods of analysis, and (7) ensure linkage to decision making. This chapter outlines the objectives, conceptual framework, monitoring questions, general methods for LSOG monitoring and provides an example of initial analyses for the Oregon Coast Range province.

Specific objectives of the LSOG monitoring approach (Hemstrom et al. 1998) include

Figure 11-1. Northwest Forest Plan area and provinces.

- Formulating effectiveness monitoring questions about the status and trends of late-successional and old-growth forest on federal lands at large landscape scales in the region of the Forest Plan.
- Providing a scientifically credible process for answering those questions, based on a conceptual model.
- Providing a process to evaluate those answers and a link to management decisions.

Hemstrom et al. (1998) list three attributes against which the quantity and quality of late-successional ecosystems may be judged: abundance and ecological diversity, process and function, and connectivity. All three attributes come directly from the scientific analysis (FEMAT 1993) that underlies the Forest Plan and from the Plan itself (FSEIS 1994).

This chapter provides an example of the application of monitoring principles discussed elsewhere in this volume. Although many potential examples exist, the following discussion draws from efforts to develop an effectiveness monitoring plan for late-successional and old-growth forests in western Oregon and Washington (Hemstrom et al. 1998). My intent is not to review the literature on monitoring old-growth forests; rather, it is to illustrate the process of translating goals in one particular land management document (FSEIS 1994) into measurable attributes, realistic analyses, and useful reports. In using this particular example, I summarize extensively from Hemstrom et al. (1998) and use their definitions and terminology. This produces some specific definitional issues, especially regarding "late-successional and old-growth forests." The land management document that establishes the basis for this particular example treats "late-successional and old-growth forests" as one vegetation stratum. Though it recognizes the variability within that stratum and the different roles of late-successional and old-growth forests, they are combined in specifying anticipated land-management outcomes. Hence, late-successional and old-growth forests are treated as one vegetation stratum below, and their definitions are based on the minimum structural attributes necessary for that stratum, as specified in the land management document. The following excerpts from the Northwest Forest Plan describe ecological characteristics (abundance and ecological diversity, process and function, and connectivity) that form the basis of the example monitoring plan.

Abundance and Ecological Diversity

"Abundance of late-successional and old-growth communities and ecosystems" refers to the total area of forest meeting structural, functional, or minimum age criteria based on ecological conditions and definitions for each physiographic province. The standards that define forests are based on the extent of three stages of late-successional and old-growth forest. The three stages are the (1) maturation, (2) transition, and (3) shifting small gap stages of late-successional and old-growth forest development. Ecological diversity is also indicated by the distribution of late-successional and old-growth communities on the landscape, and the interrelationships among a variety of geographic, climatic, elevational, topographic, and soil distributions (FSEIS 1994, chaps. 3 and 4, p. 35).

Process and Function

Process refers to ecological changes or actions that lead to the development and maintenance of late-successional and old-growth ecosystems at all spatial and temporal scales. Examples include: (1) tree establishment, maturation, and death, (2) gap formation and filling, (3) understory development, (4) small and large scale disturbances such as fire and wind, (5) decomposition, (6) nitrogen fixation, and (7) canopy interceptions of energy and matter, and (8) energy and matter transfers between the forest and atmosphere. (FSEIS 1994, chaps. 3 and 4, p. 37)

Functions refer to ecological values of the late-successional ecosystem or its components that (1) maintain or contribute to the maintenance of populations of species that use these ecosystems, and (2) contribute to the diversity and productivity of other ecosystems (such as carry over of large dead trees to early successional ecosystems, and storage of carbon in the global ecosystem). Examples of ecosystem function include habitat for organisms, climatic buffering, soil development and maintenance of soil productivity through inputs of large woody debris, nitrogen fixation, spread of biotic and abiotic disturbances through landscapes, and nutrient cycles production, storage, utilization, and decomposition. (FSEIS 1994, chaps. 3 and 4, p. 37)

Connectivity

Connectivity is a measure of the extent to which the large landscape pattern of the late-successional and old-growth ecosystem provides for biological and ecological flows that sustain late-successional and old-growth animal and plant species across the range of the northern spotted owl. Connectivity does not necessarily mean that late-successional and old-growth areas have to be physically joined in space, because many late-successional and old-growth species can move (or be carried) across areas that are not in late-successional ecosystem conditions. Large landscape features affecting connectivity are (1) distance between late-successional and old-growth areas, and (2) forest conditions in areas between late-successional and old-growth areas. (FSEIS 1994, chaps. 3 and 4, p. 38)

Old-Growth Definitions

The late-seral old-growth (LSOG) effectiveness monitoring plan (Hemstrom et al. 1998) defines LSOG from two perspectives: one from remotely sensed characteristics and one from permanent plot data at the stand scale. It establishes the statistical relation between them and evaluates trends in each (Czaplewski and Catts 1992). Remote sensing information allows answers to one set of questions (spatial features) while plot data provide information about stand-scale structure and composition. Since permanent plot methods are easier to repeat over time, they provide the most accurate estimators of LSOG amount.

LSOG forests can be described from remotely sensed information about upper canopy features, such as canopy cover, the size of tree crowns and inferences about tree diameter, canopy structure (single versus multiple layers), and, to some extent, tree species (Cohen and Spies 1992). Maps depicting LSOG using these features can be used to examine large landscape extent, patterns, and amounts. However, remote sensing cannot detect some critical features (dead wood, canopy layers below the top layer, understory vegetation, elements of the spatial distribution of these features) that determine LSOG condition at the stand scale.

LSOG forests can also be defined from stand scale, ground-based measurements of vegetation features (such as species, size, canopy complexity, and amount of dead material). A uniform grid-plot system established in the Forest Plan area measures these attributes (Max et al. 1996) and allows estimates of LSOG amounts from stand-scale characteristics. Because definitions, attributes and scale differ between remotely sensed and plot based analyses of LSOG, estimated amounts of LSOG are also likely to differ.

Remote-Sensing Definition

The LSOG effectiveness monitoring plan (Hemstrom et al. 1998) describes a classification system for forest composition and structure, based on characteristics visible in remotely sensed imagery and linked to attributes described in the Forest Plan (FSEIS 1994) and the FEMAT report (FEMAT 1993). The remote-sensing definition describes an LSOG forest as a stand dominated by trees 20 inches or more in diameter. In addition, the FEMAT report (FEMAT 1993) and FSEIS (1994) definitions distinguish stands with one canopy layer from those with two or more canopy layers. Determining canopy layers from remotely sensed information may be difficult but simple canopy structure (even upper canopies with relatively uniform crowns and few canopy breaks) can likely be distinguished from complex structure (uneven canopies with variable crown sizes and common canopy breaks) (Cohen and Spies 1992).

The LSOG effectiveness monitoring plan (Hemstrom et al. 1998) defines forest structure and composition classes that can be combined into sixteen forest vegetation mapping units (table 11-1):

1. *Potentially forested.* Land with tree canopy cover less than 10 percent. Ten percent canopy cover is a long-established standard for distinguishing forested from not-forested land. This class is restricted to lands that have been forested recently and will likely be forested again in the near future, but for which trees can not be detected in imagery.
2. *Seedling and sapling stand.* Forest with average dominant- and codominant-tree diameters between 0 and 10 inches.

Table 11-1. Forest vegetation mapping units used in LSOG effectiveness monitoring analysis for the Northwest Forest Plan area

	Composition Class		
Stand Structure Class	*Deciduous (D)*	*Mixed (M)*	*Conifer (C)*
Potentially forested (PF)	PF		
Seedling and sapling (SS)	SS-D	SS-M	SS-C
Small single-storied (SSS)	SSS-D	SSS-M	SSS-C
Medium to large single-storied (MSS)	MSS-D	MSS-M	MSS-C
Medium to large multistoried (MMS)	MMS-D	MMS-M	MMS-C
Large multistoried (LMS)	LMS-D	LMS-M	LMS-C

3. *Small single-storied stand.* Forest with average dominant- and codominant-tree diameters of 10 to 20 inches and simple canopy structure.

4. *Medium to large single-storied stand.* Forest with average dominant- and codominant-tree diameters between 20 inches and 30 inches and simple canopy structure. Forests in this category qualify as late successional from a remote-sensing perspective.

5. *Medium to large multistoried stand.* Forest with average dominant- and codominant-tree diameters between 20 and 30 inches and complex canopy structure. Forests in this category qualify as late-successional from a remote-sensing perspective.

6. *Large multistoried stand.* Forest with average dominant and codominant tree diameters of 30 inches or more and complex canopy structure. Forests in this category qualify as late-successional from a remote-sensing perspective. This class allows identification of stands that are likely to be optimum habitat for LSOG-dependent species.

7. *Deciduous/mixed/conifer.* Each forested structure class can be assigned one of three composition categories: conifer-dominated (hardwood canopy tree cover less than 20 percent of total tree cover), deciduous-dominated (conifer canopy tree cover less than 20 percent of total tree cover), and mixed (conifer and hardwood canopy tree cover are both more than 20 percent of total tree cover).

8. The FEMAT (1993) report also used broad strata of plant communities to more finely stratify LSOG by environment, composition, and structure. The LSOG effectiveness monitoring plan proposes the use

of either potential vegetation at the series level (e.g., Hall 1998) or cover type classes (Eyre 1980).

Stand-Scale Definitions

The LSOG monitoring plan (Hemstrom et al. 1998) uses three major structural elements to characterize forests at the stand-scale: live trees, standing dead trees, and fallen trees or logs. Additional important elements typically include multiple canopy layers, smaller understory trees, canopy gaps, and patchy understory. Structural characteristics of old forests vary with vegetation type, disturbance regime, developmental stage and environment (Franklin et al. 1981; Franklin and Spies 1991; Spies and Franklin 1988). The LSOG monitoring plan specifies the use of definitions of old-growth forests developed for Forest Service lands in Washington, Oregon, and California (USDA Forest Service 1992, 1993a). These definitions continue to evolve and currently cover only part of the Forest Plan area. "Old growth" is a subset of LSOG forests emphasizing those with more complex structure and, generally, larger and older trees. The LMS structure class (table 11-1) most closely represents old-growth conditions, while MSS and MMS classes represent mature forest conditions. LSOG condition includes both mature and old forests. Because site-scale permanent plots measure stand-scale features upon which LSOG definitions are based, they may provide a more reliable estimate of the total area and kinds of LSOG than does remote sensing analysis. Unfortunately, field plots cannot adequately address large-landscape spatial distribution issues.

Monitoring Questions

The LSOG monitoring plan (Hemstrom et al. 1998) poses the following effectiveness monitoring questions based on ownership (FS, BLM, other), province, plant community, and land-use allocation:

- What is the distribution and amount of structure and composition classes, including LSOG, at the large-landscape scale?

  –What is the distribution (map) and amount (area) of structure and composition classes from remote-sensing information?

  –What is the amount (area) of structure and composition classes from stand-scale (permanent-grid plot) samples?

  –What are the structure and composition characteristics (e.g., tree-diameter distribution, snags, and down woody debris) of structure and composition classes from stand-scale samples for each stratum?

–How much have the structure and composition characteristics changed since the last measurement cycle? How much are they likely to change in the near future?

–What is the error associated with these estimates?

–Are trends within expectations derived from the FSEIS (1994) and the FEMAT (1993) report?

- What is the stand-size distribution, stand interior area distribution, and interstand distance distribution of LSOG at the large-landscape scale?

  –For the region as a whole? For the region by land management allocation? For each province? For each province by land management allocation? For each province by plant community? For each province by plant community and land management allocation?

  –What is the error associated with these estimates?

  –How have these attributes changed since the last measurement cycle? How much are they likely to change in the near future?

  –Is the trend within expectations set by the FSEIS and the FEMAT report?

- What changes in distribution and amount of structure and composition classes are produced by stressors, starting with the year of the FEMAT analysis (1993), from stand-scale data?

  –What are the gains from growth and succession?

  –What are the losses from logging, fire, wind, insects, and disease?

  –What are the ramifications of these changes to future trends, especially in LSOG classes? For the region as a whole? For the region by land management allocation? For each province? For each province by land management allocation? For each province by plant community? For each province by plant community and land management allocation?

- What are the effects of silvicultural treatment and salvage on the development of LSOG structure and composition at the stand scale? Is the relationship of forest structure and composition (in stands at various ages and at multiple scales) to ecological processes and biological diversity assumed by the FEMAT report, the FSEIS, and the conceptual model (fig. 11-2) accurate?

Questions such as the last, relating to the validity of assumptions and basic science underlying the Forest Plan, are called validation monitoring issues. Hemstrom et al. (1998) included this question as a reminder to those who might develop a validation monitoring plan and to help distinguish effectiveness monitoring from validation monitoring while establishing a link between them.

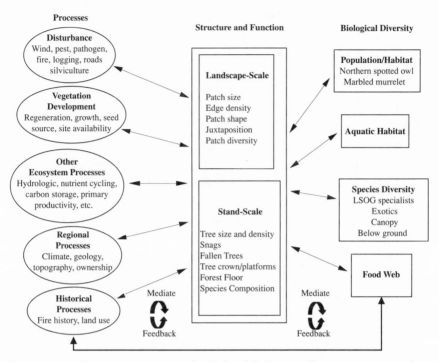

Figure 11-2. Late-successional and old-growth forest effectiveness monitoring plan conceptual model for the Northwest Forest Plan (after Hemstrom et al. 1998).

Monitoring Approach

Hemstrom et al. (1998) describe stressors as major agents of change and dynamic factors that may alter the amount, distribution, structure, and composition of LSOG forests at regional and provincial scales. Stressors include

- *Human development and use.* These can change LSOG through a wide variety of mechanisms (e.g., timber harvest, recreation, and mining). Such changes may or may not be discernible in Landsat Thematic Mapper (TM) and other remotely sensed information, but they will be detectable at the stand and site scales.
- *Wildfire and prescribed fire.* These can cause significant changes in the amount, distribution, structure, and composition of LSOG. Fire can play many roles ranging from stand replacement to slight alteration of understory structure and composition. Fire that causes complete or partial stand replacement has effects that should be visible at large-landscape scales. Underburns may not produce visible change at large-landscape scales but may generate significant changes at the stand scale

through changes in understory vegetation and the generation and composition of dead standing trees and logs.

- *Insect outbreaks and pathogen epidemics.* These can play roles similar to that of fire, though generally effects are more selective and of finer scale, especially on the west side of the Cascades. Changes range from stand replacement (rarely) to mortality of selected species and size classes within species. Effects may be hard to detect from Landsat TM data but may be visible from other remote-sensing platforms. Insects and pathogens often generate considerable dead and down woody structure in stands and may actually move stands farther along the successional continuum.
- *Wind.* Wind can have effects similar to those of fire, including partial to complete mortality of the overstory canopy. Severe blowdown often results in releases of understory vegetation from shade and competition, resulting in changed canopy structure and composition over time. Wind effects are often difficult to detect in Landsat TM data because they tend to be patchy and partial.
- *Growth, regeneration, and succession.* Although not generally considered stressors, these can produce changes in the amount, distribution, composition, and structure of LSOG forests. They tend to reduce fragmentation, interstand distance, and edge while increasing stand size, stand structure, and connection. At one extreme, in late-seral forests, succession can cause a decline in composition and structural diversity as stands lose long-lived, early seral species and move to climax condition. This circumstance is rare, however, given the time required and rates of disturbance. Analysis of early seral stages and projections of growth will be necessary to project trends.
- *Exotic species.* Some are highly competitive or pathogenic and can possibly have long-term effects on LSOG.

Conceptual Model

The LSOG monitoring conceptual model relies on the structure and composition of stands, from large landscape to stand scales, as intrinsic parts of and reasonable indicators of process, function, and biological diversity (Hemstrom et al. 1998; fig. 11-2). The model presumes that structure and composition both influence and are influenced by process and function (Spies and Franklin 1988; Edmonds et al. 1989; Franklin et al. 1989; Peterken 1996). For example, disturbance processes change stand- and landscape-scale structure and composition by killing trees, changing amounts of snags and down woody debris, changing stand size and shape

(Forman and Godron 1986; Morrison and Swanson 1990). Stand size, shape, composition, and structure influence the spread of disturbances across the landscape because some kinds of stands transmit the disturbance more readily than others (White 1979; Turner et al. 1989). Stand structure and composition also influence disturbance. Certainly, changes in composition, structure, and landscape pattern can influence ecological process and function (Turner 1989; Allen and Hoekstra 1992; Spies and Franklin 1996), and changes in structure and composition may indicate changes in underlying process and function. This conceptual model is implicitly stated in the FEMAT report (chap. 2, p. 36): "In many respects, the test of providing a functional, interacting late-successional and old-growth forest ecosystem subsumes the test of viability for the system's component species and groups of species." From this perspective, the structure and composition of vegetation at large-landscape and stand scales is a reasonable indicator of the processes, functions, and diversity of ecological systems. This conceptual framework may apply most clearly to plants. Similar models were developed for northern spotted owl (Lint et al. 1999) and marbled murrelet (*Brachyramphus marmoratus*) (Madsen et al. 1999).

Hemstrom et al. (1998) describe a relation between structure and composition and biological diversity. Species diversity is the biological foundation on which structure and composition are built. In turn, structure and composition determine habitat conditions that may support species diversity in a highly complex feedback process. Certainly, changes in structure and composition of forest stands indicate direct changes in biological diversity and may indicate habitat alteration and change in the composition of associated species. Ruggiero et al. (1991) summarize relationships of many wildlife species to habitat in Douglas-fir (*Pseudotsuga menziesii* [Mirbel] Franco) dominated forests.

Hemstrom et al. (1998) also describe two major classes of functional factors that determine the biological diversity in a given large region. The first is the amount of energy that the ecosystem can capture, which in turn is governed by latitudinal variation in solar radiation and a host of climatic and soil factors (often called *site quality* by foresters) that affect carbon fixation and primary productivity. The second major class is structure. Structure is defined as the physical arrangement of objects in three-dimensional space and their complexity. Much of the recent work on the relation of structure and biological diversity has been summarized by Bell et al. (1991). They consider structure as having three components: scale, heterogeneity, and complexity. In a general way, all of these components are related to biological diversity, but the details in any particular ecosystem can be very different.

Structure depends on spatial scale and can be described at any scale in the biological hierarchy but may have different associations and correlations with biological diversity at each scale (Hemstrom et al. 1998). For example, at the scale of biomes, mammal species richness in the United States is strongly correlated with topographic relief: more species live in mountainous areas. At the opposite end of the spatial scale, the number of lizard species on a single tree is well correlated with canopy complexity: trees with more complex canopies have more species. At the scales of stands and large landscapes, which are the focus of LSOG effectiveness monitoring, general relations exist between vegetation structural complexity and bird species diversity. Perhaps the best known of these relations is between foliage height diversity and bird species diversity first elucidated by MacArthur and MacArthur (1961). In these studies, a correlation was found between the number and evenness of layers of vegetation in a canopy and the number of bird species and the evenness of distribution of individuals among those species. A general summary of this work is that structural complexity and heterogeneity are positively correlated with diversity for many groups of organisms.

Studies of the Pacific Northwest's Douglas-fir and western hemlock (*Tsuga heterophylla* Raf. [Sarg.]) forests show that the general pattern holds both at the landscape and stand scales. Hansen et al. (1995) have shown that, at the landscape scale, greatest diversity results from a mix of successional stages each with a different canopy structure. Some bird species, for example, prefer open clearcuts while others prefer LSOG forests. Thus, a diversity of stand types including an important component of LSOG promotes diversity at the landscape scale. These different stand types and the changes in their distribution can be estimated relatively accurately from remotely sensed imagery. At the stand scale, greater canopy structural complexity is correlated with a greater abundance of some bird species. Further, the response of the abundance of some bird species is nonlinear, with a distinct threshold (Hansen et al. 1995). That is, as structural complexity increases, a point is reached where abundance for LSOG associated species increases dramatically. These results are species specific in the shape of the relationships, thresholds that may exist, and possible causal factors. Results from other groups of organisms are lacking, but all indications are that the Pacific Northwest forests roughly follow this general pattern. Thus, stand structure should be a useful indicator of biological diversity. Further, Cohen and Spies et al. (1992) have found that stand structural differences of a kind that are related to biological diversity can be estimated with remotely sensed data and in particular with Landsat TM data, which means that the detection of change of an important biological diversity indicator is logistically feasible.

Indicators

The LSOG monitoring plan (Hemstrom et al. 1998) describes indicators that hinge on the relationship of structure and composition to ecological system process and function and to biological diversity (fig. 11-3). Four kinds of indicators were considered: (1) large-landscape–scale indicators that can be addressed through analysis of remotely sensed imagery, (2) stand-scale indicators that can be addressed through analysis of field plots, (3) agents of change indicators, and (4) stand-scale silviculture and salvage-effect indicators.

Large-Landscape–Scale Indicators

Hemstrom et al. (1998) describe several indicators of LSOG conditions across large landscapes (provinces or larger). These include

- The area (amount from both plot data and remote sensing) and distribution (map) of land meeting forest-class definitions by physiographic province by plant community and land management allocation.
- The extents and sizes of stands of forest classes from remotely sensed images.
- The interior core area of LSOG stands by forest class, after account-

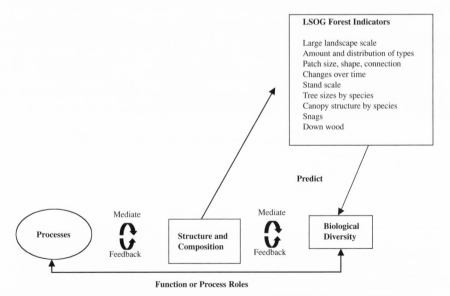

Figure 11-3. Late-successional and old-growth forest effectiveness monitoring indicators for the Northwest Forest Plan (after Hemstrom et al. 1998).

ing for edge effects. Hemstrom et al. (1998) defined interior core area of LSOG stands as the total area of individual stands from remotely sensed information minus a 100-meter buffer from nonforested and small, single-storied forest classes. Chen et al. (1992) found that edge effects may extend more than 400 meters in some cases, but most effects seem to occur within 100 meters.

- The distance between LSOG stands depicted by a frequency distribution or cumulative distribution.

Stand-Scale Indicators

Stand-scale structural and compositional attributes can be summarized from permanent-grid plots and compared with LSOG definitions for an accurate estimate of the amount of forest class by stratum (Hemstrom et al. 1998). Estimates of amounts by forest class from grid-plot data compared to those from remotely sensed information allow the development of statistical relations between remotely sensed and stand-scale definitions of LSOG (Czaplewski and Catts 1992). In addition, stratum-wide conditions of stand-scale structure and composition attributes can be summarized from plot data. Hemstrom et al. (1998) suggest the following characteristics be summarized by LSOG class:

- Area in different forest classes, including old growth according to current definitions.
- Diameter-class distribution of trees by stratum.
- Height-class distribution of trees and of selected tree species separately, by stratum.
- Height-class distribution by decay class and diameter-class distribution for standing dead trees (snags) by stratum.
- Tons, linear feet, and pieces per hectare of down wood by diameter class and decay class in each stratum.
- Other attributes useful for characterizing LSOG as they emerge from research on stand scale characteristics, ecological process and function. Stand age, where it can be determined, may be important for some species (Halpern and Spies 1995).

Ongoing research indicates that the spatial arrangement of structure at the stand scale may be important for old-growth function (J. Franklin, University of Washington, personal communication). The existing permanent grid plot design does not track spatial distribution of trees, snags, or down wood. There may be a need to add this to a subset of permanent grid plots.

Agents of Change Indicators

Drivers or stressors (agents of change) that alter remotely sensed forest classes could be tracked from both remotely sensed information and permanent-grid plot data (Hemstrom et al. 1998). Remote sensing allows detection of changes in amount and distribution of forest classes and of changes in stand size, stand interior area, and interstand distance for LSOG classes. Remotely sensed information will not generally allow determination of the agent of change, except possibly for timber harvest and large, intense fires. Most information about agents of change will likely come from analysis of field plot data. At present, each grid plot includes information on causes of tree damage (Max et al. 1996) that can be summarized to general categories of agents of change. Increases of LSOG forest classes detected in remotely sensed information might be the result of forest growth and succession. Apparent change may also result from different methods used to process remotely sensed data or from different data sources. In this sense, remotely sensed information about LSOG attributes is less statistically reliable than information from permanent-grid plot remeasurement. Hemstrom et al. (1998) describe several agents of change indicators:

- Real change in amount and distribution by forest class by stratum from both remotely sensed and field plot data.
- Real change in stand size distribution, stand interior area and interstand distance by LSOG class by stratum from remotely sensed data.
- Apparent change in amount, distribution, stand size, stand interior area, and interstand distance due solely to change in remote-sensing interpretation and analysis methods.

Stand Change Indicators from Field Plot Data

Additional stand change indicators can be gleaned from field plot data, including: the amount of forested area in each stratum that has changed structure and composition class as a result of human use (e.g., logging), fire, wind, insects, and disease; transitions among structure and composition classes by agents of change and stratum; and the amount of forested area that has changed structure and composition class because of growth or succession.

Predictive Models

Forest growth and development can be projected in the grid plot data by using stand-growth simulation models. Models can project growth trends for each forest class by stratum. Unfortunately, current growth models are oriented toward tree growth and volume (and not well-calibrated for stands beyond age one hundred) and detailed stand succession—models that may not work well for large landscapes. Models will have to be calibrated for each stratum (see Research Needs).

Stand-Scale Silviculture and Salvage-Effect Indicators

The FSEIS (1994) allows silvicultural (such as thinning, underplanting, prescribed fire, and killing trees to produce snags or down woody debris) and salvage activities in late-successional reserves as long as treatments benefit, or at least do not retard, development of late-successional forest conditions. For the most part, these activities are to be confined to younger stands within late successional reserves. A set of stand-scale attributes listed above (tree diameter class distribution, canopy structure and tree height, snag height and diameter, down woody debris amounts) could be analyzed to track changes in structure and composition in stands manipulated by silvicultural or salvage activities. Tracking these elements at the stand scale over time could test the effectiveness of silvicultural and salvage activities in maintaining or enhancing LSOG stand-scale attributes. Since analysis of change in these attributes would require a different set of plots (treated versus controlled experimental plots or case studies), which would be expensive and would require time to establish, Hemstrom et al. (1998) did not describe a sampling or analysis scheme for monitoring the effectiveness of silviculture and salvage treatments in managing or conserving LSOG.

Data Acquisition

Data for analysis of large-landscape-scale indicators (amount, distribution, stand size, stand interior area, and interstand distance by forest class) and for some agent of change indicators (large fires, evident timber harvest, etc.) most likely comes from remotely sensed imagery. Hemstrom et al. (1998) suggest a minimum polygon size of about 4 hectares for LSOG analysis, since a circular stand of this size contains a small interior core area given an edge effect of 100 meters.

The diameter of trees below the emergent canopy, tree heights, snags, and down wood, and most agents of change often cannot be reliably estimated from remotely sensed data. Data for analysis of those attributes exist in a permanent-plot grid across most forested areas of the Forest Plan region (Max et al. 1996). The sampling protocol allows estimation of measurement errors from a remeasurement of randomly selected plots.

Stand-scale attributes can be characterized from grid plots falling in each stratum. A 5.44-kilometer grid of permanent plots exists across most forested areas. Plot densities are higher in some areas (2.72- and 1.35-kilometer spacing between plot centers).

Hemstrom et al. (1998) did not develop a sampling scheme for stand-scale silviculture or salvage effects. Large local variation in silviculture and salvage treatment kind and intensity makes sampling stratification at the regional and provincial scales difficult. Sufficient numbers of permanent grid plots are unlikely to fall in areas treated by silvicultural manipulation or salvage logging to characterize effects of silviculture and salvage activities on LSOG attributes. Monitoring these effects would likely require information to classify stands into existing and emerging LSOG definitions (tree diameter class distributions, canopy structure and tree height class distributions, snag height and diameter class distributions, and down woody debris amounts).

Detecting Change and Trend

The more confidence needed in the reality of apparent change, the more costly the process. Because the state of remote-sensing science continues to evolve, analysis at each subsequent time step is likely to involve different techniques and to result in apparent change in LSOG due to changes in methods alone. In order to account for changing methods, both old (previous time step) and new methods should be evaluated at each monitoring time step, producing an estimate of change due to methods alone. The permanent-grid plot system is designed for systematic remeasurement. Plot data will provide measurements with known measurement error for LSOG elements. Change detection and rates of change can be quantified with known reliability for the stand-scale indicators.

Expected Values and Trends

The FEMAT report (1993) and the Forest Plan (1994) discuss potential or desirable changes in abundance and ecological diversity, process and function, and connectivity outcomes for LSOG in the Forest Plan area. Thresholds provided in these documents are general, regionwide, and apply only to federal lands. The preferred alternative under the Forest Plan was estimated (FSEIS 1994) to provide a 75 percent or better chance of providing abundance and diversity, process and function, and connectivity within the long-term average conditions in the moist physiographic provinces over the next one hundred years. All the alternatives considered in the Forest Plan were more likely to produce functional, interacting late-successional and old-growth forest ecosystems in the moist physiographic provinces (Olympic Peninsula, Western Washington Lowlands, Western Washington Cascades, Oregon Coast Range, Willamette Valley) than in the intermediate and dry physiographic provinces (Western Oregon Cascades, California Coast Range, Eastern Washington Cascades, Eastern Oregon Cascades, Oregon Klamath, California Klamath, California Cascades).

The FEMAT report (1993) and the FSEIS (1994) do not project reaching desired outcomes for a considerable time, since it takes decades or centuries for young stands to develop into LSOG. Changes in the first several decades should be projected for one hundred years or more to evaluate likely outcomes. The two documents do provide initial estimates of long-term trends. While the outcome elements provided in them represented the state of science at the time, several are qualitative or difficult to measure.

Reference Conditions

The FEMAT report (1993) and the FSEIS (1994) discuss long-term average reference conditions against which to judge current or future LSOG status.

> *Long term* is defined as a period of at least 200 to 1,000 years, or the time over which the full potential range of late-successional and old-growth communities can develop following severe disturbance. (FSEIS 1994, chaps. 3 and 4, p. 36)

> The long-term average regional abundance of late-successional and old-growth communities can only be approximated from a few local studies of fire history. Assuming that the average regional natural fire rotation was about 250 years for severe fires (those removing 70 percent of more of the basal area), then 60 to 70 percent of the forested area of the region was typically dominated by late-successional and old-growth forests, depending on the age at which 'mature' forest conditions develop (assume a range of eighty to one hundred years). Converting this to a single number, 65 percent, provides an estimate of the long-term average percentage of the regional landscape covered by late successional forest. This average percentage would certainly vary by physiographic province; moist, northerly provinces would have higher averages than drier provinces with higher fire frequencies.

> The total percentage of late-successional and old-growth forest would apply to a wide range of stand sizes, from less than 1 acre,

to hundreds of thousands of acres. Most of the total percentage (perhaps 80 percent or more) would probably have occurred as relatively large (greater than 1,000 acres) areas of connected forest.

The average centurial-low coverage (average of the lows that occur in one-hundred-year periods) by late-successional forest is defined as setting the lower limit of the 'typical' range. There [are] no data from which to estimate the average low from the preceding millennium. Consequently, this value was estimated based on the subjective opinions of the ecosystem experts. The Forest Ecosystem Management Assessment Team hypothesized that the average low amounts might be about 40 percent coverage by late-successional forests, with lower values expected for individual provinces. (FSEIS 1994, chaps. 3 and 4, p. 37)

Long-term averages have not been well documented for much of the Forest Plan region (Hemstrom et al. 1998). In addition, expected variation around the average is important since a single average condition may never have actually occurred.

Hemstrom et al. (1998) translated the outcomes projected in the FEMAT (1993) report and FSEIS (1994) into quantifiable attributes linked to LSOG effectiveness monitoring indicators. These translations are necessarily preliminary, since LSOG forest development takes many decades, long-term reference conditions by stratum are poorly understood, climatic conditions for the long-term average conditions may have been significantly different than they are now, and scientific understanding about forest development and succession is not well developed for many strata. They do provide an indication of the anticipated direction of change in LSOG forests as described in the Forest Plan. A description of each outcome is followed by an interpretation of that outcome in terms of LSOG indicators.

Abundance and Ecological Diversity

Outcome 1: "Late-successional and old-growth ecosystem abundance and ecological diversity on federal lands at least as high as the long-term average . . . prior to logging and extensive fire suppression. Relatively large areas (50,000 to 100,000 acres) would still contain levels of abundance and distribution of late-successional forests which are well below the regional average for long periods. However, within each physiographic province, abundance would be at least as high as province-level long-term averages, which might be higher or lower than the regional long-term average" (FSEIS 1994 chaps. 3 and 4, p. 36). This was interpreted to mean that the long-term average proportion of LSOG on

Forest Service and BLM lands would be about 65 percent (Hemstrom et al. 1998). Since this criterion is the same as LSOG cover in connectivity (below), the same number (60 percent) was used for outcome for both abundance and connectivity (table 11-2). Most (more than 80 percent) of the LSOG in the long-term average would occur in large blocks (i.e., more than 400 hectares).

Outcome 2: "Late-successional and old-growth ecosystem abundance and ecological diversity on federal lands is less than the long-term conditions (prior to logging and extensive fire suppression) but within the typical range of conditions that occurred during previous centuries" (FSEIS 1994, chaps. 3 and 4, p. 36). Hemstrom et al. (1998) interpreted this to mean that LSOG would be present in all provinces and at all elevations, but with larger gaps in distribution than those of Outcome 1. The average of the low end of the range for LSOG amount in the long-term average would be about 40 percent (FEMAT 1993). LSOG would comprise between 40 and 65 percent of Forest Service and BLM land. Less than 80 percent of the LSOG would be in stands of more than 400 hectares (Table 11-2).

Outcome 3: "LSOG abundance and ecological diversity on federal lands is considerably below the typical range of conditions that have occurred during the previous centuries, but some provinces are within the range of variability. The ecological diversity (age-class diversity) may be limited to the younger stages of late-successional ecosystems. Late-successional and old-growth communities and ecosystems may be absent from some physiographic provinces and/or occur as scattered remnant patches within provinces" (FSEIS 1994, chaps. 3 and 4, p. 36). LSOG forests would occupy 40 percent or less of Forest Service and BLM lands, or less than the average century lows from long-term baseline estimates (Hemstrom et al. 1998), but some LSOG would still exist (for example, more than 1 percent of the federal land area). Less than 80 percent of the LSOG would be in stands of more than 400 hectares (table 11-2). LSOG might be absent from some physiographic

Table 11-2. Abundance outcomes and thresholds for the Northwest Forest Plan area.

Outcome	Percentage of land covered by LSOG (%)	Percentage of land in stands of more than 400 hectares (%)	Percentage of provinces meeting both amount and stand size (%)
1	60 to 100	80 to 100	100
2	40 to 60	5 to 80	100
3	5 to 40	1 to 5	50 to 100
4	less than 5	less than 1	less than 50

provinces or some elevation zones within provinces and might occur as scattered remnant stands.

Outcome 4: "Late-successional and old-growth ecosystems are very low in abundance and may be restricted to a few physiographic provinces or elevation bands or localities within provinces. Late-successional and old-growth communities and ecosystems are absent from most physiographic provinces or occur only as small remnant patches" (FSEIS 1994, chaps. 3 and 4, p. 36). This would mean that LSOG would occupy less than 1 percent of the federal lands (Hemstrom et al. 1998). Less than 80 percent of the LSOG would be in stands of more than 400 hectares (table 11-2). LSOG forests would be absent from most provinces or occur only as small remnant forest stands.

Process and Function

The FSEIS (1994, chaps. 3 and 4, p. 38) provides descriptions of process and function outcomes that could result from implementing the Forest Plan:

Outcome 1: "The full range of natural disturbance and vegetative development processes and ecological functions are present at all spatial scales from microsite to large landscapes."

Outcome 2: "Natural disturbance and vegetative development processes and ecological functions occur across a moderately wide range of scales but are limited at large landscape scales through fire suppression and limitation of areas where late-successional ecosystems can develop."

Outcome 3: "Natural disturbance and vegetative development processes are limited in occurrence to stand and microsite scales. Many stands may be too small or not well developed enough to sustain the full range of ecological processes and functions associated with LSOG ecosystems."

Outcome 4: "Natural disturbance and vegetative development processes associated with LSOG ecosystems are extremely restricted or absent from most stands and large landscapes. Most LSOG stands are too small or not well developed enough to sustain the full range of processes and ecological functions associated with late-successional and old-growth ecosystems."

Neither the FEMAT (1993) report nor the FSEIS (1994) provide quantitative criteria for ecosystem process and function. In the near term, process and function should be maintained to the extent that ecological abundance and diversity outcomes are met (Hemstrom et al. 1998).

Connectivity

Outcome 1: "Connectivity is very strong, characterized by relatively short distances (less than 6 miles on average) between late-successional and old-growth areas. Smaller patches of late-successional and old-growth forest frequently occur. The proportion of the landscape covered by late successional and old growth conditions of all stand sizes exceeds 60 percent, a threshold when many measures of connectivity increase rapidly. At regional scales, physiographic provinces are connected by the presence of landscapes containing areas of late-successional and old-growth forests" (FSEIS 1994, chaps. 3 and 4, p. 40). Mean distances of less than 6 miles between LSOG stands of at least 400 hectares and LSOG cover greater than 60 percent indicate Outcome 1 (table 11-3, Hemstrom et al. 1998). Small stands of LSOG (riparian buffers, green tree retention in harvest units, etc.) would be common as indicated by cumulative frequency distributions of LSOG stand sizes. Large LSOG stands would connect adjacent provinces.

Outcome 2: "Connectivity is strong, characterized by moderate distances (less than 12 miles on average) between large late-successional and old-growth areas. Smaller patches of late-successional forest occur as described in Outcome 1. At regional scales, physiographic provinces are connected by the presence of landscapes containing areas of late-successional and old-growth forests. The total proportion of landscape in late-successional and old-growth conditions, including smaller patches, is at least 50 percent, so that the late-successional condition is still the dominant cover type" (FSEIS 1994, chaps. 3 and 4, p. 40). Mean distances of 6 to 12 miles between LSOG stands of at least 400 hectares and LSOG cover greater than 50 percent indicate Outcome 2 (table 11-3, Hemstrom et al. 1998). Small stands of LSOG (riparian

Table 11-3. Connectivity thresholds for the Northwest Forest Plan area.

Outcome	Mean distance between stands of more than 400 hectares (km)	Percentage of LSOG cover (%)	LSOG stands less than 400 hectares	Percentage of adjacent provinces connected with large LSOG stands (%)
1	less than 9.6	60 to 100	common	100
2	9.6 to 19.2	50 to 60	common	100
3	19.2 to 38.4	25 to 50	present	less than 100
4	more than 38.4	less than 25	absent to few	less than 100

buffers, green-tree retention in harvest units, etc.) would be common as indicated by cumulative frequency distributions of LSOG stand sizes. Large LSOG stands would connect adjacent provinces.

Outcome 3: "Connectivity is moderate, characterized by distance[s] of 12 to 24 miles between large old-growth areas. There is limited occurrence of smaller patches of late-successional forest in the matrix. The late-successional forest is at least 25 percent of the landscape, and the matrix contains some smaller areas for dispersal habitat" (FSEIS 1994, chaps. 3 and 4, p. 40). Mean distances of 12 to 24 miles between LSOG stands of at least 400 hectares and LSOG cover greater than 25 percent indicate outcome 3 (table 11-3, Hemstrom et al. 1998). Small stands of LSOG would occur in matrix lands.

Outcome 4: "Connectivity is weak, characterized by wide distances (greater then 24 miles) between old-growth areas. There is a matrix in which late-successional and old-growth conditions occur as scattered remnants or are completely absent" (FSEIS 1994, chaps. 3 and 4, p. 40). Mean distances of over 24 miles between LSOG stands of at least 400 hectares and LSOG cover less than 25 percent indicate O   utcome 4 (Table 11-3; Hemstrom et al. 1998). LSOG in matrix lands would be in small remnant stands or absent.

Baselines

Current conditions could be established from FEMAT (1993) vegetation maps. However, the vegetation maps used in the FEMAT (1993) analysis are poorly suited for use as a monitoring baseline because they were compiled from a variety of information sources, all of which used different mapping methods. Future mapping efforts cannot replicate those methods. Changes detected when comparing new maps to the FEMAT maps would contain large and unknown variation due to differing methods alone. Alternatively, a new current vegetation condition could be established from remotely sensed imagery. A stand-scale baseline could be established from permanent-grid plot data, since coverage of the region is nearly complete as of 1998. The FEMAT report (1993) and the FSEIS (1994) call for comparing existing LSOG conditions (amounts, distribution, connection, process, function) to average conditions for the period spanning several centuries to one thousand years ago. Both documents suggest developing this information through fire history studies, stratified by province and plant community, but these studies do not exist for much of the region. Initial attempts, using expert opinion and existing information, exist in the FEMAT (1993) report, the FSEIS (1994), and for national forest lands in Oregon and Washington (USDA Forest Service 1993b). However, since long-term baseline estimates do not exist for much of the area, Hemstrom et al. (1998) propose development of long-term baselines in future research.

Other Considerations

Accuracy assessments should be performed on any maps used in LSOG analysis to understand the variability associated with results (Hemstrom et al. 1998). Because the 5.44 km grid plots are independent of vegetation maps (assuming all the plots are not used for training sites in developing vegetation maps), they could be used to perform an accuracy assessment of the vegetation map as aggregated for LSOG analysis (Czaplewski and Catts 1992).

Hemstrom et al. (1998) proposed that a report be generated every five years, more often if needed, detailing the status of threshold attributes and their projected fifty- and one-hundred-year trends in comparison to FSEIS expectations. Attributes that depart significantly from projected trends (more than 10 percent) would be highlighted. Attributes below projected trends (LSOG amounts more than 10 percent below projected trends or for which the projected trends fall 10 percent below reference conditions) could trigger a variety of actions ranging from review of stand succession models and mapping methods to an examination of the Forest Plan and its implementation. Since results may require interpretation, a panel of scientists, managers, and others (as necessary) would review results and develop interpretations and recommendations for regional executives. Several other effectiveness monitoring efforts will use the same data (Lint et al. 1999; Madsen et al. 1999).

Research Needed

Several issues will likely require research investment in the near future. Among the most critical issues are

- The use of remotely sensed information to detect forest structure and composition might be enhanced with new methods or imaging systems. In addition, analysis of the sources of error in remote-sensing classifications and change detection is needed.
- Expected rates of transition among forest classes over time (succession models, disturbance models) using plant community and province strata should be developed. Existing models are probably more accurate for transitions and growth in young stands than in LSOG. These models should be calibrated by series to allow prediction that is more

accurate. Remeasured permanent grid plot data can be used to help calibrate models, which should become more accurate over time and with repeated measurements.

- Refined estimates of LSOG trends (abundance, stand core area, inter-stand distance) by province and plant community for the next one hundred years or more will be needed, requiring work on succession and disturbance models.

- Baseline, long-term conditions for LSOG by province and plant community should be documented. This might include a combination of disturbance history analysis, disturbance modeling, and stand-growth modeling. It will be important to describe ranges of conditions for LSOG attributes, not just average conditions, for the last several centuries.

- Refined definitions or indices of late-successional and old-growth forests would be helpful in assigning plots and remotely sensed stands to a position along a continuum of LSOG structure and composition conditions by stratum. This includes the need to examine within-stand spatial distribution of structural components (Franklin and Spies 1991).

- Better stand-scale projection models for LSOG stand structure and composition attributes (live trees, snags, down wood, species) could allow long-term projections (e.g., one hundred or more years) by stratum.

- Basic research on the relation between forest structure, stand age, disturbance history, and biological diversity, ecosystem processes, and ecosystem functions is needed to support (or modify) and refine the conceptual model.

Old-Growth Monitoring Analyses: Oregon Coast Range Example

A pilot test of proposed LSOG monitoring analyses for the Oregon Coast Range province provides initial examples of the kinds of results that might be expected for the Forest Plan area (Plumley et al. 1999). The Oregon Coast Range province includes about 2.3 million hectares in the coastal portion of northwestern Oregon (fig. 11-1, table 11-4). The moist, temperate environment supports extensive forests dominated by Douglas-fir, western hemlock, and several other conifer and hardwood species. The province includes the Siuslaw National Forest, BLM lands, and extensive areas of private and state-owned land.

The pilot analysis used remote-sensing data and maps from Landsat

Table 11-4. Land ownership in the Oregon Coast Range province.

Ownership	Hectares
Forest Service	252,281
BLM	311,869
Other federal	525
Total federal	564,675
State	272,305
Tribal	5,248
Private industrial	883,690
Other	582,802
Subtotal	2,308,720
Water	12,394
Total	2,321,114

TM satellite imagery. Mapped forest classes were similar to those necessary for LSOG monitoring analysis. Data from 328 permanent grid plots on the Siuslaw National Forest provided a sample from Forest Service lands, representing about 249,000 hectares. Comparisons of stand and remote-sensing perspectives cannot be made strictly according to the LSOG monitoring plan because the remote-sensing data for federal lands includes both Forest Service and BLM lands while the plot data represents only Forest Service lands. In practice, LSOG monitoring for the Forest Plan would characterize both perspectives across all lands to the extent that stand-scale data were available. Considerable progress has been made to access grid plot data from all ownerships and to develop consistent, wall-to-wall vegetation maps for the Forest Plan area (Jim Alegria, Forest Service and BLM inventory coordinator, Portland, Ore., personal communication).

LSOG forests comprise about 21 percent of the Siuslaw National Forest land (plot data, table 11-5) and about 29 percent of all Forest Service and BLM lands in the test area (remotely sensed data, table 11-5). Wildfire burned much of the Oregon Coast Range province area in the last 150 years. In addition, timber harvesting has been the dominant land use in the province for many decades. Most LSOG stands are less than about 400 hectares in total size (table 11-6). Edge effects reduce the interior area of most LSOG stands to less than 100 hectares (table 11-6). LSOG abundance and diversity conditions in the test area appear to be in Outcome 3 (table 11-2) at present. Projected trends in amount of LSOG in the test area (fig. 11-4) indicate that 60 percent or more of the Forest Service and BLM land should become LSOG by 150 years.

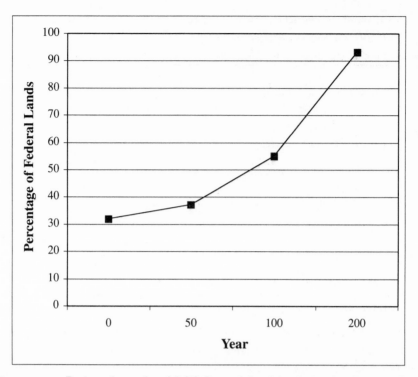

Figure 11-4. Projected trends of LSOG on federal lands in the Oregon Coast Province LSOG monitoring test area.

Table 11-5. Aerial extent and percent vegetation classes from the plot and remote-sensing perspectives for the Oregon Coast Range LSOG monitoring pilot.

	Plot data Forest Service land		Remote sensing data Forest Service and BLM land	
Vegetation class	Hectares	Percent (%)	Hectares	Percent (%)
Open	10,486	4	26,356	5
Deciduous forest	8,838	4	63,441	11
Mixed forest, sapling	10,636	4	26,235	5
Mixed forest, small	13,482	5	24,656	4
Mixed forest, medium	23,219	9	114,413	20
Mixed forest, large	5,393	2	39,717	7
Conifer forest, sapling	37,000	15	19,636	3
Conifer forest, small	35,202	14	73,158	13
Conifer forest, medium	27,113	11	110,648	20
Conifer forest, large	24,417	10	12,429	2
LSOG	53,328	21	162,794	29
Other	0	0	55,263	10

LSOG from plot data includes additional attributes (canopy layering, snags, down wood) and is a separate category. LSOG from remotely sensed data is the sum of the conifer large, conifer medium, and mixed large remotely sensed classes.

Table 11-6. Area and cumulative percent of LSOG patches in different patch size classes on Forest Service and BLM lands in the Oregon Coast Province LSOG pilot monitoring analysis area.

Patch size class	Total area		Interior area		Proportion of total	Proportion of interior total
Hectares	Hectares	Cumulative percent (%)	Hectares	Cumulative percent (%)	Percent (%)	Percent (%)
1–10	23,077	14	17,004	23	14	23
10–20	14,835	23	9,611	36	9	13
20–40	18,132	34	11,090	51	11	15
40–101	28,022	51	14,047	70	17	19
101–202	21,429	64	8,872	82	13	12
202–405	18,132	75	6,654	91	11	9
405–2024	31,319	94	6,654	100	19	9
2024–4049	3,297	96	0	100	2	0
4049–8097+	6,593	100	0	100	4	0
Total	164,835		73,931		100	100

Connectivity was not analyzed in the Oregon Coast Range province LSOG monitoring pilot, but data are available for that analysis. There are several ways to calculate interpatch distance. The LSOG monitoring plan (Hemstrom et al. 1998) did not specify which should be used. In addition, the computational process for calculating interpatch distances was beyond the computer software and hardware available at the time of the pilot analysis.

Conclusion

Several lessons have emerged from the development and implementation of the LSOG monitoring plan (Hemstrom 1998). First, it is important for those who draft management policy such as the Forest Plan to include specific, measurable characteristics that can be used to track the effectiveness of policy in achieving its goals. It is then imperative that monitoring plans reflect those same characteristics and specify methods to measure them. Although it is inevitably difficult, from a policy development perspective, to be specific about expectations, avoiding specificity will only make monitoring and establishment of public faith more difficult for future managers.

Second, ecological change occurs across large landscapes involving millions of hectares and involves ecological processes that require time. Monitoring plans that only track current conditions and compare to expected results without extrapolating likely future conditions may suffer time-lag effects. Although projecting likely future conditions given recent trends may be difficult, it can help managers recognize areas where management may not produce desired results and allow for proactive change or increased attention to monitoring specific characteristics.

In addition, a monitoring plan based on an explicit conceptual model, as discussed in this book, allows those who implement effectiveness monitoring plans to understand the conceptual framework and derivation of monitoring indicators. Since monitoring plans may be implemented over many years or decades, this is an important consideration given that those who develop monitoring plans may not be those who implement them.

It may be difficult to acquire the necessary data and implement the analysis needed to support effectiveness monitoring across large landscapes, especially where the budgets and personnel of several organizations must be considered. New or novel approaches to data collection and analysis may require more resources or time than estimated, budget emphases may change, and new issues may arise (see chap. 8). These difficulties indicate the need for continuing commitment to a monitoring program and the necessary resources.

ACKNOWLEDGMENTS

The author gratefully acknowledges the work done by Jane Kertis, Harriet Plumley, Thomas Spies, and others for the Coast Pilot monitoring analysis.

LITERATURE CITED

Allen, T. F. H., and T. W. Hoekstra. 1992. *Toward a unified ecology*. New York: Columbia University Press.

Bell S. S., E. D. McCoy, and H. R. Mushinsky. 1991. *Habitat Structure: The physical arrangements of objects in space*. New York: Chapman and Hall.

Chen, J., J. F. Franklin, and T. A. Spies. 1992. Vegetation responses to edge environments in old-growth Douglas-fir forests. *Ecological Applications* 2:387–396.

Cohen, W. B., and T. A. Spies. 1992. Estimating structural attributes of Douglas-fir/western hemlock forest stands from Landsat and SPOT imagery. *Remote Sensing of the Environment* 41:1–17.

Czaplewski, R. L., and G. P. Catts. 1992. Calibration of remotely sensed proportion or area estimates for misclassification error. *Remote Sensing of the Environment* 39:29–43.

Edmonds, R. L., D. Binkley, M. C. Feller, P. Sollins, A. Abee, and D. D. Myrold. 1989. Nutrient cycling: Effects on productivity of Northwest forests. Pp. 17– in *Maintaining the long-term productivity of Pacific Northwest forest ecosystems*, edited by D. A. Perry, R. Meurisse, B. Thomas, R. Miller, J. Boyle, J. Means, C. R. Perry, and R. F . Powers. Portland, Ore.: Timber Press.

Eyre, F. H., ed. 1980. *Forest cover types of the United States and Canada*. Washington, D.C.: Society of American Foresters.

FEMAT. 1993. *Forest ecosystem management: An ecological, economic and social assessment.* Report of the Forest Ecosystem Management Assessment Team. USDA Forest Service, USDC National Marine Fisheries Service, USDI Bureau of Land Management, USDI National Park Service, and U.S. Environmental Protection Agency. Portland, Ore.

Forman, R. T. T., and M. Godron. 1986. *Landscape ecology.* New York: John Wiley and Sons.

Franklin, J. F., K. Cromack Jr., W. Dennison, A. McKee, C. Maser, J. Sedell, F. Swanson, and G. Juday. 1981. *Ecological characteristics of old-growth Douglas-fir forests.* General Technical Report PNW-118, U.S. Department of Agriculture, Forest Service, Pacific Northwest Research Station, Portland, Ore.

Franklin, J. F., D. A. Perry, T. D. Schowalter, M. E. Harmon, A. McKee, and T. A. Spies. 1989. Importance of ecological diversity in maintaining long-term site productivity. Pp. 82–97 in *Maintaining the long-term productivity of Pacific Northwest forest ecosystems,* edited by D. A. Perry, R. Meurisse, B. Thomas, R. Miller, J. Boyle, J. Means, C. R. Perry, and R. F. Powers. Portland, Ore.: Timber Press.

Franklin, J. F., and T. A. Spies. 1991. Ecological definitions of old-growth Douglas-fir forests. Pp. 61–69 in *Wildlife and vegetation of unmanaged Douglas-fir forests,* coordinated by L. F. Ruggiero, K. B. Aubry, A. B. Carey, and M. Huff. General Technical Report PNW-GTR-285, U.S. Department of Agriculture, Forest Service, Pacific Northwest Research Station, Portland, Ore.

FSEIS. 1994. *Final supplemental environmental impact statement on management of habitat for late-successional and old-growth forest related species within the range of the northern spotted owl.* U.S. Department of Agriculture, Forest Service, and U.S. Department of the Interior, Bureau of Land Management. Portland, Ore.

Hall, F. C. 1998. *Pacific Northwest ecoclass codes for seral and potential natural plant communities.* General Technical Report PNW-418, U.S. Department of Agriculture, Forest Service, Pacific Northwest Research Station, Portland, Ore.

Halpern, C. B., and T. A. Spies. 1995. Plant species diversity in natural and managed forests of the Pacific Northwest. *Ecological Applications* 5:913–934.

Hansen, A. J., W. C. McComb, R. Vega, M. G. Raphael, and M. Hunter. 1995. Bird habitat relations in natural and managed forests in the west Cascades of Oregon. *Ecological Applications* 5:555–569.

Hemstrom, M. A., T. Spies, C. Palmer, R. Kiester, J. Teply, P. McDonald, and R. Warbington. 1998. *Late-successional and old growth forest effectiveness monitoring plan for the Northwest Forest Plan.* General Technical Report PNW-GTR-438, U.S. Department of Agriculture, Forest Service, Pacific Northwest Research Station, Portland, Ore.

Lint, J., B. Noon, R. Anthony, E. Forsman, M. Raphael, M. Collopy, and E. Starkey. *Northern spotted owl effectiveness monitoring plan for the Northwest Forest Plan.* General Technical Report PNW-GTR-438, U.S. Department of Agriculture, Forest Service, Pacific Northwest Research Station, Portland, Ore.

MacArthur, R. H., and J. W. MacArthur. 1961. On bird species diversity. *Ecology* 42:594–598.

Madsen, S., D. Evans, T. Hamer, P. Hensen, S. Miller, K. Nelson, D. Roby, and M. Stepanian. *Marbled murrelet effectiveness monitoring plan for the Northwest Forest Plan.* General Technical Report PNW-GTR-439, U.S. Department of Agriculture, Forest Service, Pacific Northwest Research Station, Portland, Ore.

Max, T. A., H. T. Schreuder, J. W. Hazard, D. D. Oswald, J. Teply, and J. Alegria. 1996. *The Pacific Northwest Region vegetation and inventory monitoring system.* Research Paper PNW-RP-493, U.S. Department of Agriculture, Forest Service, Pacific Northwest Forest and Range Experiment Station, Portland, Ore.

Morrison, P. H., and F. J. Swanson. 1990. *Fire history and pattern in a Cascade Range landscape.* General Technical Report PNW GTR-254, U.S. Department of Agriculture, Forest Service, Pacific Northwest Research Station, Portland, Ore.

NRC (National Research Council) 2000. *Environmental issues in Pacific Northwest forest management.* National Research Council. Washington, D.C.: National Academy Press.

Peterken, G. F. 1996. *Natural woodland: Ecology and conservation in northern temperate regions.* Cambridge, U.K.: Cambridge University Press.

Plumley, H., T. Spies, C. Hawkins, B. Daniels, C. Frounfelker, J. Kertis, C. Murdock, D. Rainsford, and C. Wettstein. 1999. Oregon Coast Range effectiveness monitoring pilot study. Document on file, Siuslaw National Forest, Corvallis, Ore.

Ruggiero, L. F., K. B. Aubry, A. B. Carey, and M. H. Huff. 1991. *Wildlife and vegetation of unmanaged Douglas-fir forests.* General Technical Report PNW GTR-285, U.S. Department of Agriculture, Forest Service, Pacific Northwest Research Station, Portland, Ore.

Spies, T. A., and J. F. Franklin. 1988. Old growth and forest dynamics in the Douglas-fir region of western Oregon and Washington. *Natural Areas Journal* 8:190–201.

———. 1996. The diversity and maintenance of old-growth forests. Pp. 296–314 in *Biodiversity in managed landscapes: Theory and practice,* edited by R. C. Szaro, and D. W. Johnston, New York: Oxford University Press.

Spurr, S. H., and B. V. Barnes. 1973. *Forest ecology.* 2d ed. New York: Ronald Press.

Turner, M. G. 1989. Landscape ecology: The effect of pattern on process. *Annual Review of Ecology and Systematics* 20:171–197.

Turner, M. G., R. H. Gardner, V. H. Dale, and R. V. O'Neil. 1989. Predicting the spread of disturbance across heterogeneous landscapes. *Oikos* 55:121–129.

USDA Forest Service. 1992. Old growth definitions/descriptions for forest cover types. Internal memo dated 19 June 1992, U.S. Department of Agriculture, Forest Service, Pacific Southwest Region, San Francisco, Calif.

———. 1993a. *Region 6 interim old growth definition[s] [for the] Douglas-fir series, grand fir/white fir series, interior Douglas-fir series, lodgepole pine series, Pacific silver fir series, ponderosa pine series, Port Orford cedar series, tanoak (redwood) series, western hemlock series.* U.S. Department of Agriculture, Forest Service, Pacific Northwest Region, Portland, Ore.

———. 1993b. *A first approximation of ecosystem health.* U.S. Department of Agriculture, Forest Service, Pacific Northwest Region, Portland, Ore.

White, P. S. 1979. Pattern, process and natural disturbance in vegetation. *Botanical Review* 45:229–299.

CHAPTER 12

Monitoring Wetland Ecosystems Using Avian Populations: Seventy Years of Surveys in the Everglades

Peter Frederick and John C. Ogden

Birds have frequently been suggested as bioindicators of environmental conditions (Custer and Osborn 1977; Ogden 1993; Erwin et al. 1996; Erwin and Custer 2000). This idea is intuitively attractive because birds are often strong bioaccumulators in the energetic sense, are uniquely mobile, feed at various levels in local food webs, and are usually easy to locate and census. However, considerable criticism has been aimed at the overuse of animals in general, and birds specifically, as bioindicators (Morrison 1986; Temple and Wiens 1989; Niemi et al. 1997). This skepticism is well placed, and stems from several sources. The use of any organism as a reliable bioindicator demands that a cause-and-effect relationship between environmental variation and response in the organism be documented—a criterion that is often lacking in field monitoring studies. For instance, changes in the size of avian breeding populations are often touted as an indicator of the abundance and availability of food sources in the area being monitored. Yet variation in bird populations can be due to various local and distant factors, such as habitat conditions elsewhere, overwinter survival rate, or fecundity in previous years. Thus, it is clear that the monitored response of the bioindicator species must be linked specifically and mechanistically to some measurable aspect of environmental variability or to the end result of some key environmental process.

The species or attribute serving as an indicator must also be matched in grain and scale with the environment or environmental attribute being monitored (Drury 1980). Birds have been regularly monitored at extremely large geographic scales; the U.S. Breeding Bird Survey, the National Audubon Christmas Bird Count (McCrimmon et al. 1997), and the European Seabirds at Sea program (Stone et al. 1995) are excellent examples. Although these programs produce data that allow an understanding of animal movements and population change over very large time and spatial scales, the results are often spatially and temporally too coarse for detecting the effects of environmental change at the scale of ecosystems or the temporal scale of many management actions. Similarly, there are many bird monitoring programs designed specifically to answer questions about the reproductive ecology and demography of particular species at large scales but which are not designed or able to use bird populations to understand ecosystem processes (Robinson et al. 1995).

Despite these caveats, there are numerous examples of the successful use of birds as bioindicators in a variety of types of ecosystems. Long-term trends of penguin-breeding aggregations have been used to detect changes in fishery resources in the Antarctic (Lemaho et al. 1993), physiological stress indicators in birds have been linked with habitat quality (e.g., Marra and Holberton 1998), and historical records of irruption patterns of passerine birds have been used to infer past climatic events (Kinzelbach et al. 1997).

In this chapter, we describe a long-term monitoring program in the Florida Everglades, which has used various response variables (nesting population size, reproductive success, contaminant loads, location of nesting, and habitat-use patterns) in long-legged wading birds (herons, egrets, ibises, storks, and spoonbills, order Ciconiiformes) as a means to understand ecosystem processes and linkages between the Everglades and other ecosystems. The resulting picture of past and present ecosystem behavior has been used to help set goals and monitoring guidelines for the restoration of critical ecosystem functions (Davis and Ogden 1994; Ogden et al., chap. 5).

The utility of wading birds as a monitoring tool in this ecosystem has been examined previously in some detail (Kushlan and Frohring 1986; Ogden 1993; Walters et al. 1992; Ogden 1994; Bancroft 1989). Our goal in this paper is to critically evaluate what this monitoring program has (and has not) told us about environmental attributes and functions within, and external to, the Everglades ecosystem. We also hope to identify the factors that have made this program successful, in an attempt to find attributes that might identify similar programs in other ecosystems.

Wading Birds as Biological Indicators

Like many birds chosen to be bioindicators, long-legged wading birds are generally large, highly mobile, top-level consumers in the aquatic food web and have high energetic needs. In addition, many species of wading birds are strongly social and often breed and feed in highly aggregated groups. This, combined with the white or light-colored plumage of many species, makes the finding, counting, and monitoring of these animals in a large ecosystem relatively efficient and accurate. Indeed, it is difficult to imagine any other vertebrate that can be monitored with any accuracy or without extreme cost within the approximately 4,000-square-kilometer landscape of the Everglades (Ogden 1993). Wading birds are also known to forage and breed almost exclusively in wetlands, and, when breeding, to forage within a fairly well-defined range surrounding the colony (Bancroft et al. 1994; Smith 1995). This implies that some aspects of reproduction might be profitably used to reflect local environmental differences within the ecosystem. Breeding-site fidelity is highly variable among species, ranging from storks that return annually to the same colony, to ibises that are extremely nomadic (Frederick and Ogden 1997). Nonetheless, most species seem capable of moving their breeding sites in response to consistently unfavorable conditions.

There is also a large but somewhat diffuse body of evidence that links various aspects of wading bird reproduction to the availability of food. In most large wetland ecosystems in the world, the timing of breeding of wading birds usually coincides with the greatest availability of food. In South Carolina, Bildstein et al. (1990) demonstrated that annual numbers of nesting white ibises (*Eudocimus albus*) were in direct proportion to the availability of crayfishes in freshwater marshes. In the Everglades, Kushlan (1976c) showed that white ibises shifted their timing of nesting to coincide with the time at which available food energy was at a maximum. Similarly, in the Everglades, the Llanos of Venezuela, the Pantanal of Brazil, and the Usamacinta Delta and Yucatan of Mexico, wood storks (*Mycteria americana*) breed only during the dry season, when fishes are trapped in high densities in pools and depressions as a result of rapidly receding waters (Kushlan et al. 1975; Leber 1980; Ogden et al. 1988; Ramo and Busto 1992; Gonzalez 1999; Bouton 1999).

More mechanistic studies have also demonstrated links between the availability of food and reproductive success. For example, Powell (1983) found that food-supplemented great white herons (*Ardea herodias*) in Florida Bay had significantly higher clutch and brood size than did unsupplemented birds. Hafner et al. (1993) found that increases in productivity of little egrets (*Egretta garzetta*) were associated with increased food availability. In the Everglades, interruptions in food supply have been closely correlated with mass nesting abandonments, whether the interruptions are brought about as a result of drought (Bancroft et al. 1990), cold weather (Frederick and Loftus 1993), or flooding (Kushlan et al. 1975; Frederick and Collopy 1989a; Smith and Collopy 1995; Frederick and Ogden 1997). Growth rates of nestling herons are directly related to food intake rates (Salatas 2000), and growth rates in snowy egrets have been correlated with survival rates of fledglings during the first month of life (Erwin et al. 1996).

Within the Everglades, the relative effects of other potential causes of variation in reproductive success have been investigated in some detail. Losses of nest contents to predation has been found to be surprisingly rare in the central Everglades in most years (Frederick and Collopy 1989b), and effects of both researcher disturbance (Frederick and Collopy 1989c) and availability of nesting habitat (Frederick and Spalding 1994) have been found to be negligible. In a large-scale survey of the importance of disease in Everglades wading birds, only one parasitic disease was found to have any effect on reproduction (Spalding and Forrester 1991). Although this disease (eustronglylidosis, caused by parasitic nematodes of the genus *Eustronglyides*) can cause very high mortality of nestlings in some colonies, the disease seems associated only with the relatively uncommon sites of high nutrient deposition within the Everglades (Spalding et al. 1993).

Thus food availability seems to be strongly linked to nesting success in wading birds in general, and variation in food availability explains much of the variation in nesting success specifically within the Everglades. Studies linking choice of nesting site and timing of nesting with availability of food are less well established for the Everglades, but the evidence (above) suggests that location and timing of nesting may also be used as indicators of prey availability and abundance in wetland ecosystems. This information collectively suggests that the cueing and success of nesting are driven largely by the availability of prey, and that variation in reproductive effort and productivity can, within some limits, be interpreted as an indicator of those ecological and physical features that affect the abundance and availability of prey.

The conditions affecting availability of wetland prey to wading birds are probably numerous, but the density of prey animals and depth of water have often been found to be primary components. Wading birds take many types of aquatic prey, using a wide variety of foraging tactics and behaviors (Kushlan 1976b, 1978). Nearly all foraging is in shallowly flooded wetlands, and foraging success is highly dependent upon appropriate conditions. Variation in foraging success may be dependent on a variety of characteristics of the foraging site, including prey density (Renfrow 1993; Surdick 1998; Gawlik 2002), water depth (Powell 1987; Renfrow 1993; Gawlik 2002), water temperature (Frederick and Loftus 1993), dissolved oxygen (Hafner et al. 1993), and vegetative density (Surdick 1998). Of these variables, dissolved oxygen probably plays a minor role, since the wetlands of the Everglades marshes are shallow and poorly stratified. Similarly, water temperature is only an important factor in the Everglades during relatively brief periods of cold. Within the Everglades, Surdick (1998) found that water depth, prey density, and vegetative density were the factors most commonly affecting foraging success and choice of foraging sites of four species of wading birds, and that these factors often interacted.

Historical and Contemporary Wading Bird Monitoring

The Everglades is nearly unique in having an exceptionally long written record of wading bird nesting. Perhaps most importantly, a well-documented time series of nesting events was begun prior to the period of the most extreme hydrological manipulations, allowing some sense of wading bird dynamics in the ecosystem before much of the natural hydrological variability of the pre-drainage Everglades had been altered.

The history and nature of this nesting record has been reviewed in detail by Robertson and Kushlan (1974), Ogden (1978), Kushlan and Frohring (1986), and Ogden (1994), and we present an overview here. Although there are several written accounts of wading bird nesting from the latter part of the nineteenth century (Scott 1887, 1889, 1890), the first extended series of annual estimates of the nesting aggregations came from wardens and biologists of the National Audubon Society, who regularly checked breeding colonies and roosts in the coastal areas of what is now Everglades National Park (ENP) between the early 1930s and mid-1940s.

Between the late 1940s and the mid-1960s, there was, with the exception of occasional surveys of wood storks, a hiatus in systematic monitoring of wading bird reproduction. Regular estimates of estuarine breeding colonies were resumed by ENP staff during the late 1960s and were intensified and expanded with the help of National Audubon Society biologists to irregularly include parts of the current Water Conservation Areas (WCAs) of the Everglades through the mid-1980s (see fig. 12-1).

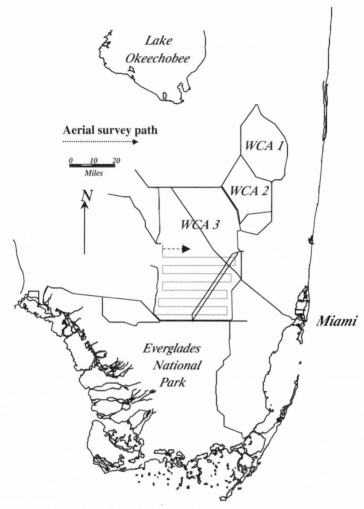

Figure 12-1. Map of the Everglades area showing boundaries of water conservation areas, major urban areas, and landscape features. The approximate flight path of a portion of the systematic aerial surveys for wading bird nesting colonies is shown as a dotted line. Transects are oriented east-west, and spaced 2.6 kilometers apart.

During the same period, the National Audubon Society Research Department (NAS) began annual surveys of wood stork colonies in peninsular Florida and in so doing documented many previously unknown colonies in the central Everglades. These surveys by ENP and NAS were mainly aimed at estimating numbers of birds in large, previously active colonies, and search patterns were not systematic.

Systematic Surveys of Breeding Colonies

Systematic aerial surveys were initiated in the WCAs of the Everglades in 1986 and were adopted in ENP shortly thereafter. These surveys are flown in east-west transects spaced 2.6 kilometers apart (see fig. 12-1). This spacing was determined empirically by flying naive observers near known colonies at decreasing horizontal distances until the colonies were routinely recognized by observers. The surveys are flown once monthly between February and June, and cover the entire mainland Everglades from Florida Bay to northern end of WCA 1. In the WCAs, systematic ground searches are also performed between April and late May of each year by airboat, and each tree island is individually checked for nesting activity. Although the aerial surveys are an efficient means to locate and count large colonies (those with more than one hundred pairs) dominated by white-plumaged birds, they are inefficient for discovering small colonies, particularly those of dark-colored species. Using intensive ground surveys by airboat as a standard, aerial surveys in the Everglades have been shown to miss over 79 percent of all colonies (most of which are very small), and over 30 percent of nesting wading birds (Frederick et al. 1996a). The ground survey program has thus led to the first reliable estimates of dark-colored wading birds in the ecosystem, as well as information on the dispersion of small colonies and solitary nests.

Systematic Surveys of Distribution

Beginning in 1986, systematic aerial estimates of numbers of all wading birds on the marsh surface were initiated. This Systematic Reconnaissance Flight (SRF) program was designed to estimate the geographic distribution and total numbers of all wading birds (breeding, non-breeding, adult, juvenile) within the Everglades ecosystem. Quite distinct from the surveys of breeding colonies, the SRFs are flown at low altitude, and numbers of birds are estimated using strip-transect methodology (Norton-Griffiths 1987; see fig. 12-2). Estimates of total numbers and densities are then derived from the strip counts, which constitute approximately 16 percent of the total area. These surveys have been flown monthly between January and June of 1986 through the present in ENP and the WCAs, and sporadically in Big Cypress National Preserve.

Reproductive Success Studies

Measures of reproductive success have been collected in Everglades colonies at various times and in various manners. Up until the early 1960s, reproductive success was scored only by whether an apparently

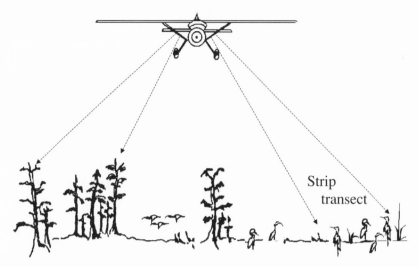

Figure 12-2. Schematic diagram of Systematic Reconnaissance Flight methodology. Flights are conducted at 61 meters in altitude, and observers count birds in strip transects on either side of the aircraft. Estimates of total numbers of birds on the marsh surface are then derived from these strip-counts by taking into account the percentage of the marsh surveyed.

large proportion of the nesting pairs in a colony succeeded or failed. During the late 1960s and 1970s, wood stork productivity was measured as the numbers of young raised per nest start (Ogden 1994). Between 1986 and 1991, repeated checks of marked nests of various species were used to estimate reproductive success and productivity in a large proportion (10–20 percent) of the nests in the southern and central Everglades (Bancroft et al. 1990; Frederick 1995); measurement of these parameters continued in the central Everglades through 1994 and sporadically thereafter (Frederick 1995).

The more recent studies of reproductive success have been linked with studies of foraging movements by reproductive adults (Ogden 1977; Bancroft et al. 1994; Frederick 1995), food habits and energetic requirements of young birds (Bancroft et al. 1994; Williams 1997; Frederick et al. 1999; Salatas 2000), and foraging behavior and foraging habitat (Surdick 1998; Gawlik 2002) to produce a picture of foraging distribution, reproductive success, and reproductive energetics of breeding birds.

Studies of Contaminants

Contaminant studies of wading birds have been sporadically undertaken in the Everglades, beginning with Ogden et al.'s (1974) reports of heavy metals and organochlorines in eggs, and followed by Spalding and Forrester's (1991) survey of diseases and contaminants. During the 1990s, an intensive investigation of the sources, accumulation points, and effects of methylmercury in the ecosystem's biota led to ongoing annual collections of feather samples from young great egrets and investigations of effects of this contamination (Spalding and Forrester 1991; Sunlof et al. 1994; Beyer et al. 1997; Williams 1997; Bouton et al. 1999; Frederick et al. 1999; Sepulveda et al. 1999; Spalding et al. 2000a,b; Frederick et al. 2000).

What Has Monitoring Revealed about the Ecosystem?

Ogden (1994) summarized a comparison of recent (1974–1989) wading bird nesting with that of the 1930s and 1940s. Between these two periods, the numbers of nesting birds declined by over 90 percent (fig. 12-3). The comparison between the two periods was biased toward finding more birds in the later period, since the more recent survey methodology was systematic and more efficient at finding birds than were the mostly ground-based estimates of the 1930s. The more recent surveys also covered vastly more area—the 1930s estimates were only of coastal colonies and did not penetrate the interior marshes. Thus the 90 percent reduction between the two periods seems conservative. During the period 1930–1946, Ogden (1994) suggested that sixty-nine thousand to eighty-nine thousand birds nested in many years, with peaks of two hundred thousand. The current nesting numbers rarely exceed thirty thousand birds, with peaks of no more than sixty thousand (Frederick and Collopy 1989a; Frederick 1995; see fig. 12-3).

The Importance of Hydrological Variability

One of the most profound puzzles of the Everglades has been that a vast, nutrient-poor wetland system should be capable of supporting a large, concentrated biomass of wildlife. The wading bird monitoring programs have provided evidence of several ecological relationships that explain how the large numbers of wading birds could be supported by a system with such low energy density. The most obvious way is through annual fluctuations in water level, which serve to concentrate aquatic prey animals to the point that they are energetically profitable to consume. This process may create a broad "drying front," which progresses across the landscape as surface water dries during the spring nesting season, allowing wading birds (and other vertebrates) to exploit a moving wave of protein. This annual process has been well documented through aerial

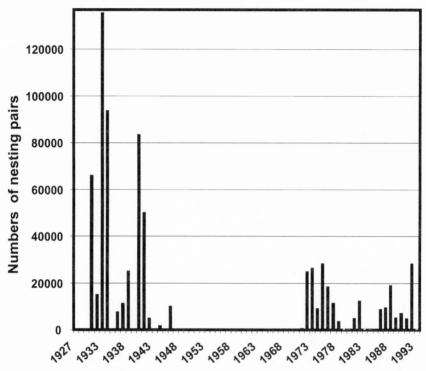

Figure 12-3. Numbers of nesting pairs of all wading bird species surveyed, 1930–1998. Although surveys of wood storks were regularly conducted during the 1950s and 1960s, there were no regular surveys of all wading birds during this time.

surveys designed to monitor numbers of birds (Kushlan 1977; Hoffman et al. 1994).

Other cycles of food availability may be less regular, with return intervals on the order of several to many years. Ogden (1994) noted that during the period of the 1930s, there was far greater interannual variability in colony presence and size than there has been during the most recent thirty years. For instance, during the 1930s and early 1940s, alternate years often showed severe drought with little or no nesting followed by one or more years with extremely large nestings (to over two hundred thousand birds). Ogden suggested that the large nestings of the 1930s and early 1940s were in part dependent upon the alternation of flood and drought. This suggested that there was something about multiyear patterns of hydrological variability that strongly affected nesting.

One prediction from this general observation is that abnormally large nesting events may be more likely following severe droughts than at other times (Frederick and Ogden 2001). Using the entire nesting record, we statistically identified eight abnormally large nesting events; all but one of these occurred within two years after severe droughts. Similarly, all but two of the severe droughts during the same period were followed by abnormally large nesting events. The biotic mechanisms behind this statistically significant association are unknown, but at least three processes have been suggested: (1) prey are temporarily superabundant following droughts as a result of reliberation of nutrients, (2) prey are superabundant following droughts because predatory fish have been killed off through desiccation (Kushlan 1976a; Walters et al. 1992), and (3) prey are more available following droughts due to more open vegetation (Surdick 1998). These hypotheses have predictions that are specific enough to test with further monitoring of fishes and aquatic macroinvertebrates.

The idea that both prey animals and wading birds depend strongly on nonannual natural hydrological fluctuations for their pulses of productivity derives support from studies of other Everglades biota as well as from more general examples of riverine systems (Junk et al. 1989). For example, long-term stable water conditions in the Water Conservation Areas have been shown to be detrimental to emergent vegetation and to both nesting and hunting success of snail kites (Bennetts and Kitchens 1997), and rivers with flood pulses are known to have more productive fisheries than those that do not (Junk et al. 1989).

Thus the evidence from wading bird monitoring in the Everglades has led directly to recommendations that natural hydrological fluctuation become a priority in water management and has served as part of a growing body of evidence that hydrological fluctuation is necessary to the normal functioning of many types of wetland systems. These examples also highlight the need to understand the mechanisms involved in creating pulses of productivity. This gives further justification for monitoring populations of prey animals at a large-enough scale to enable linkages with the wading bird studies.

Estuarine Productivity

The record of early twentieth-century colony locations also has demonstrated that there has been a major shift in the geographic location of nesting (Ogden 1978, 1994). During the latter part of the nineteenth century, and during the 1930s, all of the major colonies described for the Everglades were in the coastal zone or along the mangrove/freshwater marsh interface. In contrast, by the period 1986–1999, an annual average of 83 percent of the wading bird nests were located in freshwater areas of the Everglades (fig. 12-4). Although turnover in use of colony locations

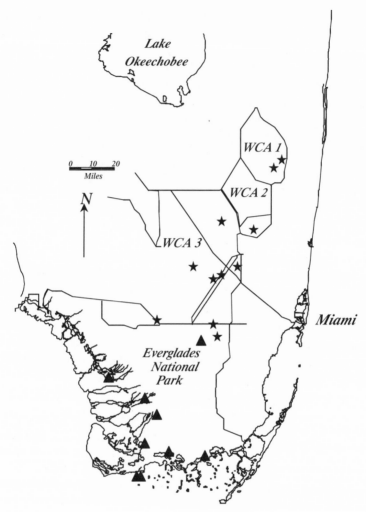

Figure 12-4. Map of the Everglades showing the locations of all colonies more than five hundred pairs. Colonies during the period 1930–1960 are shown as solid triangles and those from 1970–1998 are shown as solid stars. Although the inland colonies may simply have gone undetected during the early surveys, it seems clear that the coastal colonies have disappeared during the recent period.

is common in wading birds (Bancroft et al. 1988; Frederick 1995; Smith and Collopy 1995; Frederick and Ogden 2001), the loss of the entire region of formerly productive coastal colonies was so complete and has been maintained for such a long time (more than thirty years) that the abandonment of the area seemed indicative of a profound change in coastal ecosystem function.

The loss of coastal colonies implied that the coastal foraging habitat or prey base had become degraded in some fashion, and studies initiated in the 1980s and 1990s on other aspects of the coastal ecosystem have borne this hypothesis out (McIvor et al. 1994). The most profound changes in the coastal zone have probably been driven by an extreme reduction in surface freshwater flows to the region (Smith et al. 1989; Fennema et al. 1994) and have included increases in the salinity of the estuary, decreases in shrimp production (Browder 1985), decreases in sport-fish catches, and decreases in the standing stocks and densities of small "forage" fishes in the coastal marshes and mangrove ecotone (Lorenz 1997). What is most significant about the use of wading bird monitoring information in this story is that the decreases in coastal wading bird colonies were recognized as much as twenty years before evidence from other sources suggested that the productivity of the coastal ecosystem had collapsed (Ogden 1978).

The estuarine zone may also have been productive because it offered wading birds a variety of habitat and wetland types in which to forage that would not be available in the more homogeneous freshwater marshes. The importance of heterogeneous foraging habitats for buffering birds against unpredictable water-level fluctuations has been observed in modern-day studies of coastal-nesting wading birds in the Everglades (Bancroft et al. 1990, 1994) and points to the need to understand fluctuations in prey animal populations in those habitats.

Defining Healthy Hydrological Patterns

Wading bird monitoring studies have also demonstrated important differences in the conditions that have led to productive nesting during historical versus modern periods, and ultimately these observations have led to fundamental changes in surface water management. Ogden (1994) noted that during the 1930s and 1940s large nesting events occurred in years with either wet or dry conditions but that during the modern period, productive nesting occurred only during the drier years. This pattern of nesting in dryer years has been linked to a direct relationship between annual nesting effort and rate of surface water recession for both wood storks (Kushlan et al. 1975) and white ibises (Frederick and Collopy 1989b). An analysis of nest failures in the Everglades during the recent period has similarly shown that abandonment is common whenever a drying water regime is reversed by high rainfall (Frederick and Collopy 1989b).

This difference between historical and modern nesting responses was something of a puzzle. However, monitoring of marsh hydrology and aquatic biota eventually showed that overdrainage of the freshwater marsh resulted in marked decreases in the abundance and standing stock of small fishes and invertebrates (Loftus and Eklund 1994; Loftus et al.

1992). Thus drainage practices have led to short hydroperiod in the freshwater marsh, resulting in depauperate prey animal communities. It then made sense that the wading birds would show progressively greater dependence on drying events through time, because they were using them as a mechanism to concentrate the few prey present. In this case, the combination of changing bird-nesting responses and fish population dynamics were required to fully realize the ramifications of long-term marsh drainage (Walters et al. 1992; Fennema et al. 1994).

It is important to realize that in the absence of the long-term bird monitoring effort, our impression of suitable wading bird foraging conditions would probably be narrowly (and incorrectly) focused on the "beneficial" effects of rapidly falling water for stimulating nesting and increasing nesting success. In fact, this latter impression has been widely held, and has led to management policies that routinely dried much of the marsh surface during the dry season. Thus a negative feedback mechanism probably existed between management that favored annual drying events and consequent increasing dependence of wading birds on drying events.

Tracking Changes in Aquatic Contamination Levels

Wading birds are empirically good accumulators of contaminants for a number of reasons. Wading birds are known to feed at or close to the top of the aquatic food web and show high bioaccumulation potential (Custer and Osborn 1977; Jurczyk 1993; Erwin and Custer 2000). Since the majority of the food gathered by adult wading birds (e.g., white ibises and great egrets) is know to come from distances of 10 kilometers or less from the colony (summarized by Bancroft et al. 1994; Smith 1995), the tissues of young wading birds are known to be composed largely of resources from within this area. The use of young birds therefore largely avoids contamination signals that might come from other parts of the range of these migratory birds. The sampling unit here is the colony site, and although this grain for sampling (20-kilometer-diameter circles) may seem large, it is probably appropriate for monitoring contamination in the aquatic food web of the Everglades ecosystem (rough dimensions 60 by 180 kilometers).

To date, only mercury has been shown to consistently occur in high concentrations in wading birds in the Everglades, though there have been no systematic investigations to date of PCBs or dioxins. The Everglades aquatic food web is highly contaminated with mercury (Frederick 2000), and levels in great egret chicks are in some years higher than has so far been demonstrated for young of any fish-eating bird (Sunlof et al. 1994; Sepulveda et al. 1999). These levels have been shown to be high enough to result in reduced health and altered hunting behavior of juvenile birds

(Spalding et al. 1994; Bouton et al. 1999; Spalding et al. 2000a,b), and there may also be effects on reproduction. Concentrations of mercury in feathers from nestlings have differed markedly and consistently among colonies, and these differences track geographic variation in mercury sampled from mosquitofish (*Gambusia holbrooki*, Stober et al. 1996). In addition, laboratory studies have demonstrated a clear and predictable relationship between mercury consumed and mercury concentration in feathers of young great egrets (Spalding et al. 2000a). Thus, there is empirical evidence from lab and field that feather tissue concentrations are a good indicator of mercury concentration in prey.

In this monitoring program, geographic variation in contamination among colonies has been largely swamped by differences among years, and we have measured a significant decline in mercury concentrations between 1994 and 2000 (73 percent reduction, Frederick et al. 2002). This monitoring system has alerted those studying mercury mass transport and dynamics to look for processes that may have led to this significant decrease in contamination.

Wading birds have thus served an important sentinel role for contaminants in the Everglades (as much by what they have not accumulated as by what they have) and have demonstrated both geographic and temporal differences in concentrations of mercury. Further understanding of the dynamics of mercury at the ecosystem scale is likely to rely in large part on monitoring studies of top-level consumers like wading birds. The success of the mercury monitoring efforts suggests that the birds should also be monitored for other contaminants, since south Florida currently applies more pesticides per hectare, and uses a wider array of pesticides and herbicides, than any other area of the United States.

Linkages between Wetland Ecosystems

Relatively few animals move between the Everglades and other ecosystems. Those that do include the Florida panther (*Felis concolor coryi*), West Indian manatee (*Trichechus manatus*), snail kite (*Rostrhamus sociabilis*), migratory birds, and numerous euryhaline fishes and penaeid shrimps (Browder 1985). Of these, wading birds are probably the longest-monitored both within and outside of the Everglades and in many cases have been the subject of more ecological research. Thus, wading birds are one of the few well-studied Everglades animal groups that have the capability of leaving the ecosystem or of being attracted there from other locations in response to favorable ecological conditions. In this sense, wading birds can function as a true bellwether of environmental conditions among ecosystems in a way that few other species can.

A history of statewide surveys, banding records, and other sources of information indicates that the wading birds that utilize the Everglades are panmictic with, and demographically linked to, wading birds in other wetland areas of the southeast and the eastern Caribbean (Stangel et al. 1990, 1991; Frederick et al. 1996). The Everglades "population" then, is actually not distinct in its genetics or social organization and instead belongs to a loose grouping of animals that occur in a space perhaps as large as the eastern and southeastern United States and the eastern Caribbean. It is important to remember, then, that the dynamics of birds in the Everglades ecosystem may be strongly influenced by conditions in other regions, and, conversely, that management in the Everglades may well influence patterns of distribution and abundance elsewhere.

For example, the Everglades probably served as an important source for the restoration of wading birds in the southeastern United States during the period immediately following the decline of the plume trade (Ogden 1978). The very large nesting aggregations of wading birds documented during the period 1930–1946 were not reported elsewhere in the United States, and several of the species (wood storks, white ibises) were not known to breed outside of Florida at the time. Similarly, very large colonies of great egrets, white ibises, and snowy egrets in Florida during the 1930s immediately preceded the rapid expansion of nesting by these species in the Atlantic coastal plain and Mississippi Valley.

During the last two decades, however, interregional monitoring indicates that this situation may have reversed itself, and the Everglades may now have become a demographic sink rather than source for many species. Wood stork reproductive parameters in the Everglades during the recent period are exceptionally low for the species (Ogden 1994), and it is very unlikely that Everglades wood storks are replacing themselves. Between 1976 and the present, the percentage of the U.S. wood stork population nesting in south Florida dropped from 70 to 13 percent (Coulter et al. 1999). Through a comparison of breeding and Systematic Reconnaissance Flight (SRF) surveys, we have found that during the 1990s an average of 70 percent of adult great egrets, white ibises, and wood storks annually remained as nonbreeders in the Everglades during the breeding season. Although the breeding population appears to be stable or slightly increasing, it is very likely that "sitting out" reproduction in this manner will lead to a sharp decline in recruitment. This suggests that birds breeding in the Everglades may not be reproducing fast enough to cause local recruitment, and that stable populations are maintained instead by new birds from elsewhere. Thus, even an apparently stable population may not be a healthy or self-sustaining one (Temple and Wiens 1989; Sadoul 1997). This example illustrates one of the ways that local monitoring, even at the ecosystem level, can be extremely misleading if not considered in the context of dynamics in other ecosystems.

The Everglades is in a geographically key position for migrating and wintering wading birds (Byrd 1978; Root 1988), and the SRF surveys have documented especially large numbers of birds using the Everglades during the winter pre-breeding season. These studies have shown that in some years the Everglades may host a substantial proportion of regional populations of some species (Bancroft et al. 1992). The high usage of the area suggests that large numbers of birds are regularly able to assess conditions in the Everglades during the prebreeding period, and the large interannual variance in breeding effort suggests that these migratory birds may include the Everglades as a potential site when deciding where to nest. This scenario bolsters the notion that the numbers of birds nesting in the Everglades is indicative of conditions there, and that wading bird reproduction is a useful bioindicator of restoration.

Events in other ecosystems may also affect wading bird use of the Everglades. For example, white ibises are known to make large-scale shifts in breeding location depending in part on comparing breeding conditions and food resources in past and prospective breeding sites (Ogden 1978; Frederick et al. 1996b; Frederick and Ogden 1997). During the period 1980–1995, the numbers of white ibises in Louisiana increased dramatically, probably in response to a large increase in impoundment acreage devoted to commercial production of crayfishes (*Procambarus* spp., Fleury and Sherry 1992). During the same period, a 50 percent reduction was documented in the total numbers of ibises nesting in Florida (Runde 1991). This suggests strongly that the increase in aquaculture in Louisiana was directly related to decreases in ibises nesting in Florida and in the Everglades.

It therefore seems very important to extend the monitoring of wading birds to sites outside the boundaries of the Everglades ecosystem. This is particularly critical for interpreting changes in wading bird numbers and, eventually, for evaluating the potential for and success of restoration efforts within the Everglades. Monitoring of wading birds throughout their southeastern United States and Caribbean range is far from complete, but useful information can (as above) be gleaned through cooperative efforts with other states and countries.

Assessment of the Everglades Monitoring Programs

We believe that the information presented so far demonstrates that careful, long-term monitoring of an avian species assemblage can be extremely useful in understanding both the past and present dynamics of a large aquatic ecosystem like the Everglades. Of the Everglades ecosystem features that Davis and Ogden (1994) felt were key to the design of restoration plans, our understanding of nearly all of them has been revealed in part by our understanding of wading bird dynamics. Monitoring of wading birds has demonstrated or helped to demonstrate the importance of large-spatial-extent, habitat mosaics, refugia, sheet flow into the estuaries, seasonal energy pulses, seasonal depth patterns, and long uninterrupted hydroperiods to the functioning of the Everglades. By this standard, we feel that the bird monitoring programs have been an unqualified success. However, there are a number of areas in which these efforts could be improved, and a critical evaluation here may serve to aid in the design of monitoring programs in other ecosystems.

Perhaps the single most obvious weakness of the Everglades avian monitoring program is that there is a poor understanding of processes affecting birds outside the ecosystem. It seems clear that most of the wading bird species in the Everglades wander widely during the nonbreeding season and may frequently be attracted to nest in other locations. Thus the ability to interpret reproductive or population changes in the group of birds using the Everglades may be strongly dependent upon the changing attributes of other wetlands throughout the Caribbean and the southeastern United States. Our ability to distinguish between population changes resulting from factors endogenous or exogenous to the Everglades is poorly developed in large part because comparable monitoring systems elsewhere either do not exist or are incomplete. For example, statewide monitoring of wading bird colonies is occasionally accomplished in Florida, Georgia, South Carolina, Texas, and Louisiana but is conducted infrequently (often five to fifteen years between surveys) and is never performed in all states in the same year. The opportunity to compare responses at these sites with annual fluctuations in the Everglades nesting numbers is therefore extremely rare.

Similarly, the movements of birds to the Caribbean, and in particular to Cuba, are very poorly documented, yet changes in conditions in the Caribbean could strongly influence our interpretation of the dynamics of birds in the Everglades. In Europe, Hafner et al. (1994) found that annual fluctuations in little egrets breeding in the Camargue were partly explained by variation in rainfall in the African wintering grounds for this species. It is probably unrealistic to ever expect the development of simultaneous annual surveys for wading birds throughout the potential range of the Everglades birds, but regular, consistent monitoring in many key areas could greatly expand the ability to interpret dynamics in any one of the wetland locations. The Everglades also lacks any kind of banding or marking program, which could help identify regular wintering habitats and alternative breeding sites.

The wading bird monitoring programs of the Everglades have been used intensively and extensively as a restoration and research tool in large part because the historical record enabled the establishment of target restoration goals (numbers of wading birds nesting, locations of colonies, and attributes of the ecosystem that nesting is known to depend upon, see Ogden et al. 1997). However, it is not clear that even the "benchmark" period (1931–1946) was representative of a natural, pre-drainage ecosystem. Considerable drainage of the ecosystem had occurred prior to the 1930s (Smith et al. 1989), and some years during the 1930s may have been particularly dry (Frohring et al. 1988). Thus, there is the danger that the use of numbers of wading birds from that period as a restoration target could be misleading. Nonetheless, the period of the 1930s was probably broadly representative of some important characteristics of the pre-drainage Everglades, such as a consistently nesting population, a concentration of nesting in the coastal zone, and high interannual variability in nesting. To avoid the overuse of the 1930s benchmark, restoration targets for wading birds have deemphasized target breeding numbers and focused instead on attributes of productive nesting (timing, location, levels of productivity) and on the ecological functions that are thought to have supported coastal nesting (Ogden et al. 1997).

One of the most common criticisms of waterbird monitoring in general is that censuses are inaccurate, perhaps to the point that change in nesting or total numbers are overshadowed by error resulting from inaccuracies in finding colonies and estimating numbers of birds (e.g., Drury 1980; Frohring et al. 1988; Rodgers et al. 1995). To some extent, the Everglades survey record is buffered from this criticism simply because the change in breeding numbers between early and recent time periods is probably too large (well over one hundred thousand birds) to be the result of inaccuracies in estimation. However, there are important problems that remain with the estimation of large numbers of birds in a large landscape. Within the various Everglades management units, nesting birds are variously estimated by ground surveys, aerial surveys, or both. Given the inaccuracies of using either one of these techniques alone (Frederick et al. 1996a), there is a very real need to standardize estimation procedures throughout the ecosystem.

An evaluation of the utility of the SRF survey system is somewhat premature since much of the data from those systematic surveys are still being analyzed at the time of writing. However, it is true that the majority of insights gained from monitoring wading birds have come directly from

monitoring nesting in various ways, and comparatively little has been derived from the studies of distributional patterns. At the time that the SRF
flights were initiated in the mid-1980s, there was great hope that the resulting distribution maps of birds in the Everglades landscape could be used
both to determine the ecological attributes necessary for wading bird foraging and nesting, and to assay the success of hydrological restoration efforts.
In our view, there has been, and continues to be, only mild progress on the
first goal. Although the SRF survey information has been used to identify
the fact that birds follow patterns of shallow water and avoid areas of dense
vegetation, the grain of the information (2-x-2-kilometer cells) is too coarse
to be of much value in finer distinctions of habitat quality. This problem results from an obvious trade-off between geographic extent and detail of surveys, and is one that is likely to be repeated in other extensive ecosystems.
This example illustrates that researchers should be very careful about
matching the grain of surveys with the question at hand (Begg et al. 1997).

The SRF surveys may have successfully accomplished the second goal—
evaluating success of restoration through changes in distributions. One of
the most obvious signals of restoration success would be a shift in distribution of birds from the WCAs areas to the coastal zone (Ogden et al. 1997).
During the period of SRF surveys (1985–present), no large and consistent
shifts of wading bird density distributions have been documented within
the Everglades. This reveals that the goal of shifting birds back to the
estuarine zone has not been realized, which is not surprising since no large-
scale restoration actions have been taken to date. Nonetheless, it is also
quite possible that the detection of such large-scale shifts could be accomplished with a far less intensive and expensive program.

The real value of the SRF surveys may lie in their ability to estimate
total numbers of birds using the marsh habitat. While the estimates do
contain some important biases as a result of the characteristically clumped
distribution of birds, there are several robust conclusions that have arisen
from this series of population estimates. First, it seems inescapable that
the Everglades functions in some years as an important concentration
point for wintering wading birds. Second, in conjunction with the breeding bird surveys the SRF counts have resulted in the tentative conclusion
that a large proportion of the wading birds remaining in the Everglades
throughout the breeding season do not actually attempt to breed. One obvious extension of this finding is that the Everglades may function as a
reproductive sink.

One disappointment of the Everglades monitoring program is that
there has been little success in matching spatial information about birds
with spatial attributes of their prey animals. Such linkages could elucidate
the relative importance of changes in prey density to the wading bird de-

clines. The basis of the problem is that it is inherently more costly and logistically difficult to sample fish and invertebrates over a large landscape than it has been to sample birds. For example, during the unanticipated large nesting event of 1992, fish and macroinvertebrate samplings were available from only three locations in the Everglades, and none of them were in the areas where birds nested or fed in high densities. Thus it was impossible to tell whether the large nesting event was related to elevated prey densities. The Everglades fish monitoring program has since been expanded to include over twenty-five sites spread throughout the Water Conservation Areas and Everglades National Park, and it is hoped that this larger network of observation sites will yield a better match with the wading bird distributional studies.

Conclusion

The factors that have led to the overall success of the wading bird monitoring programs may be general, or some cases unique, to the Everglades. To assess the general applicability of the Everglades avian monitoring program to other ecosystems, it seems important to identify the attributes that were essential to the success of the program.

Existence of Historical Baseline Data

The existence of a run of reasonably accurate historical records during a period prior to intensive water management has been the single most valuable attribute of the wading bird program. It is striking, for example, that without the information from the 1930–1946 period there would have been little hint from the birds of the pervasive changes that have taken place in the estuarine zone of the Everglades. It is also sobering to realize that on the basis of the relationships between hydrology and nesting from the recent past (1960–1996), we would have assumed that rapid drying of the marsh surface every year was the key to stimulating nesting. With the inference of the historical period, it became clear that this relationship is probably an artifact of a prey community degraded by years of overdrainage. This example suggests that ecological relationships derived from a current, disturbed ecosystem may be extremely misleading for establishing restoration targets.

The historical data also have compelled us to realize that the natural hydrological variability of the Everglades was in fact key to the ability of the ecosystem to produce the enormous pulses of secondary productivity that fueled the fish and wildlife populations for which the ecosystem was once famous. The historical information has also been central to the development of restoration goals, both for target patterns of nesting by wading birds and for ecosystem attributes (timing and minimum flows of water, restoration of flows to the estuary, reconnection of wetland compartments).

The importance of the historical data in the Everglades example speaks very strongly to the need to begin monitoring studies in other ecoregions, particularly those systems that are so far relatively unimpacted by anthropogenic actions. Although the Everglades has been truly and perhaps uniquely blessed with the quality and quantity of historical data on local wading bird population dynamics, similarly useful records are available for many ecosystems, especially for commercially harvested species (e.g., blue crabs and oysters in Chesapeake Bay, salmon in the Columbia River drainage). Even short runs of historical data may be of value in identifying ecosystem attributes and conditions.

Long, Continuous Time Series

Although the historical information has been a critical tool, the sheer length and relative continuity of the Everglades wading bird data set have been of key importance in discerning recurring and changing patterns. The need for long time-series of data for developing trend analyses seems obvious. What is less obvious is the emerging importance of long-term monitoring for detecting rare, trend-setting ecological events in ecosystems. For instance, a minimum of twenty-five years of information has been necessary to detect the powerful effect of severe winters on wetland bird populations in the French Camargue (Johnson 1997; Hafner and Fasola 1997). In the Everglades, the apparent relationship between severe droughts and ensuing bursts of secondary productivity and wading bird nesting relies on evidence of drought events spread out over seven decades of monitoring. Similarly, the differences in patterns shown by nesting birds during the historical and recent periods in the Everglades have demanded at least a decade of continuous data during each era to distinguish the effect of recurrent periods of drought and flood. It is also true that the sheer length of the record has been a strong argument for continuing the program in the face of many other competing funding priorities.

Strong Public Appeal

The initiation and continued public support for monitoring wading birds is directly traceable to their popular appeal. Indeed, the earliest inquiries into the status of breeding wading birds during the late 1800s were the result of public interest in the fate of these animals during the plume-hunting era. The protection of the birds through a network of wardens and the resulting record of nesting during the 1930s and 1940s was supported directly through funds generated by the public's interest in these programs. This interest continues to the present, and the wading bird monitoring programs continue to enjoy wide public interest as evidenced by the numerous newspaper articles and radio spots that are aired every year on this subject. By comparison, it has been much harder to garner support for monitoring of fishes, invertebrates, and many other aquatic creatures in the Everglades.

Besides being exceptionally beautiful, visible, and evocative animals, these birds are also easy for the public to envision as an indicator of the less-visible and less aesthetically appealing aspects of a wetland ecosystem. In the public eye, they have become symbolic of the health of the Everglades; large numbers of wading birds equates with a healthy ecosystem. Thus the ability to stimulate public support for long-term monitoring may rely in large part on picking a species or group that is inherently popular with humans.

Integration over an Appropriate Spatial Scale

Wading birds are sufficiently mobile to be capable of including the entire Everglades ecosystem when making foraging and nesting decisions. They are also able to collect food and energy over large areas and are tertiary consumers, which is of interest for monitoring both contaminants and prey animal populations. At the same time, the restricted range of foraging during breeding ensures that samples from eggs and young are representative of a specific region of the ecosystem.

Wading Birds Are Biologically Well Understood

Much is known about relationships between wading birds and their environment from studies within and outside the Everglades. This includes a solid linkage between food availability and reproduction, the relative importance of predation to reproductive success, the effects of contaminants, and the responses of birds to various kinds of natural and anthropogenic perturbations. This information is essential to the interpretation of changes in avian population sizes and breeding parameters and their eventual use as indicators of environmental change.

Monitoring Represents a Wide Range of Ecological Needs

Differences in habitat preferences, food habits, and behavior among the dozen or so species of wading birds that are commonly monitored in south

Florida can result in differences among species in responses to environmental conditions (Gawlik 2002). These differences in responses can help to reveal the specific nature of the ecological effects of water management impacts. The use of any single species could not accomplish this goal.

A Range of Spatial and Temporal Scales Is Represented

Mobility and poor site fidelity are characteristic of many species of wading birds that make them good indicators of all levels of environmental conditions. At small time and spatial scales, foraging wading birds may relocate within hours following changes in water depths caused by management practices or sudden rainfall. Mid-scale responses include changes between consecutive years in the location and timing of nesting activity, again in response to regional differences in hydropatterns caused by management practices and annual differences in rainfall. Long-term changes over years or decades are best illustrated by the ecosystem-wide population trends reported for most Everglades wading birds, but may also track changes in contaminants, such as the Everglades mercury example.

Wading Birds Are Easily Monitored

The white species of wading birds have been shown to be easily located, identified, and counted or estimated from low-flying aircraft and as a group are probably far cheaper to monitor than any vertebrate.

LITERATURE CITED

Bancroft, G. T. 1989. Status and conservation of wading birds in the Everglades. *American Birds* 43:1258–1265.

Bancroft, G. T., W. Hoffman, and R. Sawicki. 1992. The importance of the water conservation areas in the Everglades to the endangered wood stork (*Mycteria americana*). *Conservation Biology* 6:392–398.

Bancroft, G. T., S. D. Jewell, and A. M. Strong. 1990. *Foraging and nesting ecology of herons in the lower Everglades relative to water conditions.* South Florida Water Management District, West Palm Beach, Fla.

Bancroft, G. T., J. C. Ogden, and B. W. Patty. 1988. Colony formation and turnover relative to rainfall in the Corkscrew Swamp area of Florida during 1982 through 1985. *Wilson Bulletin* 100:50–59.

Bancroft, G. T., A. M. Strong, R. J. Sawicki, W. Hoffman, and S. D. Jewell. 1994. Relationships among wading bird foraging patterns, colony locations, and hydrology in the Everglades. Pp. 615–687 in *Everglades: The ecosystem and its restoration*, edited by S. M. Davis and J. C. Ogden, Delray Beach, Fla.: St. Lucie Press.

Begg, G. S., J. B. Reid, M. L. Tasker, and A. Webb. 1997. Assessing the vulnerability of seabirds to oil pollution: Sensitivity to spatial scale. *Colonial Waterbirds* 20:339–352.

Bennetts, R. E., and W. M. Kitchens. 1997. Population dynamics and conservation of snail kites in Florida: The importance of spatial and temporal scale. *Colonial Waterbirds* 20:324–329.

Beyer, W. N., M. Spalding, and D. Morrison. 1997. Mercury concentrations in feathers of wading birds from Florida. *Ambio* 26:97–100.

Bildstein, K. L., J. Johnston, W. Post, and P. C. Frederick. 1990. Freshwater wetlands, rainfall, and the breeding ecology of white ibises in coastal South Carolina. *Wilson Bulletin* 102:84–98.

Bouton, S. N. 1999. Ecotourism in wading bird colonies in the Brazilian Pantanal: Biological and socioeconomic implications. Master's thesis, University of Florida, Gainesville.

Bouton, S. N., P. C. Frederick, M. G. Spalding, and H. Lynch. 1999. The effects of chronic, low concentrations of dietary methylmercury on appetite and hunting behavior in juvenile great egrets (*Ardea albus*). *Environmental Toxicology and Chemistry* 18:1934–1939.

Browder, J. 1985. Relationship between pink shrimp production on the Tortugas and water flow patterns in the Florida Everglades. *Bulletin of Marine Science* 37:839–856.

Byrd, M. A. 1978. Dispersal and movements of six North American ciconiiforms. Pp. 161–186 in *Wading birds*, edited by A. Sprunt IV, J. C. Ogden, and S. Winckler, National Audubon Society Research Report No. 7. National Audubon Society, New York.

Coulter, M. C., J. A. Rodgers, J. C. Ogden, and F. C. Depkin. 1999. Wood stork (*Mycteria americana*). In *The birds of North America*, edited by A. Poole and F. Gill. No. 409, the Academy of Natural Sciences and the American Ornithologists Union, Philadelphia and New York. 28 pp.

Custer, T. W., and R. G. Osborn. 1977. *Wading birds as biological indicators: 1975 colony survey*. U.S. Fish and Wildlife Service Special Science Report No. 206, U.S. Fish and Wildlife Service, Washington, D.C.

Davis, S., and J. C. Ogden, 1994. Everglades: The ecosystem and its restoration. Delray Beach, Fla.: St. Lucie Press.

Drury, W. H. 1980. Coastal surveys—northeast and northwest. *Transactions of the Linnean Society of New York* 9:57–75.

Erwin, R. M., and T. W. Custer. 2000. Herons as indicators. Pp. 311–330 in *Heron conservation*, edited by J. Kushlan and H. Hafner, London: Academic Press.

Erwin, R. M., J. G. Haig, D. B. Stotts, and J. S. Hatfield. 1996. Reproductive success, growth and survival of black-crowned night-heron (*Nycticorax nycticorax*) and snowy egret (*Egretta thula*) chicks in coastal Virginia. *Auk* 113:119–130.

Fennema, R. J., C. J. Neidrauer, R. A. Johnson, T. K. MacVicar, and W. A. Perkins. 1994. A computer model to simulate natural Everglades hydrology. Pp. 249–289 in *Everglades: The ecosystem and its restoration*, edited by S. M. Davis and J. C. Ogden, Delray Beach, Fla.: St. Lucie Press.

Fleury, B. E., and T. W. Sherry. 1994. Long term population trends of colonial wading birds and the impact of crawfish aquaculture on Louisiana wading birds. *Auk* 112:612–632.

Frederick, P. C. 1995. *Wading bird nesting success studies in the Water Conservation Areas of the Everglades, 1992–1995*. Final report to South Florida Water Management District, West Palm Beach, Fla.

———. 2000. Mercury and its effects in the Everglades ecosystem. *Reviews in Toxicology* 3:213–255.

Frederick, P. C., K. L. Bildstein, B. Fleury, and J. C. Ogden. 1996b. Conservation of large, nomadic populations of white ibises (*Eudocimus albus*) in the United States. *Conservation Biology* 10:203–216.

Frederick, P. C., and M. W. Collopy. 1989a. Nesting success of five ciconiiform species in relation to water conditions in the Florida Everglades. *Auk* 106:625–634.

———. 1989b. The role of predation in determining reproductive success of colonially nesting wading birds in the Florida Everglades. *Condor* 91:860–867.

———. 1989c. Researcher disturbance in colonies of wading birds: Effects of frequency of visit and egg-marking on reproductive parameters. *Colonial Waterbirds* 12:152–157.

Frederick P. C., and W. F. Loftus. 1993. Responses of marsh fishes and breeding wading birds to low temperatures: A possible behavioral link between predator and prey. *Estuaries* 16:216–222.

Frederick, P. C., and J. C. Ogden. 1997. Philopatry and nomadism: Contrasting long-term movement behavior and population dynamics of white ibises and wood storks. *Colonial Waterbirds* 20:316–223.

———. 2001. Pulsed breeding of long-legged wading birds and the importance of infrequent severe drought conditions in the Florida Everglades. *Wetlands* 21:484–491.

Frederick, P. C., and M. G. Spalding 1994. Factors affecting reproductive success of wading birds (Ciconiiformes) in the Everglades. Pp. 659–691 in *Everglades: The ecosystem and its restoration*, edited by S. Ogden and J. C. Davis. Delray Beach, Fla.: St. Lucie Press.

Frederick, P. C., M. G. Spalding, and R. Dusek. 2002. Wading birds as bioindicators of mercury contamination in Florida: Annual and geographic variation. *Environmental Toxicology and Chemistry* 21:163–167.

Frederick, P. C., M. G. Spalding, M. S. Sepulveda, G. E. Williams Jr., L. Nico and R. Robbins. 1999. Exposure of great egret nestlings to mercury through diet in the Everglades of Florida. *Environmental Toxicology and Chemistry* 18:1940–1947.

Frederick, P. C., T. Towles, R. J. Sawicki, and G. T. Bancroft. 1996a. Comparison of aerial and ground techniques for discovery and census of wading bird (Ciconiiformes) nesting colonies. *Condor* 98:837–840.

Frohring, P. C., D. P. Voorhees, and J. A. Kushlan. 1988. History of wading bird populations in the Florida Everglades: A lesson in the use of historical information. *Colonial Waterbirds* 11:328–335.

Gawlik, D. E. 2002. The effects of prey availability on the numerical response of wading birds. *Ecological Monographs* 72:329–346.

Gonzalez, J. A. 1999. Nesting success in two wood stork colonies in Venezuela. *Journal of Field Ornithology* 70:18–27.

Hafner, H., P. J. Dugan, M. Kersten, O. Pineau, and J. P. Wallace. 1993. Flock feeding and food intake in little egrets *Egretta garzetta* and their effects on food provisioning and reproductive success. *Ibis* 135:25–32.

Hafner, H., and M. Fasola. 1997. Long-term monitoring and conservation of herons in France and Italy. *Colonial Waterbirds* 20:298–305.

Hafner, H., O. Pineau, and Y. Kayser. 1994. Ecological determinants of annual fluctuations in numbers of breeding little egrets (*Egretta garzetta*) in the Camargue, France. *Revue D'Ecologie, La Terre et La Vie* 49:53–62.

Hoffman, W., G. T. Bancroft, and R. J. Sawicki. 1994. Foraging habitat of wading birds in the Water Conservation Areas of the Everglades. Pp. 585–614 in *Everglades: The ecosystem and its restoration*, edited by S. M. Davis and J. C. Ogden, Delray Beach, Fla.: St. Lucie Press.

Johnson, A. R. 1997. Long-term studies and conservation of greater flamingos in the Camargue and Mediterranean. *Colonial Waterbirds* 20:306–315.

Junk, W. J. B., B. Bayley, and R. E. Sparks. 1989. The flood pulse concept in river-floodplain systems. *Canadian Special Publications in Fisheries and Aquatic Sciences* 106:110–127.

Jurczyk, N. U. 1993. An ecological risk assessment of the impact of mercury contamination in the Florida Everglades. Master's thesis, University of Florida, Gainesville.

Kinzelbach, R., B. Nicolai, and R. Schlenker. 1997. The bee-eater *Merops apiaster* as an indicator for climatic change: Notes on the invasion in the year 1644 to Bavaria, Switzerland, and Baden. *Journal Fur Ornithologie* 138:297–308.

Kushlan, J. A. 1976a. Environmental stability and fish community diversity. *Ecology* 57:821–825.

———. 1976b. Feeding behavior of North American herons. *Auk* 93:86–94.

———. 1976c. Site selection for nesting colonies by the American white ibis (*Eudocimus albus*) in Florida. *Ibis* 118:590–593.

———. 1977. Population energetics of the American white ibis. *Auk* 94:114–122.

———. 1978. Feeding ecology of wading birds. Pp. 249–298 in *Wading birds*, National Audubon Society Research Report No. 7, edited by A. Sprunt IV, J. C. Ogden, and S. Winckler. National Audubon Society, New York.

Kushlan, J. A., and P. C. Frohring. 1986. The history and status of the southern Florida wood stork population. *Wilson Bulletin* 98:368–386.

Kushlan, J. A., J. C. Ogden, and A. L. Higer. 1975. *Relation of water level and fish availability to wood stork reproduction in the southern Everglades.* U.S. Geological Survey Open-file Report, U.S. Geological Survey, Tallahassee, Fla.

Leber, K. K. 1980. Habitat utilization in a tropical heronry. *Bresnia* 17:97–136.

Lemaho, Y. J., P. Gendner, E. Challet, C. A. Bost, J. Gilles, C. Verdon, C. Plumere, J. P. Robin, and Y. Handrich. 1993. Undisturbed breeding penguins as indicators of changes in marine resources. *Marine Ecology Progress Series* 95:1–6.

Loftus, W. F., and A. M. Eklund. 1994. Long-term dynamics of an Everglades small-fish assemblage. Pp. 461–484 in *Everglades: The ecosystem and its conservation*, edited by S. M. Davis and J. C. Ogden, Delray Beach, Fla.: St. Lucie Press.

Loftus, W. F., R. A. Johnson, and G. H. Anderson. 1992. Ecological impacts of the reduction of groundwater levels in short-hydroperiod marshes of the Everglades. In *American Water Resources Association Proceedings: First International Symposium on Groundwater Ecology.*

Lorenz, J. J. 1997. *The effects of hydrology on resident fishes of the Everglades mangrove zone.* Homestead, Fla.: Everglades National Park and National Audubon Society.

Marra, P. P., and R. L. Holberton. 1998. Corticosterone levels as indicators of habitat quality: Effects of habitat segregation in a migratory bird during the non-breeding season. *Oecologia* 116:284–292.

McCrimmon, D. A., S. T. Fryaska, J. C. Ogden, and G. S. Butcher. 1997. Nonlinear population dynamics of six species of Florida Ciconiiformes assessed by Christmas bird counts. *Ecological Applications* 7:581–592.

McIvor, C. C., J. A Ley, and R. D. Bjork. 1994. Changes in freshwater inflow from the Everglades to Florida Bay including effects on biota and biotic processes: A review. Pp. 117–148 in *Everglades: The ecosystem and its restoration*, edited by S. M. Davis and J. C. Ogden, Delray Beach, Fla.: St. Lucie Press.

Morrison, M. L. 1986. *Bird populations as indicators of environmental change. Current Ornithology 6*, edited by R. L. Johnston, Chicago: Plenum Press.

Niemi, G. J., J. M. Hanowski, A. R. Lima, T. Nicholls, and N. Weiland. 1997. A critical analysis of the use of indicator species in management. *Journal of Wildlife Management* 61:1240–1252.

Norton-Griffiths, M. 1987. *Counting Animals.* Vol. 1. *Handbooks on techniques currently used in African wildlife ecology.* Nairobi, Kenya: African Wildlife Leadership Foundation.

Ogden, J. C. 1977. An evaluation of interspecific information exchange by waders on feeding flights from colonies. *Colonial Waterbirds* 1:155–162.

———. 1978. Recent population trends of wading birds on the Atlantic and Gulf Coastal plains. Pp. 137–154 in *Wading birds*, edited by A. Sprunt Jr., J. C. Ogden, and S. Winckler. National Audubon Society Research Report No. 7, National Audubon Society, New York.

———. The Everglades story: The role of waterbirds in integrated management. 1993. Pp. 240–244 in *Waterfowl and wetland conservation in the 1990s: A global perspective*, edited by M. Moser, R. C. Prentice, and J. van Vessem. International Waterfowl Research Bureau Special Publication No. 26. Slimbridge, U.K.

———. 1994. A comparison of wading bird nesting dynamics, 1931–1946 and 1974–1989 as an indication of changes in ecosystem conditions in the southern Everglades. Pp. 533–570 in *Everglades: The ecosystem and its restoration*, edited by S. M. Davis and J. C. Ogden, Delray Beach, Fla.: St. Lucie Press.

Ogden, J. C., W. B. Robertson, G. E. Davis, and T. W. Schmidt. 1974. Pesticides, polychlorinated biphenyls and heavy metals in upper food chain levels, Everglades National Park and vicinity. Department of the Interior, National Technical Information Service, Washington, D.C.

Ogden, J. C., C. E. Knoder, and A. Sprunt IV. 1988. Colonial wading bird populations in the Usamacinta Delta, Mexico. Pp. 595–605 in *Ecologia de los rios Usamacinta y Grijalva*, Instituto Nacional de Investigaciones sobre Recursos Bioticos, Tabasco, Mexico.

Ogden, J. C., G. T. Bancroft, and P. C. Frederick. 1997. Reestablishment of healthy wading bird populations. In *Ecological and precursor success criteria for south Florida ecosystem restoration: A science sub-group report to the working group of the South Florida Ecosystem Restoration Task Force*. South Florida Ecosystem Restoration Task Force, Florida International University, Miami.

Powell, G. V. N. 1983. Food availability and reproduction by great white herons (*Ardea herodias*): A food addition study. *Colonial Waterbirds* 6:139–147.

———. 1987. Habitat use by wading birds in a subtropical estuary: Implications of hydrography. *Auk* 104:740–749.

Ramo, C., and B. Busto. 1992. Nesting failure of the wood stork in a Neotropical wetland. *Condor* 94:777–781.

Renfrow, D. H. 1993. The effects of fish density on wading bird use of sediment ponds on an east Texas coal mine. Master's thesis, Texas A&M University, College Station.

Robertson, W. B., and J. A. Kushlan. 1974. The southern Florida avifauna. *Miami Geological Society Memoirs* 2:414–452.

Robinson S. K., F. R. Thompson, T. M. Donovan, D. R. Whitehead, and J. Faaborg. 1995. Regional forest fragmentation and the nesting success of migratory birds. *Science* 267:1987–1990.

Rodgers, J. A., S. B. Linda, and S. A. Nesbitt. 1995. Comparing aerial estimates with ground counts of nests in wood stork colonies. *Journal of Wildlife Management* 59:656–666.

Root, T. 1988. *Atlas of wintering North American birds*. Chicago: University of Chicago Press.

Runde, D. E. 1991. *Trends in wading bird nesting populations in Florida, 1976–1978 and 1986–1989*. Final Performance Report, Nongame Section, Florida Game and Freshwater Fish Commission, Tallahassee.

Sadoul, N. 1997. The importance of spatial scales in long-term monitoring of colonial Charadriiforms in southern France. *Colonial Waterbirds* 20:330–338.

Salatas, J. 2000. The relationships among food consumption, growth and survival in nestling herons. Master's thesis, University of Florida, Gainesville.

Scott, W. E. D. 1887. The present condition of some of the bird rookeries of the Gulf coast of Florida. *Auk* 4:135–144, 213–222, 273–284.

———. 1889. A summary of observations on the birds of the Gulf coast of Florida. *Auk* 6:13–18.

———. 1890. A summary of observations on the birds of the gulf coast of Florida. *Auk* 7:14–23.

Sepulveda, M. S., P. C. Frederick, M. G. Spalding, and G. E. Williams Jr. 1999. Mercury contamination in free-ranging great egret nestlings (*Ardea alba*) from southern Florida. *Environmental Toxicology and Chemistry* 18:985–992.

Smith, J. P. 1995. Foraging flights and habitat use of nesting wading birds (Ciconiiformes) at Lake Okeechobee, Florida. *Colonial Waterbirds* 18:139–158.

Smith, J. P., and M. W. Collopy. 1995. Colony turnover, nest success and productivity, and causes of nest failure among wading birds (Ciconiiformes) at Lake Okeechobee, Florida (1989–1992). Archives of Hydrobiologie. *Special Issues in the Advance of Limnology* 45:287–316.

Smith, T. J., III, J. H. Hudson, G. V. N. Powell, M. B. Robblee, and P. J. Isdale. 1989. Freshwater flow from the Everglades to Florida Bay: An historical reconstruction based on fluorescent banding in the coral *Solenastrea bournoni*. *Bulletin of Marine Science* 44:274–282.

Spalding, M. G., and D. J. Forrester. 1991. *Effects of parasitism and disease on the reproductive success of colonial wading birds (Ciconiiformes) in southern Florida*. Florida Game and Fresh Water Fish Commission, Final Report. Tallahassee.

Spalding, M. G., G. T. Bancroft, and D. J. Forrester. 1993. The epizootiology of eustrongylidosis in wading birds (Ciconiiformes) in Florida. *Journal of Wildlife Diseases* 29:237–249.

Spalding, M. G., R. D. Bjork, G. V. N. Powell, and S. F. Sundlof. 1994. Mercury and cause of death in great white herons. *Journal of Wildlife Management* 58:735–739.

Spalding, M. G., P. C. Frederick, H. C. McGill, S. N. Bouton, and L. R. McDowell. 2000a. Methylmercury accumulation in tissues and effects on growth and appetite in captive great egrets. *Journal of Wildlife Diseases* 36:411–422.

Spalding, M. G., P. C. Frederick, H. C. McGill, S. N. Bouton, I. Schumacher, C. G. M. Blackmore, L. Richey, and J. Harrison. 2000b. Histologic, neurologic, and immunologic effects of methylmercury in captive great egrets. *Journal of Wildlife Diseases* 36:423–435.

Stangel, P. W., Jr., J. A. Rodgers, and A. L. Bryan. 1990. Genetic variation and population structure of the Florida wood stork. *Auk* 107:614–619.

———. 1991. Low genetic differentiation between two disjunct white ibis colonies. *Colonial Waterbirds* 14:13–16.

Stober, Q. J., J. D. Scheidt, R. Jones, K. Thornton, R. Ambrose, and D. France. 1996. *South Florida ecosystem assessment. Monitoring for adaptive management: Implication for ecosystem restoration (Interim report)*. Environmental Protection Agency 904-R-96-008: 26. Washington, D.C.

Stone, C. J., A. Webb, C. Barton, N. Ratcliffe, T. C. Reed, M. L. Tasker, C. J. Camphuysen, and M. W. Pienkowski. 1995. *An atlas of seabird distribution in northwest European waters*. Peterborough, U.K.: Joint Nature Conservation Committee.

Sunlof, S. F., M. G. Spalding, J. D. Wentworth, and C. K. Steible. 1994. Mercury in livers of wading birds (Ciconiiformes) in southern Florida. *Archives of Environmental Contamination and Toxicology* 27:299–305.

Surdick, J. A. 1998. Biotic and abiotic indicators of foraging site selection and foraging success of four Cicioniiform species in the freshwater Everglades of Florida. Master's thesis. University of Florida, Gainesville.

Temple, S. A., and J. A. Wiens. 1989. Bird populations and environmental changes: Can birds be bio-indicators? *American Birds* 1989:260–270.

Walters, C. J., L. Gunderson, and C. S. Holling. 1992. Experimental policies for water management in the Everglades. *Ecological Applications* 2:189–202.

Ware, F. J., H. Royals, and Ted Lange. 1990. Mercury contamination in Florida largemouth bass. *Proceedings of the Annual Conference of the Southeast Association of Fish and Wildlife Agencies* 44:5–12.

Williams, G. E., Jr. 1997. The effects of methylmercury on the growth and food consumption of great egret nestlings in the central Everglades. Master's thesis, University of Florida, Gainesville.

CHAPTER 13

Setting and Monitoring Restoration Goals in the Absence of Historical Data: The Case of Fishes in the Florida Everglades

Joel C. Trexler, William F. Loftus, and John H. Chick

In designing and conducting ecosystem monitoring, one must ask to what end are we monitoring? The usual answer, to document the status of trends of biological or physical indicators toward some restoration goal, presupposes that well-defined goals have been established. However, setting goals and assessing success of ecological restoration are among the greatest challenges of the restoration enterprise (Yount and Neimi 1990). By definition, restoration entails regaining some aspects of the ecological system that have been lost (Gore 1985; Bradshaw 1987). However, quantitative biological data on historical ecosystems are often limited in quantity or quality, or are completely absent (Cairns et al. 1977). Compounding the challenge, ecological restoration is at times controversial, and goals and assessments may come under intense scrutiny, even taking center stage in court proceedings (e.g., Rizzardi 2001). Thus, a substantial burden may rest on ecologists to justify assumptions and methods used in establishing and monitoring restoration targets.

In many ecological management and restoration cases, recapturing historical or "natural system" conditions for an entire ecosystem is neither practical nor desirable. In the Florida Everglades, for example, endangered species management and natural system management at times appear to be

351

in conflict (e.g., management for snail kites versus wood storks [Banner 1990]). Targeting portions of a regional ecosystem for restoration of one or more endangered species or some ecosystem function, while other portions are identified for "natural-system" management, may be the most common circumstance facing managers.

Ideally, before beginning a restoration action, historical ecosystem conditions should be delineated by data collected prior to human intervention in the ecosystem. Such data are rare, and generally occur in a narrative rather than quantitative framework (Egan and Howell 2001). Use of contemporary reference areas for comparison begs the question of whether those areas are themselves unaltered. Human alterations of ecosystems are pervasive and often act over large spatial scales affecting habitats not directly manipulated. For example, extirpation of large predators and herbivores by humans has probably dramatically altered most, if not all, natural ecosystems (e.g., Jackson 2001). Such effects may extend back into ancient times when human activities such as burning grasslands or overharvesting animals undoubtedly altered ecosystems dramatically (Jackson et al. 2001). Thus, historical conditions inferred from techniques of paleoecology may still reflect the effects of early human cultures on ecosystem composition and processes. Identification of targets must be explicit with regard to what time period of ecosystem history is to be restored.

Scientists are forced to recreate historical conditions by analogy through description of reference areas or by analytical tools such as simulation models because little or no historical data are typically available in planning restoration (Clark et al. 2001). Reference or background sites are a preferred approach in restoration efforts, but true historical references seldom exist. More critical, it is generally questionable that nearby regional ecosystems are truly comparable, even if relatively pristine sites exist (see Hargrove and Pickering 1992). As Marjory Stoneman Douglas wrote in her book *River of Grass*, "[t]here is but one Everglades." To which we might add, "Or Colorado River or Konza Prairie, or Olympic Peninsula." Thus, the premise of our chapter is that regional ecosystem initiatives must depend heavily on developing a predictive ecological understanding (or perhaps "post-dictive") of the ecosystem being restored to set restoration targets. Most importantly, a critical part of developing this understanding generally includes data gathered by retrospective monitoring (Noon, chap. 2). The process of moving from retrospective monitoring to prospective monitoring should include development of assessment tools to recreate "natural system" conditions that can serve as targets. We will illustrate this premise through our monitoring effort of fishes inhabiting the Florida Everglades.

Performance Measures: Targets for Monitoring

Assessing the progress of ecosystem restoration requires measurement of ecological parameters (physical or biological) that both are indicative of contemporary ecosystem conditions and trends, and can be projected to the idealized end points of the restoration as targets. A diversity of these so-called "performance measures" must be identified to allow assessment of ecosystem function or to measure progress toward specific targets, possibly for societal goals. For example, the number of nesting pairs of wading birds in the Florida Everglades serves both goals: as top carnivores they are indicative of food-web integrity *and* their presence and status is of great aesthetic interest to the public (Frederick and Ogden, chap. 12). Also, performance measures should be selected to reflect ecosystem responses across a range of time scales because important ecosystem processes may operate with multiyear time lags. For example, large predatory fishes require several years to reach sexual maturity and may induce important regulatory effects on communities of their prey. The exclusive use of ecosystem metrics that respond rapidly, such as periphyton or bacterial communities, would not indicate whether top trophic levels have been restored in an aquatic community. Finally, to be of greatest utility, it must be possible to make defensible projections of values for each performance measure in a restored ecosystem. Attainment of the projected values for monitoring indicators is the target for judging restoration success.

Several approaches can be employed to validate performance measures for a natural-system scenario. When time and monetary resources are limited, managers may be forced to rely on the opinion of experts deemed most familiar with the ecosystem—opinions usually termed "best professional judgment." At its best, this approach is an informal form of more-sophisticated model building; however, it suffers from obvious limitations of subjectivity and repeatability. Most significant for managers, the prospects of this approach withstanding an aggressive legal challenge are dubious. At the other extreme, application of quantitative historical data could provide the strongest basis for choice of restoration targets. Unfortunately, historical data are often lacking, and historical data collected by past generations of scientists often prove difficult to apply because of changing techniques and expectations of data-quality control (e.g., wading birds: Frohring et al. 1988, Ogden 1994; fishes: Miller 1990, Trexler 1995).

A third approach for setting restoration targets is the use of reference or background sites as benchmarks for desirable ecosystem management. This approach uses existing areas for identification, analysis, and documentation of performance measures and their targets. There are limitations to this approach, in addition to concerns discussed earlier that no "pristine" habitats remain to provide baseline conditions, particularly with removal of large grazers or carnivores, or deposition of airborne toxics, such as mercury (Zillioux et al. 1993). Karr and Chu (1999) make a compelling case for the use of baseline knowledge of geophysical setting and undisturbed biological condition to identify metrics for monitoring. A presumption of this view is that extensive baseline knowledge has already been established to permit the development of relatively generic measures of ecosystem health or integrity. Unfortunately, many regional ecosystem initiatives are in unique habitats that lack such detailed knowledge and may include all remaining examples of the endangered habitat, all of which may have experienced some level of human disturbance. Thus, setting targets for indices based on modern conditions in habitats perceived as pristine may undervalue the resources of the historical ecosystem (Jackson et al. 2001)—Daniel Pauly's (1995) "shifting baseline syndrome."

Ideally, development of simulation models for exploration of management scenarios (Fitz et al. 1996; DeAngelis et al. 1998; Curnutt et al. 2000; Sklar et al. 2001) brings together all three previous approaches and makes explicit the assumptions and data leading to restoration targets (DeAngelis et al., chap. 6). In doing so, simulation models formalize best professional judgment and extrapolate historical and contemporary knowledge of ecosystem state and function to a common framework. Models can also explore the implications of hypothesized species extinctions and anthropogenic effects through scenario building. Of course, validation of those scenarios may be possible only indirectly, but, short of time travel, they provide a powerful tool to peer into historical ecosystems. In the remainder of this chapter, we will use our studies of fish communities in the Florida Everglades to illustrate the linkage of ecological research, monitoring, and model building in a process leading to the development of restoration targets.

Monitoring and Model Development

Over the past century, goals for the Everglades have changed in terms of their relative degree of public support. Throughout much of this period, monitoring of the ecosystem's physical and biotic factors has contributed to the dialogue about such goals.

History of Monitoring

Alteration of the natural southerly flow of water through the Everglades, begun in the 1890s, was well underway by the early 1900s (reviewed in Blake 1980). Drainage was accelerated after two major hurricanes flooded

Lake Okeechobee in 1926 and 1928, killing thousands of people. The major canals and levees blocking natural flow patterns from the north were in place by 1930. Although formation of a national park in the Everglades region was suggested as early as 1905, the Everglades National Park (ENP), located at the southern end of the ecosystem, was not officially dedicated until 1947. The Tamiami Trail and Canal, completed in 1928, marked the northern boundary of the park and limited the southerly flow of water into it until large plugs were opened in the mid-1960s (S-12 structures; fig. 13-1). By that time, there was widespread agreement that the Park was not receiving adequate water supplies to fulfill its conservation mission. A federal commitment to re-think the south Florida water-delivery and drainage system, developed over the preceding one hundred years, was codified in a report from the U.S. Army Corps of Engineers in 1968 (Blake 1980, 191). This culminated in the Comprehensive Everglades Restoration Plan (CERP), a major restoration initiative passed by Congress in 2000 (Anonymous 2000).

Quantitative collections of marsh fishes and aquatic invertebrates from the Everglades are relatively recent and, prior to the 1950s studies, were

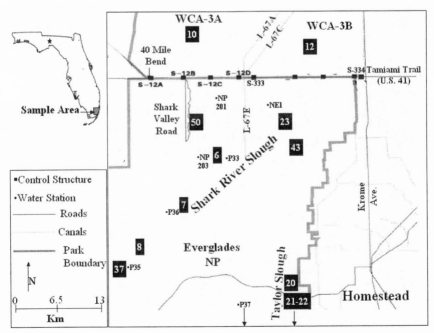

Figure 13-1. Map of the study sites in the Shark River Slough. Numbers in black squares indicates study sites. Sites 6, 23, and 50 are discussed in the text. Three plots, not shown, separated by approximately 1 kilometer of marsh habitat, were present at each study site.

entirely qualitative. The first quantitative studies of southern Florida freshwater animals were conducted in the early 1950s–1960s (Hunt 1952; Reark 1960, 1961; reviewed in Loftus and Eklund 1994). Reark (1960, 1961) was the first to collect fish and invertebrate density and biomass data with relation to vegetation cover. He compiled the only database, albeit very limited, from Shark River Slough (the primary drainage of the southern Everglades) before the operation of the S-12 water-control structures (fig. 13-1). Those flood gates placed control of southern marsh hydrological patterns in the hands of water managers, exacerbating departures in the timing, quantity, and spatial distribution from natural water flow in the Everglades National Park brought about by the Tamiami Trail. Thus, all data on aquatic animals from the Everglades south of the Tamiami Canal, and probably all data on the Everglades generally, were collected long after marsh hydrology was disturbed by the construction of the drainage system.

From 1965 to 1972, the U.S. Geological Survey (USGS) was contracted by Everglades National Park to collect data on aquatic-animal community composition and population variability related to hydrology in Shark River Slough (Higer and Kolipinski 1967; Kolipinski and Higer 1969). Those data all appeared in the peer-reviewed literature over a period of years (fishes: Kushlan 1976, 1980; apple snails [*Pomacea paludosa*]: Kushlan 1975; crayfish [*Procambarus* spp.] Kushlan and Kushlan 1979; and freshwater prawns [*Palaemonetes paludosus*] Kushlan and Kushlan 1980). In 1976, the research division of Everglades National Park began a long-term commitment to study the aquatic ecosystem. ENP supported an eight-year investigation of fish dynamics with relation to hydrology, using improved techniques (a 1-square-meter throw trap) developed and tested by Kushlan (1981; Loftus and Eklund 1994). Information generated by that eight-year study was used to design the next study that ran from October 1985 to 1992 (Loftus et al. 1990). In that study, a new sampling design was initiated to test hypotheses about the effect of marsh hydroperiod on community parameters. Collections at two of the long-term sites (6 and 23) were continued in this study, and a new site (50) added (fig. 13-1).

Following Hurricane Andrew in 1992, the original ENP-funded program was supplemented with support from the U.S. Army Corps of Engineers and the South Florida Water Management District. The new funding sources related to Everglades restoration projects, and the transfer of program personnel to the newly formed National Biological Service in 1993, resulted in major changes to the monitoring program. The increased funding also provided the first opportunity to build a larger regional sampling program by expanding the ENP program (fig. 13-1). That program expanded the system of monitoring sites in Everglades

National Park and Water Conservation Area 3A and 3B to a total of twenty sites. A collaboration of Florida International University, ENP, and USGS personnel now share the sampling responsibilities. The most important aspect of the expanded program is that it employs a standardized and consistent sampling design to enable data comparisons across the region (Jordan et al. 1997a). Several reports and papers resulting from this long-term program have been published (Trexler et al. 1996a,b, 2000, Trexler and Loftus 2001; Loftus et. al. 1990, 1997; Turner et al. 1999).

Scope of Existing Monitoring

Monitoring aquatic animals in the southern Everglades has focused on the ridge-and-slough habitat that dominates the central Everglades ecosystem (Gunderson 1994). Sampling depends heavily on use of a 1-square-meter throw trap and is limited to spikerush-dominated (primarily *Eleocharis cellulosa*) prairie and slough habitats. The throw trap was described in Kushlan (1981), and the method has been well supported by a literature on sampling bias and effectiveness through the range of conditions where it is employed for Everglades monitoring (Kushlan 1981; Jacobsen and Kushlan 1987; Chick et al. 1992; Loftus and Eklund 1994; Jordan et al. 1997a). Limitations of the technique include that it does not sample large fishes (longer than 8-centimeter standard length) or small macroinvertebrates (able to pass through a 2-millimeter sieve) well. Large fishes are not enumerated well both because of trap avoidance and because they are too sparse to be collected effectively at the 1-square-meter scale (though sampling juveniles of these species is not biased by avoidance). Although crustaceans (crayfish and grass shrimp), dragonfly naiads, and several groups of aquatic bugs (e.g., Belostomatidae and Naucoridae) are sampled effectively, the majority of Everglades macroinvertebrates are too small for effective enumeration by field sorting (Turner and Trexler 1997). Since 1997, throw-trap sampling has been supplemented by boat electrofishing for larger species (Chick et al. 1999) to provide a more comprehensive monitoring of fish communities. Smaller macroinvertebrates have largely been ignored in landscape-scale monitoring of the Everglades (McCormick et al. 2001), but an aquatic invertebrate monitoring program has been started in the Everglades National Park. A separate fish-monitoring effort more than ten years in duration is ongoing in the mangrove zone at the southern edge of the freshwater Everglades. That study uses permanently installed drop-nets because of the challenges presented for sampling fishes among the prop roots of red mangrove tress (Lorenz et al. 1997; Lorenz 1999). Fishes are also sampled by ENP and USGS personnel in the Rocky Glades and other short-hydroperiod areas of Park by the use of wire minnow traps and throw traps (see also Kobza 2001; Trexler et al. 2001).

Special monitoring initiatives are in place for documentation of mercury in fishes, though long-term funding has not been identified. Eastern mosquitofish (*Gambusia holbrooki*) and largemouth bass (*Micropterus salmoides*) have been the targets of those efforts. Biologists from the U.S. Environmental Protection Agency and Florida International University have estimated total mercury burden in mosquitofish sampled from over seven hundred locations selected by a stratified random design to cover the entire Everglades marsh ecosystem. Their collections were made from 1995 to 1999 (Stober et al.1998, and subsequent reports). The sampling was completed in four synoptic sampling events in April 1995, May 1996, May 1999, and September 1999.

Clearly, not all ecosystem elements can be selected for monitoring. Each choice must be justified based on its responsiveness to management actions and Everglades restoration goals. Fishes, as well as aquatic invertebrates and frogs, have been chosen because of their critical roles in the Everglades food web and because they are important in the diets of charismatic species of wading birds and alligators. For example, most wading bird species found in the Everglades consume fishes as a portion of their diet, and many species, such as white ibis, also consume large quantities of crayfish. Ornithologists believe that wading bird nesting success is largely limited by food availability (Frederick and Spalding 1994; Ogden 1994). Apple snails serve as food for a number of birds, notably the limpkin and endangered snail kite. Fishes, crayfish, frogs, and apple snails are major diet items of alligators in the Everglades (Fogarty 1984; Barr 1997). In addition, pig frogs are economically important in the Water Conservation Areas, where they are harvested for commercial sale. Finally, the species or groups selected for monitoring are known to be sensitive to changing patterns of hydrology and water quality, responding to management actions over a range of time periods from months to years.

Performance Measures and Targets

The Comprehensive Everglades Restoration Plan is in its early phases and work on its monitoring program primarily involves development of restoration targets for performance measures (see Busch and Trexler, chap. 1), tools for assessment of those measures, and assimilation of data needed to complete assessments (Ogden et al., chap. 5). Performance measures and restoration targets are being set separately for several different habitats in the ecosystem. Targets for fish communities in ridge-and-slough habitats illustrate this process.

Four types of performance measures have been proposed for fishes that inhabit the ridge-and-slough habitat from a series of workshops employing best professional judgment. These four metrics are abundance, size distribution, relative abundance, and contaminants (Ogden et al., chap. 5). It is proposed that the recovery targets for these ridge-and-slough performance measures will be set by reference sites in Shark River Slough or patterns predicted from hydrological information from a natural system model (NSM). The hydrological NSM estimates water depth for 6.4-square-kilometer grid cells based on actual rainfall data across the freshwater ecosystem at a daily time step under natural landscape patterns, including the removal of all canals and levees (Fennema et al. 1994). Performance measures and their targets that are presently under consideration for fishes include (1) *Abundance.* In response to restoration of lengthened hydroperiods, patterns of fish population dynamics should include increases in marsh fish numbers (measured as density) and biomass. (2) *Size distribution.* In response to lengthened hydroperiods, the range of biomass and body length of marsh fishes should increase by increasing the frequency of large-bodied species. (3) *Relative abundance.* In response to lengthened hydroperiods, the relative abundance of centrarchids and chubsuckers should increase. (4) *Nonnative species.* An additional goal is to maintain the low frequency of nonnative fishes currently observed in the interior of the ridge-and-slough system. (5) *Contaminants.* Levels of mercury and other toxins in marsh fishes should be reduced.

Several challenges to the process leading to these performance measures have emerged. Criticisms of this process include that a rigorous conceptual and empirical basis for the performance measures has yet to be developed. Also, forcing the ecosystem into discrete habitat categories, such as ridge and slough or marl prairies, creates problems because the habitats are not uniform and distinctions emerge gradually along hydroperiod gradients. Spatially referenced simulation models can overcome these difficulties, but spatially explicit conceptual models become confusingly complex as the basis for discussion among agency personnel. Thus, the current conceptual models have emerged as an expedient mechanism to facilitate communication among groups with wide-ranging experience and interests. It seems reasonable that developing targets will be an evolutionary process, with modeling becoming critical to establish an objective foundation for targets in the long run.

Application of Performance Measures

Assessing restoration targets is a four-step process: (1) collect data indicative of fish communities under the range of present-day field conditions,

(2) develop analytical tools to predict values of performance measures from hydrological parameters (stressors), (3) use NSM hydrological data and analytical tools to derive restoration targets, and (4) once restoration begins, compare observed values for performance measures to targets. This process need not be complex if adequate data are available, as illustrated by our work with monitoring data from Everglades fishes. We have used statistical analysis of long-term data records from ridge-and-slough communities to develop functions that predict fish density given two hydrological parameters: days since an area last dried (water depth less than 5 centimeters) and average water depth during the thirty days prior to the assessment (fig. 13-2). The density of small fishes generally increases as time passes after a dry-down event because of population growth and re-colonization of the dried area. However, some species are always most abundant at short-hydroperiod sites (illustrated by flagfish in fig. 13-2), and these temporarily increase in relative abundance at long-hydroperiod sites following a dry-down event. The eastern mosquitofish is a unique species in re-colonizing the marsh rapidly following a dry-down event, and displaying little or no long-term trend with respect to time since dry down (fig. 13-2). Its relative abundance drops as time passes after a dry-down event because species characteristic of long-hydroperiods (illustrated by bluefin killifish in fig. 13-2) increase in density over time, while mosquitofish density remains relatively constant.

We estimated the functional relationship of fish density and days since marsh drying at twenty existing monitoring locations from WCA-3A, 3B, Shark River Slough, and Taylor Slough. The quality of the estimates varied among those sites related to the length of the existing data record and whether the site dried during the period when data were collected. Fortunately, the functions estimated from most locations were fairly similar in shape, so the results appear robust with some important exceptions. Data from one of our monitoring sites located near natural creeks in southern Shark River Slough (site 37, fig. 13-1) indicated a much more rapid recovery from dry-down events than for central marsh sites. This illustrates that assessment models must be spatially explicit and incorporate landscape features such as dry-season aquatic refuges (DeAngelis et al. 1997). It also provides a caveat for simplistic application of habitat characterizations.

The existence of long-term records for fishes in the Everglades makes the development of predictive models of performance measures possible. ENP personnel have collected records of fish density for over twenty years at two locations in Shark River Slough, one in the main slough south of the S-12C structure (Site 6; fig. 13-1) and one in the northeast Shark River Slough east of the L67E levee (Site 23; fig. 13-1). Time since the

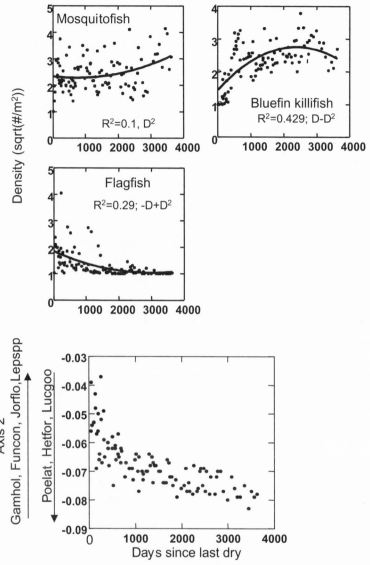

Figure 13-2. Patterns of fish density and relative abundance relative to days since marsh site was last dry for Site 6 between 1977 and 1999. (Top) Square-root transformed density is plotted for three characteristic species: mosquitofish, bluefin killifish, and flagfish. The coefficient of determination and parameters in a regression model are reported on each panel: D = days since last dry. (Bottom) Results from an ordination of relative abundance patterns using nonmetric multidimensional scaling plotted against days since last dry. Gamhol = *Gambusia holbrooki* (mosquitofish); Funcon = *Fundulus confluentus* (marsh killifish); Lepspp = *Lepomis* spp. (sunfish); Poelat = *Poecilia latipinna* (sailfin mollies); Hetfor = *Heterandria formosa* (least killifish); Lucgoo = *Lucania goodei* (bluefin killifish).

last dry-down event and water depth yielded a good fit to those data (fig. 13-3) and the results are presented here to illustrate the approach. Those sites also provide insight into the effect of manipulating hydrology as an ecological restoration tool for the Everglades. Management of northeast Shark River Slough (NESRS), east of the L-67E levee, was modified in

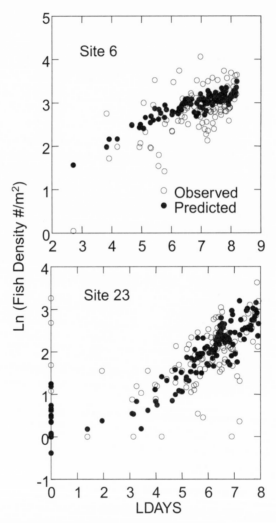

Figure 13-3. Observed and predicted densities of fishes (all species) plotted relative to days since last dry down. The predicted values were derived from a Poisson regression of the model: fish density = (average water depth the month samples were collected) + (natural log transformed number of days since the study site was last dry) + (natural log transformed number of days since the study site was last dry)^2. Site 6 R^2 = 58.4 %; Site 23 R^2 = 78.2%.

1985 under a federal program to lengthen the hydroperiod to better reflect historical conditions there. Management at the SRS site was not modified at that time, and we considered it to be a reference area for this "experiment" in management.

We obtained NSM-derived estimates of water depth for each study site under "natural system" conditions from 1971 to 1995 and used our statistical model to estimate fish density over that time period. We then compared the NSM-predicted density to the density of fishes observed from monitoring. The results indicate that the SRS site was much more similar to NSM-predicted conditions from 1971 to 1995 than was the NESRS site (fig. 13-4). From 1981 to 1986, the SRS site actually had more fishes than expected by the natural system model (fig. 13-4), because managers kept the site from drying out in 1981 when NSM predicted it should have dried. In contrast, the northeast Shark River Slough had fewer fishes than predicted by the natural system model throughout this period. Unfortunately, much of the perceived recovery of this study area occurred after the end of the NSM data record (1995), so further assessment of the success of the management experiment will have to wait for additional NSM calculations. By 1999, the density of fishes observed in the northeast Shark River Slough was in the range of that predicted previously for NSM conditions (approximately twenty fish per square meter), but this corresponded to a series of very wet years. We do not know if the natural system model will predict an even higher density of fishes than in previous years. The density of fishes in the northeast Shark River Slough in 1999 was not different from the reference site in SRS, providing some support for the hypothesis that restoration goals were met by this experimental water delivery.

This exercise illustrates several goals for further improvement in setting ecoregional assessment targets. First, we need to complete calibration of the hydrologically based statistical model to the observed data from our other eighteen study sites. Next, those equations must be incorporated into a landscape-based model that produces output in maps as is used in assessment models like the Across Trophic Level System Simulation (ATLSS; DeAngelis et al. 1998) and the Everglades Landscape Model (ELM; Fitz et al. 1996). Finally, we need access to NSM output that is updated routinely. Our efforts illustrate the importance of uninterrupted gathering of monitoring data to improve the statistical models and provide tests of status and trends in the ecosystem. Gaps in the long-term data greatly hamper this effort. For example, data collection was stopped by budget constraints for most of 1988 in the Shark River Slough, and those data available suggest an unusually high density of fishes (fig. 13-4). Finally, we need to continue to improve the assessment models, most

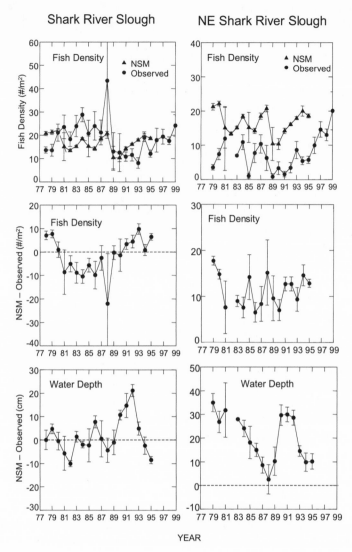

Figure 13-4. Results from analysis of fish density in Shark River Slough and northeast (NE) Shark River Slough. Top Panel: Plot of observed fish density from 1977 and 1999, and density predicted from the natural system model (NSM). NSM data were limited to the years from 1978 to 1995. Middle Panel: The difference between NSM-predicted fish density and observed fish density from 1978 to 1995. Positive values indicate fewer fishes were recorded than predicted under NSM conditions and negative values indicate more fishes than expected under NSM conditions. Bottom Panel: The difference between NSM predicted water depth (centimeters) and observed water depth in central Shark River Slough from 1978 to 1995. Positive values indicate shallower water was recorded than predicted under NSM conditions and negative values indicate deeper water than expected under NSM conditions.

clearly by incorporating information on the nutrient status of areas in the Everglades (see next section). It is expected that changing hydrology will change the landscape pattern of system productivity in subtle, but potentially important, ways that should be incorporated into natural-system scenarios. Current work with the ELM model is addressing this effort, and future development of fish-assessment tools will benefit from this work.

Monitoring Large-Fish Communities

Beginning in 1997, we used airboat electrofishing to monitor the distribution and species composition of large fishes (more than 8 centimeters standard length) in freshwater marshes of the Everglades. From October 1997 to December 1999, we sampled large fishes four times per year at sites within three regions of the Everglades: (1) Taylor Slough and (2) Shark River Slough in Everglades National Park, and (3) Water Conservation Area 3A (WCA-3A) (fig. 13-1). These three regions fall along a gradient of hydroperiod (the proportion of the year when marshes are flooded) with the shortest hydroperiod marshes found in Taylor Slough and the longest hydroperiod marshes in WCA-3A (Trexler et al. 2001). While total catch-per-unit-effort (CPUE) varied significantly through time, among regions, and among sites within regions, regional variation was the greatest. CPUE was consistently greater in both WCA-3A and Shark River Slough compared to Taylor Slough. Large-bodied predators, such as Florida gar (*Lepisosteus platyrhincus*) and largemouth bass (*Micropterus salmoides*), dominated the large-fish community of WCA-3A, whereas the Shark River Slough community was composed primarily of smaller species, such as lake chubsuckers (*Erimyzon sucetta*) and spotted sunfish (*Lepomis punctatus*). Composition of the Taylor Slough large-fish community was highly variable through time. As seen with fishes collected by throw trap, CPUE of large fishes was positively correlated to hydroperiod and soil total phosphorus. Phosphorus is the limiting element for plant growth in the Everglades (Davis 1994), and its availability in soil is positively correlated with hydroperiod.

A simulation model suggested that encounter rates between prey fishes and a large predatory fish is positively related to both predator density and hydroperiod. The potential influence of large-predatory fishes is greatest in the longest hydroperiod and least in short hydroperiod marshes. Interestingly, the density of forage species, estimated by throw trapping at the same study sites, was greatest in Shark River Slough where the large-fish community was dominated by the omnivorous lake chubsucker (diet in SRS approximately 33 percent detritus; 65 percent invertebrates; Loftus

2000). These data illustrate that fish communities vary regionally within habitat types such as ridge and slough. Spatially explicit performance measures, rather than simply habitat-based ones, are needed to incorporate regional variation of this type.

Monitoring Nonnative Fishes

Fish monitoring has provided fortuitous information on the invasion and status of nonnative fishes in the Everglades (Trexler et al. 2000). The monitoring data indicate that introduced fishes are present in the ridge-and-slough community at very low numbers. The long-term nature of the monitoring data has illustrated a boom-bust pattern of invasion by several species entering Shark River Slough. For example, the dry years of 1989 and 1990 corresponded to the range expansion and boom in abundance of at least one nonnative fish, the pike killifish (*Belonesox belizanus*). Monitoring in 2001 and beyond should be focused on identifying similar patterns following this dry period, should they occur. Nonnative species, especially the Mayan cichlid (*Cichlasoma uropthalmus*), are known to be more abundant in habitats surrounding the Everglades ridge and slough. Long-term monitoring in the mangrove zone indicated great fluctuation in the abundance of that species, possibly related to infrequent freeze events (Trexler et al. 2000). Fish monitoring in short-hydroperiod habitats suggests that hydrological management may affect the success of nonnative species there through its influence on their demography and dispersal ecology.

Monitoring Contaminants

The REMAP study revealed "hot spots" of mercury contamination in WCA-3A and Everglades National Park that have led to the development of models of mercury deposition and bioaccumulation (Stober et al. 1998). Also, the temporal data indicate that the concentration of total mercury in Everglades mosquitofish declined by as much as 24 percent from the 1995–1996 sampling events to those in 1999. Similar results have been obtained by monitoring of largemouth bass by the biologists of the Florida Fish and Wildlife Conservation Commission (Lange et al. 2000), a trend that may have reversed in the following drier years of 2000 and 2001. The long hydroperiods of the unusually wet years 1997–1999 have been implicated in this decline, a hypothesis with implications for management that can be tested from future monitoring results. Fish mercury contamination has complex origins in food web ecology that is mediated by both community ecology (food-chain length) and mercury biogeochemistry (availability of methyl mercury to the food web) (Stober et al. 1998). Thus, while mercury should be monitored because of its ecosystem and human health effects, further work is needed before restoration targets can be generated for this parameter.

Assumptions for Natural-System Simulations

Setting monitoring targets by any technique requires that assumptions be made about ecosystem function in general, and specifically in the interpretation of data on which the targets are based. For example, determining whether the Comprehensive Everglades Restoration Plan fulfills the goals of restoring Everglades ridge-and-slough communities requires that targets be identified against which contemporary communities can be compared as restoration progresses. We have illustrated a statistical and simulation approach for establishing NSM targets for fishes, based on ecological studies at a diversity of sites. Those sites represent the spectrum of hydroperiods that we believe also characterized the historical Everglades ridge-and-slough system. Many of those sampling sites may also be considered "reference sites," although they have experienced variable degrees of hydrological alteration. Given the absence of "pristine" Everglades habitats, we have identified nine assumptions inherent in producing and interpreting our natural-system scenario.

1. *Wading bird predation did not regulate numbers of fishes in the historical Everglades.* There is strong evidence that the number of wading birds nesting in southern Florida has decreased significantly in the past century. In the contemporary ecosystem, there is no evidence that wading birds regulate the number of fishes at the landscape scale. The predatory effects of wading birds appear to be very local and limited in time to periods of dry-down when fishes become available. These events are uncommon and generally lead to situations in which most fishes would have perished had they not been consumed. In the unmanaged ecosystem, such events would have been confined to the edges of the system in most years because systemwide dry-downs occurred less frequently than at present.

2. *Recent patterns of wading bird nesting have not grossly altered landscape patterns of fish productivity.* Frederick and Powell (1994) indicated that wading birds could transport and concentrate phosphorus across the ecosystem. Contemporary wading bird populations may be consuming 4.9 tons (dry mass) less prey than historical populations (Frederick and Powell 1994). Thus nutrient transport could have been substantial under historical conditions when bird rookeries of the southern Everglades included thousands of nests. This probably created local

conditions of high primary and secondary production in the vicinity of rookeries. Historically, rookeries were concentrated in the mangrove zone of the extreme southern Everglades, but nesting birds have moved north into areas now in the Water Conservation Areas. Thus, this landscape pattern of avian-supported productivity has changed. At present, these effects are limited to the immediate vicinity of rookeries where they may be important, but they probably have little effect on system-level production.

3. *The Everglades is a smaller ecosystem than it was historically, and the remaining edge habitats have been diminished in area and quality.* Marl prairies and rockland habitats have been reduced in area and in their ability to support aquatic species. Transverse glades draining east to Biscayne Bay have been developed and channelized. Also, the Everglades agricultural area south of Lake Okeechobee was formerly a large area of the historical Everglades ecosystem. The implications of the loss of these habitats, some of which were never well described in their pre-drainage condition, are unknown but are unlikely to have been beneficial.

4. *Drainage has altered soil formation and led to loss of soil in much of the Everglades.* The loss of deep peats and changes to shallow peat and marl substrates has had unknown implications for aquatic community dynamics.

5. *Mercury contamination, which has reached high levels in Everglades fishes, has had little ecological effect on these communities.* While there are no historical data to determine if this is a new phenomenon, studies have demonstrated links to anthropogenic environmental changes (Stober et al. 1998). Mercury contamination may arise from recent mercury deposition and/or biogeochemical processes that methylate elemental mercury naturally present in Everglades sediments. The health effects of chronic low-level mercury exposure to fishes, and implications for their population dynamics, have not been explored for any fish that inhabits the Everglades. Other heavy metals, such as selenium, arsenic, and cadmium are also potential contaminants that may affect fish health and can be transported throughout the ecosystem by the system of canals.

6. *Our results can be generalized across the ridge-and-slough habitat, although our fish data have mainly been collected from one habitat (spikerush-dominated wet-prairie or sloughs).* Wet prairies are one of the two predominant habitats of the central Everglades (Gunderson 1994). Limiting our efforts to this habitat could bias our conclusions:

First, there is some evidence, largely anecdotal, that historical Everglades wet-prairie sloughs were deeper than those studied today, with abrupt edges. Related to this is the proposal that the relative spatial coverage of sawgrass-dominated ridges versus wet-prairie sloughs was different in the historical ecosystem as a function of more-rapid velocity of downstream flow (Chris McVoy, personal communication).

Second, applying data from wet-prairie samples may overestimate fish density in the total ridge-and-slough habitat. Data from sampling sawgrass-dominated ridges adjacent to wet-prairie sloughs indicate slightly fewer fishes on the ridges, which can be so densely vegetated that little interstitial aquatic habitat remains. The ridges also dry before the adjacent sloughs (Jordan et al. 1997b; Trexler et al. 2001).

7. *The presence of levees and water-control gates, their effect on hydrology, and their proximity to our study sites do not greatly influence our results.* Preliminary sampling data indicate that the electrofishing CPUE of large fishes diminishes with distance from canals for approximately 1 kilometer. Also, preliminary results from radio tracking Florida gar in WCA-3A suggest that large fishes move several kilometers from canals into the marsh to feed. All of our monitoring sites are at least 3 kilometers from canals, and generally much farther. Canals may function as large, albeit artificial, dry-season refuges. Implications of the large area of such deep-water refuges that were absent under historical conditions are not well understood (but see Loftus 2000; Trexler et al. 2000). Of course, the function of artificial refuges must be contrasted with the historical ecosystem, which held much more water in its central flow-way and seldom dried compared to that habitat today.

8. *Nonnative species, including several species of fishes, have not altered the community dynamics as revealed by our contemporary monitoring and experimental studies.* At least for the ridge-and-slough communities of the southern Everglades, there is reason to accept this assumption (Trexler et al. 2000). New threats from introduced fishes arise routinely in the region (e.g., the recently introduced Asian swamp eel [*Monopterus* sp., probably multiple species] has a life history unique among current invaders, with unpredictable implications [Collins et al. 2002]).

9. *The hydrological Natural System Model provides adequate estimates of historical water depths that can serve as a basis for simulating historical communities.* There are few hydrological data for validation of historical water depths predicted by the NSM. The model's predictions have received extensive evaluation by hydrologists and biologists, who agree with them in general (a process of best professional judgment), but who are often critical of its predictions in local areas. Continued

development of the hydrological Natural System Model is a linchpin to further modeling of biological restoration scenarios.

Model Evaluation, Uncertainty, and Prospects for Monitoring Targets

The advent of societal pressure for science-based management has led to use of simulation models to predict the outcome of ecological management (Sarewitz and Pielke 1999). However, a burgeoning literature emphasizes that it is at best challenging, if not impossible, to verify or validate natural-system or predictive models (e.g., Oreskes et al. 1994, but see Rykiel 1996). In place of validation, it may more appropriate to discuss model evaluation under a diversity of scenarios (Oreskes 1998) leading to delineation of the uncertainties that emerge (Snowling and Kramer 2001). It may be possible to identify and minimize uncertainties (Rotmans and van Asselt 2001), even when predictions per se are impossible.

Despite the reality that we can never validate our NSM predictions, we believe that alternative approaches for goal setting are even more problematic and subject to whims of politics, policy, and expediency. Of course, policy makers and managers appropriately set the final direction of any public works project such as the Comprehensive Everglades Restoration Plan, but our NSM goals provide guidance for benchmarks that seek to minimize subjectivity as a starting point for management action.

The restoration enterprise must include a sustained effort to improve natural-system models and scenarios as new technology and information become available. A key future step, in addition to improving the natural-system model, is to delineate and perhaps quantify uncertainties in the output. Philosophers of science point out that uncertainty cannot be eliminated, and recommend that managers include it explicitly in the decision process (Reynolds and Ford 1999; Rotmans and van Asselt 2001; Reynolds and Reeves, chap. 7). Techniques permitting explicit estimation of uncertainty need to be incorporated into the NSM process to target priority areas for improvement, and minimize the potential of negative impacts of unknowns.

Ecological systems are not static. While the chemical properties of the limiting nutrient, phosphorus, may have been relatively constant throughout the brief history of the Everglades, the same cannot be said for the reactions that determine the abundance of organisms that inhabit it. New species enter ecosystems and others go extinct (naturally or with human assistance), with potential to dramatically change the regulation of biological populations and communities. Species themselves are not stable over evolutionary time, and examples can be cited in which characteristics influencing population dynamics have changed over ecological time (e.g., Reznick et al. 1997; Grether et al. 2001). We have attempted to delineate the assumptions necessary to develop an Everglades natural-system model based on modern ecological interactions, but uncertainty will remain an inevitable partner in this exercise. Because of this, the use of reference sites must remain an important complement to NSM predictions, even when their comparability is imperfect. For example, the Everglades is an oligotrophic, hydrologically pulsed ecosystem (Turner et al. 1999) with few counterparts. Wetlands in Mexico's Yucatan peninsula and nearby Belize may provide reference ecosystems for examination of key ecological processes that they may have in common. However, in a more practical vein, we have illustrated an example where locations within the restoration served as references for local subprojects within the larger effort. This must be done with circumspection, however, because it precludes documentation of landscape-scale effects and other forms of ecological "surprise."

The Everglades restoration initiative may be relatively well endowed with monitoring data that will assist in assessment of ecosystem alterations as the project proceeds. However, even for this project, many areas are poorly documented and require uncomfortable extrapolation beyond existing data to build NSM scenarios. This hodgepodge of monitoring at the outset of a regional initiative is to be expected (see Hemstrom, chap. 11). Further, we have identified nine challenging assumptions that must be considered for the application of fishery data in assessment of ecosystem restoration. The challenge for the Comprehensive Everglades Restoration Plan, and all ecoregional initiatives, is to maximize the use of existing historical and contemporary data in assessment programs that address both landscape-scale (ecoregion-wide) restoration and short-term local project needs.

Acknowledgements

We wish to acknowledge and thank the many ENP biologists who have contributed to the collection of the long-term data used in this report, especially Oron Bass and Sue Perry. Eric Nelson and Victoria Foster provided data and data management assistance in preparing this report, and Jennifer Barnes and the South Florida Water Management District supplied us with NSM output for our study sites. Work reported here was funded by a variety of sources, including cooperative agreements CA5280-6-9011 and CA5280-8-9003 from Everglades National Park and National Science Foundation grant no. 9910514 to J. C. Trexler, and base funding from Everglades National Park and USGS-BRD to W. F. Loftus.

LITERATURE CITED

Anonymous. 2000. *Comprehensive Everglades restoration plan*. Vol. 1. *Master program management plan, final, August 2000*. South Florida Water Management District and U.S. Army Corps of Engineers. [Online at http://208.60.32.205/pm/mpmp_08182000.htm]

Banner, A. 1990. *Biological opinion: Modified water delivery to Everglades National Park*. U.S. Department of Interior, U.S. Fish and Wildlife Service Coordination Act Report, Vero Beach, Fla.

Barr, B. 1997. Food habits of the American alligator, *Alligator mississippiensis*, in the southern Everglades. Ph.D. diss., University of Miami, Coral Gables, Fla.

Blake, N. M. 1980. *Land into water—water into Land: A history of water management in Florida*. Tallahassee: University Press of Florida.

Bradshaw, A. D. 1987. Restoration: An acid test for ecology. Pp. 23–29 in *Restoration ecology*, edited by W. R. Jordan III, M. E. Gilpin, and J. D. Aber. Cambridge, U.K.: Cambridge University Press.

Cairns, J., Jr., K. L. Dickson, and E. E. Herricks, eds. 1977. *Recovery and restoration of damaged ecosystems*. Charlottesville: University Press of Virginia.

Chick, J. H., S. Coyne, and J. C. Trexler. 1999. Effectiveness of airboat electrofishing for sampling fishes in shallow vegetated habitats. *North American Journal of Fisheries Management* 19:957–967.

Chick, J. H., F. Jordan, J. P. Smith, and C. C. McIvor. 1992. A comparison of four enclosure traps and methods used to sample fishes in aquatic macrophytes. *Journal of Freshwater Ecology* 7:353–361.

Clark, J. S., S. R. Carpenter, M. Barber, S. Collins, A. Dobson, J. A. Foley, D. M. Lodge, M. Pascual, R. Pielke Jr., W. Pizer, C. Pringle, W. V. Reid, K. A. Rose, O. Sala, W. H. Schlesinger, D. H. Wall, and D. Wear. 2001. Ecological forecasts: An emerging imperative. *Science* 293:657–660.

Collins, T., J. C. Trexler, L. G. Nico, and T. Rawlings. 2002. Genetic diversity in a morphologically conservative invasive taxon: Multiple swamp eel introductions in the southeastern United States. *Conservation Biology* 16:1024–1035.

Curnutt, J. L., J. Comiskey, M. P. Nott, and L. J. Gross. 2000. Landscape-based spatially explicit species index models for Everglades restoration. *Ecological Applications* 10:1849–1860.

Davis, S. M. 1994. Phosphorus inputs and vegetation sensitivity in the Everglades, Pp. 357–378 in *Everglades: The system and its restoration*, edited by S. M. Davis and J. C. Ogden. Boca Raton, Fla.: St. Lucie Press.

DeAngelis, D. L., L. J. Gross, M. A. Huston, W. F. Wolff, D. M. Fleming, E. J. Comiskey, and S. M. Sylvester. 1998. Landscape modeling for Everglades ecosystem restoration. *Ecosystems* 1:64–75.

DeAngelis, D. L., W. F. Loftus, J. C. Trexler, and R. E. Ulanowicz. 1997. Modeling fish dynamics in a hydrologically pulsed ecosystem. *Journal of Aquatic Ecosystem Stress and Recovery* 6:1–13.

Egan, D., and E. A. Howell, eds. 2001. *The historical ecology handbook: A restorationist's guide to reference ecosystems*. Washington, D.C.: Island Press.

Fennema, R. J., C. J. Neidrauer, R. A. Johnson, T. K. MacVicar, and W. A. Perkins. 1994. A computer model to simulate natural Everglades hydrology, Pp. 249–289 in *Everglades: The system and its restoration*, edited by S. M. Davis and J. C. Ogden. Boca Raton, Fla.: St. Lucie Press.

Fitz, H. C., E. B. DeBellevue, R. Costanza, R. Boumans, T. Maxwell, L. Wainger, and F. H. Sklar. 1996. Development of a general ecosystem model for a range of scale and ecosystems. *Ecological Modelling* 88:263–295.

Fogarty, M. J. 1984. The ecology of the Everglades alligator. Pp. 211–218 in *Environments of south Florida present and past II*, edited by P. J. Gleason. Coral Gables, Fla.: Miami Geological Society.

Frederick, P. C., and G. V. N. Powell. 1994. Nutrient transport by wading birds in the Everglades. Pp. 571–584 in *Everglades: The system and its restoration*, edited by S. M. Davis and J. C. Ogden. Boca Raton, Fla.: St. Lucie Press.

Frederick, P. C., and M. G. Spalding. 1994. Factors affecting reproductive success of wading birds (Ciconiiformes) in the Everglades ecosystem. Pp. 659–691 in *Everglades: The system and its restoration*, edited by S. M. Davis and J. C. Ogden. Boca Raton, Fla.: St. Lucie Press.

Frohring, P. C., D. P. Voorhees, and J. A. Kushlan. 1988. History of wading bird populations in the Florida Everglades: A lesson in the use of historical information. *Colonial Waterbirds* 11:328–335.

Gore, J. A. 1985. Introduction. Pp. vii–xii in *The restoration of rivers and streams: Theory and experience*, edited by J. A. Gore. Stoneham, Mass.: Butterworth Publishers.

Grether G. F., D. F. Millie, M. J. Bryant, D. N. Reznick, and W. Mayea. 2001. Rain forest canopy cover, resource availability, and life history evolution in guppies. *Ecology* 82:1546–1559.

Gunderson, L. H. 1994. Vegetation of the Everglades: Determinants of community composition, Pp. 323–340 in *Everglades: The system and its restoration*, edited by S. M. Davis and J. C. Ogden. Boca Raton, Fla.: St. Lucie Press.

Hargrove, W. W., and J. Pickering. 1992. Pseudoreplication: A sine qua non for regional ecology. *Landscape Ecology* 6:251–258.

Higer, A. L., and M. C. Kolipinski. 1967. Pull-up trap: A quantitative device for sampling shallow-water animals. *Ecology* 48:1008–1009.

Hunt, B. P. 1952. Food relationships between Florida spotted gar and other organisms in the Tamiami Canal, Dade County, Florida. *Transactions of the American Fisheries Society* 82:13–33.

Jackson, J. B. C. 2001. What was natural in the coastal oceans? *Proceedings of the National Academy of Sciences of the United States of America* 98:5411–5418.

Jackson, J. B. C., M. X. Kirby, W. H. Berger, K. A. Bjorndal, L. W. Botsford, B. J. Bourque, R. H. Bradbury, R. Cooke, J. Erlandson, J. A. Estes, T. P. Hughes, S. Kidwell, C. B. Lange, H. S. Lenihan, J. M. Pandolfi, C. H. Peterson, R. S. Steneck, M. J. Tegner, and R. R. Warner. 2001. Historical overfishing and the recent collapse of coastal ecosystems. *Science* 293:629–637.

Jacobsen, T., and J. A. Kushlan. 1987. Sources of sampling bias in enclosure fish trapping: Effects on estimates of density and diversity. *Fishery Research* 5:401–412.

Jordan, F., S. Coyne, and J. C. Trexler. 1997a. Sampling fishes in vegetated habitats: Effects of habitat structure on sampling characteristics of the 1-square-meter throw trap. *Transactions of the American Fisheries Society* 126:1012–1020.

Jordan, F., H. L. Jelks, and W. M. Kitchens. 1997b. Habitat structure and plant species composition in a northern Everglades wetland landscape. *Wetlands* 17:275–283.

Karr, J. R., and E. W. Chu. 1999. *Restoring life in running waters.* Washington, D.C.: Island Press.

Kobza, R. M. 2001. Community structure of fishes in habiting hydrological refuges in a threatened karstic habitat. Master's thesis, Florida International University, Miami.

Kolipinski, M. C., and A. L. Higer. 1969. Some aspects of the effects of the quantity and quality of water on biological communities in Everglades National Park. Open File Report 69-007, U.S. Geological Survey, Tallahassee, Fla.

Kushlan, J. A. 1975. Population changes in the apple snail (*Pomacea paludosa*) in the southern Everglades. *Nautilus* 89:21–23.

———. 1976. Environmental stability and fish community diversity. *Ecology* 57:821–825.

———. 1980. Population fluctuations of Everglades fishes. *Copeia* 1980:870–874.

———. 1981. Sampling characteristics of enclosure fish traps. *Transactions of the American Fisheries Society* 110:557–562.

Kushlan, J. A., and M. S. Kushlan. 1979. Observations on crayfish in the Everglades, Florida, USA. *Crustaceana*, Supplement 5:115–119.

———. 1980. Population fluctuations of the prawn, *Palaemonetes paludosus*, in the Everglades. *American Midland Naturalist* 103:401–403.

Lange, T., D. Richard, and H. Royals. 2000. Abstract. Long-term trends of mercury bioaccumulation in Florida's largemouth bass. Annual Meeting South Florida Mercury Science Program, Tarpon Springs, Fla. 8–11 May.

Loftus, W. F. 2000. Accumulation and fate of mercury in an Everglades aquatic food web. Ph.D. diss., Florida International University, Miami.

Loftus, W. F., O. L. Bass, and J. C. Trexler. 1997. Long-term fish monitoring in the Everglades: Looking beyond the park boundary. Pp. 389–392 *Making protection work*, edited by D. Harmon. March. Ninth Conference on Research and Resources Management in Parks and On Public Lands, The George Wright Society, Albuquerque, N.M.

Loftus, W. F., J. D. Chapman, and R. Conrow. 1990. Hydroperiod effects on Everglades marsh food webs, with relation to marsh restoration efforts, Pp. 1–22 in *Fisheries and coastal wetlands research*. Vol. 6. Edited by G. Larson and M. Soukup. Proceedings of the 1986 Conference on Science in National Parks. Tucson: University of Arizona Press.

Loftus, W. F., and A. M. Eklund. 1994. Long-term dynamics of an Everglades fish community. Pp. 461–483 in *Everglades: The system and its restoration*, edited by S. M. Davis and J. C. Ogden. Boca Raton, Fla.: St. Lucie Press.

Lorenz, J. J. 1999. The response of fishes to physicochemical changes in the mangroves of northeast Florida Bay. *Estuaries* 22:500–517.

Lorenz J. J., C. C. McIvor, G. V. N. Powell, and P. C. Frederick. 1997. A drop net and removable walkway used to quantitatively sample fishes over wetland surfaces in the dwarf mangroves of the southern Everglades. *Wetlands* 17:346–359.

McCormick, P. V., S. Newman, S. Miao, D. E. Gawlik, D. Marley, K. R. Reddy, and T. D. Fontaine. 2001. Effects of anthropogenic phosphorus inputs on the Everglades. Pp. 83–126 in *The Everglades, Florida Bay, and coral reefs of the Florida Keys: An ecosystem sourcebook*, edited by J. W. Porter and K. G. Porter. Boca Raton, Fla.: CRC Press.

Miller, S. J. 1990. Kissimmee River fishes: A historical perspective. Pp. 31–42 in *Proceedings of the Kissimmee River Restoration Symposium, Orlando, Florida, October 1988*, edited by M. K. Loftin, L. A. Toth, and J. Obeysekera. South Florida Water Management District, West Palm Beach, Fla.

Ogden, J. C. 1994. A comparison of wading bird nesting colony dynamics (1931–1946 and 1974–1989) as an indication of ecosystem conditions in the southern Everglades. Pp. 533–570 in *Everglades: The system and its restoration*, edited by S. M. Davis and J. C. Ogden. Boca Raton, Fla.: St. Lucie Press.

Oreskes, N. 1998. Evaluation (not validation) of quantitative models. *Environmental Health Perspectives* 106:1453–1460.

Oreskes, N., K. Shrader-Frechette, and K. Belitz. 1994. Verification, validation, and confirmation of numerical models in the earth sciences. *Science* 263:641–646.

Pauly, D. 1995. Anecdotes and the shifting baseline syndrome of fisheries. *Trends in Ecology and Evolution* 10:430.

Reark, J. B. 1960. *Ecological investigations in the Everglades*. First annual report to Everglades National Park. University of Miami, Coral Gables, Fla.

————. 1961. Ecological investigations in the Everglades. Second annual report to Everglades National Park. University of Miami, Coral Gables, Fla.

Reynolds, J. H., and E. D. Ford. 1999. Multi-criteria assessment of ecological process models. *Ecology* 80:538–553.

Reznick D. N, F. H. Shaw, F. H. Rodd, and R. G. Shaw. 1997. Evaluation of the rate of evolution in natural populations of guppies (*Poecilia reticulata*). *Science* 275:1934–1937.

Rizzardi, K. W. 2001. Alligators and litigators: A recent history of Everglades regulation and litigation. *Florida Bar Journal* 75:18–27.

Rotmans, J., and M. B. A. van Asselt. 2001. Uncertainty management in integrated assessment modeling: Towards a pluralistic approach. *Environmental Monitoring and Assessment* 69:101–130.

Rykiel, E. J., Jr. 1996. Testing ecological models: The meaning of validation. *Ecological Modelling* 90:229–244.

Sarewitz, D., and R. Pielke. 1999. Prediction in science and policy. *Technology in Society* 21:121–133.

Sklar, F. H., H. C. Fitz, Y. Wu, R. Van Zee, and C. McVoy. 2001. The design of ecological landscape models for Everglades restoration. *Ecological Economics* 37:379–410.

Snowling, S. D., and J. R. Kramer. 2001. Evaluating modeling uncertainty for model selection. *Ecological Modelling* 138:17–30.

Stober, Q. J., D. Scheidt, R. Jones, K. W. Thornton, L. M. Gandy, D. Stevens, J. Trexler, and S. Rathbun. 1998. *South Florida ecosystem assessment: Monitoring for ecosystem restoration.* Final Technical Report, Phase I. EPA 904-R-98-002. United States EPA Region 4 Science and Ecosystem Support Division and Office of Research and Development, Athens, Ga. [Online at www.epa.gov/region4/sesd/reports/epa904r98002.html]

Trexler, J. C. 1995. Restoration of the Kissimmee River: A conceptual model of past and present fish communities and its consequences for evaluating restoration success. *Restoration Ecology* 3:195–210.

Trexler, J. C., and W. F. Loftus. 2001. *Analysis of relationships of Everglades fish with hydrology using long-term databases from the Everglades National Park.* Cooperative Agreement CA5280-8-9003.

Trexler, J. C., W. F. Loftus, and O. L. Bass. 1996a. Documenting the effects of Hurricane Andrew on Everglades aquatic communities, Part 3. Pp. 6–136 in *Effects of Hurricane Andrew on the structure and function of Everglades aquatic communities*, edited by J. C. Trexler, L. Richardson, and K. Spitze. Final Report CA5280-3-9014 to Everglades National Park, Homestead, Fla.

Trexler, J. C., W. F. Loftus, O. L. Bass Jr., and C. F. Jordan. 1996b. High-water assessment: The consequences of hydroperiod on marsh fish communities, Pp. 103–124 in *Proceedings of the 1996 Conference: Ecological Assessment of the 1994–1995 High Water Conditions in the southern Everglades*, edited by T. V. Armentano. 22–23 August. Florida International University, Miami.

Trexler, J. C., W. F. Loftus, C. F. Jordan, J. Chick, K. L. Kandl, T. C. McElroy, and O. L. Bass. 2001. Ecological scale and its implications for freshwater fishes in the Florida Everglades. Pp. 153–181 in *The Everglades, Florida Bay, and coral reefs of the Florida Keys: An Ecosystem Sourcebook*, edited by J. W. Porter and K. G. Porter. Boca Raton, Fla.: CRC Press.

Trexler, J. C., W. F. Loftus, F. Jordan, J. Lorenz, J. Chick, and R. M. Kobza. 2000. Empirical assessment of fish introductions in a subtropical wetland: An evaluation of contrasting views. *Biological Invasions* 2:265–277.

Turner, A., and J. C. Trexler. 1997. Sampling invertebrates from the Florida Everglades: A comparison of alternative methods. *Journal of the North American Benthological Society* 16:694–709.

Turner, A. M., J. C. Trexler, F. Jordan, S. J. Slack, P. Geddes, and W. F. Loftus. 1999. Targeting ecosystem features for conservation: Standing crops in the Florida Everglades. *Conservation Biology* 13:898–911.

Yount, J. D., and G. J. Niemi. 1990. Recovery of lotic communities and ecosystems from disturbance: A narrative review of case studies. *Environmental Management* 14:547–569.

Zillioux, E. J., D. B. Porcella, and J. M. Benoit. 1993. Mercury cycling and effects in freshwater wetland ecosystems. *Environmental Toxicology and Chemistry* 12:2245–2264.

CHAPTER 14

Monitoring Biodiversity for Ecoregional Initiatives

William L. Gaines, Richy J. Harrod, and John F. Lehmkuhl

The conservation of biodiversity has become an important issue receiving national and international attention (Noss 1991; Noss and Cooperrider 1994; Wilson 1992). Biodiversity has been defined in various ways (Harrod et al. 1996), leading to confusion amongst the general public about what biodiversity conservation means (Brussard et al. 1992). For example, Volland (1980) and Gast et al. (1991) define biodiversity as "the variety, distribution, and structure of plant and animal communities, including all vegetative stages, arranged in space over time that support self-sustaining populations of all natural and desirable naturalized plants and wild animals." We use biodiversity to describe the variety of life forms, the ecological roles they perform, and the genetic diversity they contain (Wilcox 1984).

Biodiversity hierarchy theory suggests that what happens at the higher levels of ecological organization, such as the landscape or ecosystem level, influences the lower levels, such as the species or genetic level (Allen and Star 1982; Noss 1990). Biodiversity hierarchy is composed of the genetic, species-population, community-ecosystem, and landscape or regional levels (Grumbine 1992; Harrod et al. 1996). Although these levels are convenient for the sake of discussion, they are not completely isolated or

377

strictly independent (King 1997). Rather, the members of the system at one level are themselves systems of elements of the next lower level and are at the same time components of the systems that occupy the next higher level (King 1997). In addition, biodiversity can also be influenced by scale, both temporal and spatial.

Regional-scale ecosystem initiatives, like the Northwest Forest Plan for federal lands within the range of the northern spotted owl (*Strix occidentalis caurina*) (USDA and USDI 1994), frequently involve the conservation of well-known rare or endangered species, and generally incorporate the conservation of overall biodiversity, at least implicitly. With these important goals in ecoregional initiatives, it is similarly important to assure that key species, as well as system biodiversity, are monitored to determine if goals and objectives of the initiative are achieved (Ringold et al. 1999). A decision to commit resources to monitor biodiversity could also be based on legal mandates such as the National Forest Management Act (NFMA 1976) or the Endangered Species Act (ESA 1973 as amended) (Harrod et al. 1996). In addition, monitoring is an important element of ecosystem management and an adaptive management approach (Woodley 1993; Everett et al. 1994; Noss and Cooperrider 1994; Ringold et al. 1999). We hope that by providing some "real world" examples of how biodiversity can be monitored, we will encourage land management agencies to develop comprehensive and scientifically credible monitoring programs.

Biodiversity Monitoring Approach

The initial phase in biodiversity surveys is estimating diversity at one point in time and location (in other words, knowing what species or communities are present). The second phase, monitoring biodiversity, involves estimating diversity at the same location, for multiple time periods, in order to make inferences about change (Wilson et al. 1996).

Wilson et al. (1996) identified attributes of biodiversity that can be assessed at each level of ecological organization. At the landscape level, attributes that could be monitored include the identity, distribution, and proportions of each type of habitat, and the distribution of species within those habitats. At the ecosystem level, richness, evenness, and diversity of species, guilds, and communities are important. At the species level, abundance, density, and biomass of each population may be of interest. And, at the genetic level, genetic diversity of individual organisms within a population is important. It is best to assess and interpret biodiversity across all these levels of organization by using various approaches at several spatial and temporal scales (Noss 1990; Noss and Cooperrider 1994; Niemela 2000).

We propose a three-phase approach to monitoring biodiversity: identifying monitoring questions; identifying monitoring methods; and analysis and interpretation of information for integration into management strategies (fig. 14-1). Each of these phases are described in greater detail below.

Identifying Monitoring Questions

This phase includes identifying and refining biodiversity monitoring questions, determining data needs to address the questions, and prioritizing monitoring questions and data needs. Examples of monitoring questions appropriate to each biodiversity level are shown in table 14-1. Prioritizing monitoring questions is important because the resources available to accomplish monitoring are likely to be limited.

Identifying monitoring questions is a critical and difficult step. It could be accomplished through an interdisciplinary process with experts knowledgeable of the issues at the appropriate level (e.g., landscape, ecosystem, species, genetic) and should be considered an iterative process that is adapted as new information becomes available.

In implementing the Northwest Forest Plan, monitoring questions could be derived from information available in watershed analysis, late-successional reserve (LSR) assessments, or could come from other regional assessments. For example, in the Wenatchee National Forest LSR assessment (U.S. Forest Service 1997), the assessment team developed monitoring questions, proposed data collection methods, and identified appropriate expertise needed to accomplish the monitoring. The monitoring questions were ranked as low, moderate, and high priority.

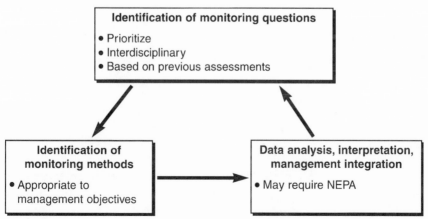

Figure 14-1. Three phases of biodiversity monitoring.

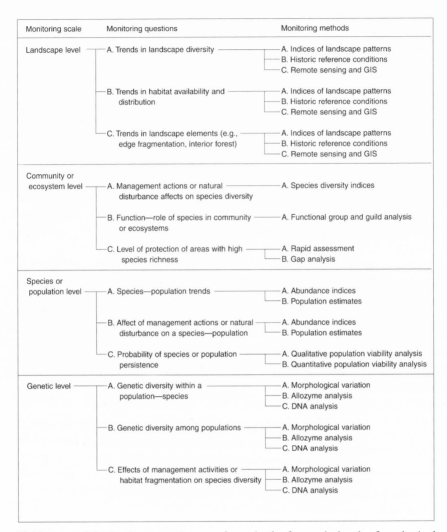

Monitoring scale	Monitoring questions	Monitoring methods
Landscape level	A. Trends in landscape diversity	A. Indices of landscape patterns B. Historic reference conditions C. Remote sensing and GIS
	B. Trends in habitat availability and distribution	A. Indices of landscape patterns B. Historic reference conditions C. Remote sensing and GIS
	C. Trends in landscape elements (e.g., edge fragmentation, interior forest)	A. Indices of landscape patterns B. Historic reference conditions C. Remote sensing and GIS
Community or ecosystem level	A. Management actions or natural disturbance affects on species diversity	A. Species diversity indices
	B. Function—role of species in community or ecosystems	A. Functional group and guild analysis
	C. Level of protection of areas with high species richness	A. Rapid assessment B. Gap analysis
Species or population level	A. Species—population trends	A. Abundance indices B. Population estimates
	B. Affect of management actions or natural disturbance on a species—population	A. Abundance indices B. Population estimates
	C. Probability of species or population persistence	A. Qualitative population viability analysis B. Quantitative population viability analysis
Genetic level	A. Genetic diversity within a population—species	A. Morphological variation B. Allozyme analysis C. DNA analysis
	B. Genetic diversity among populations	A. Morphological variation B. Allozyme analysis C. DNA analysis
	C. Effects of management activities or habitat fragmentation on species diversity	A. Morphological variation B. Allozyme analysis C. DNA analysis

Table 14-1. Monitoring questions and methods for each level of ecological organization.

Ultimately, management must determine which monitoring questions should be addressed.

Identifying Monitoring Methods

Methods selected for monitoring biodiversity depend on management objectives. A management objective of maintaining species viability would involve different monitoring methods than those of an objective of restoring inherent disturbance regimes.

Selecting the appropriate biodiversity monitoring approach includes identifying methods that will provide answers to specific monitoring questions (table 14-1). A wide range of methods is available, and the selection of the appropriate methods would be made based on costs, available resources, and statistical constraints. A critical part of this step is the determination of sample sizes, sampling strategies, and desired statistical power (Nyberg 1998).

Data Analysis, Interpretation, and Management Integration

Periodically, data collected from monitoring would be analyzed and integrated into management strategies based on the knowledge gained. If monitoring reveals that adjustments need to be made in management strategies, then this becomes a decision step for managers or policy makers.

Examples of Monitoring at Each Biodiversity Level

Presented below are several examples of monitoring biological diversity at each level of ecological organization, beginning with the landscape level. For each level, we provide definitions and background on the relevance of monitoring to land managers; examples of monitoring questions; a summary of available monitoring methods and references to determine appropriate methods; and case studies of the actual application of these methods, the results they can produce, and interpretations of these results. Some methods have not been adequately developed due to our limited understanding of certain aspects of biodiversity (Harrod et al. 1996). Other methods have serious limitations, which we attempt to point out.

Landscape Monitoring

A landscape has been defined as a land area with groups of plant communities or ecosystems forming an ecological unit with distinguishable structure, function, geomorphology, and disturbance regimes (Forman and Godron 1986; Noss 1983; Romme and Knight 1982). Landscape diversity is the number of ecosystems, or combinations of ecosystems, and types of interactions and disturbances present within a given landscape.

The relevance of landscape structure to biodiversity has been established in the scientific literature (e.g., Forman and Godron 1986). Landscape features such as patch size, heterogeneity, and connectivity have major implications for ecosystem processes and species composition, distribution, and viability (Noss and Harris 1986). Because of this, it may be important that managers monitor elements of biodiversity at the landscape scale to meet species viability requirements or objectives of an ecoregional initiative.

Examples of monitoring questions at this level could include the following: What is the current level of landscape diversity and how does it compare with historic or sustainable levels? What are the trends in habitats or populations of a particular species? What are the trends in landscape features, such as the amount of edge, patch size, and forest interior?

Presented below are two approaches to assessing biological diversity at a landscape scale: measuring landscape patterns and comparing current conditions to historic reference conditions. Each of these approaches relies on the use of geographic information systems (GISs) and requires mapped vegetation and other layers that can be analyzed with GIS technologies.

INDICES OF LANDSCAPE PATTERNS

Landscape pattern measurements, or metrics, can be classified into three categories: patch, class, and landscape (McGarigal and Marks 1993). Patch metrics describe the attributes of individual patches of vegetation. The size, shape, edge, or nearest-neighbor relations of individual patches are measured. Class metrics describe those same patch attributes as the mean, minimum, maximum, or variance for a class of mapped landscape attributes (e.g., late-successional forest). Landscape pattern metrics describe these and other attributes for all landscape classes combined without distinction between different classes. For example, mean patch size might be measured for all patches in a landscape as a landscape metric rather than for just one vegetation type (class). In addition to the diversity and evenness indices that will be discussed later, there are many other landscape metrics available to index specific aspects of landscape spatial pattern (O'Neill et al. 1999).

O'Neill et al. (1988) and Turner (1990) proposed indices of dominance, contagion, fractal complexity, and disturbance as standard measures of landscape pattern. Dominance is an index of vegetation type composition and equitability. Contagion measures the extent to which a single class of vegetation or all types combined are clumped (high contagion) or patterned in a fine-scale (low-contagion) mosaic. Fractal indices describe patch shape and boundary (edge) complexity. The perimeter-area fractal is the most commonly used fractal index, and the one described by O'Neill et al. (1988); however, other forms of the fractal index have been developed. For example, Korcak's fractal incorporates the number-area relation of vegetation patches to describe patch-size variability and fragmentation (Burrough 1986; Milne 1988). Most fractal indices can be easily calculated with simple regression methods (Burrough 1986). A disturbance index is calculated as a ratio of the area of disturbed vegetation to that of undisturbed vegetation.

Other class or landscape pattern indices continue to be developed by landscape ecologists and can be found in the literature. For example, Lehmkuhl et al. (1991) and Lehmkuhl and Raphael (1993) used a form of the contagion index, as suggested by Turner et al. (1989), to measure fragmentation of late-successional forest. Ripple et al. (1991) and Lehmkuhl and Raphael (1993) calculated an index of late-successional forest isolation that incorporates information on both the total area and the spatial distribution of late-successional forest patches by using a standard GIS proximity function. Schumaker (1996) developed a patch cohesion index of habitat connectivity, and Metzger and Muller (1996) developed metrics to describe boundary characteristics. O'Neill et al. (1999) review recent developments of pattern indices.

A typical application of these indices to forest landscapes in western Washington can be found in Lehmkuhl et al. (1991). They found that forest patch shape is not complex as measured by the fractal index (maximum value of 1.27 within a possible range of 1 to 2). The dominance indices showed that no single vegetation type was dominant in most landscapes (mean value of 0.93 compared to a maximum value of 3). Their landscapes showed high vegetation type contagion (mean of 1.7 out of a possible 20). The disturbance index showed a relatively low level of logging disturbance on the landscape (mean of 20 percent disturbed). Many other examples are found in the literature.

Calculation of landscape pattern metrics is only meaningful if interpreted in terms of specific environmental end points, such as wildlife or water quality (O'Neill et al. 1999). The bias or limitation of each index also needs to be considered. Lehmkuhl et al. (1991) found that contagion of all vegetation types proved to be a variable closely associated with differences among vertebrate assemblages in forests of western Washington and Oregon. O'Neill et al. (1988) and Turner (1990) caution users, however, that the index can be sensitive to the number of types in the landscape so that comparisons of different areas with different numbers of vegetation types could be misleading if the index is not normalized to the number of types. Lehmkuhl et al. (1991) also found their isolation index is another attribute of late-successional forest pattern associated with differences in vertebrate assemblages. The isolation index, however, is correlated with the total area of late-successional forest, which would indicate that simply measuring late-successional forest area might be an effective and simple method. Lehmkuhl and Raphael (1993) also concluded that simply measuring the area of habitat is the single best indicator of habitat pattern distinguishing northern spotted owl areas on the Olympic

Peninsula of Washington. Others, similarly, are finding that a simple measure of habitat area in many cases may be the best indicator of habitat pattern, or fragmentation, for areas where suitable habitat is more than 40 percent of the landscape (Andren 1994; With and Crist 1995; Meyer et al. 1998).

Landscape edge complexity, measured by the perimeter-area fractal and Korcak's fractal, is also associated with some differences in vertebrate occurrence in western Washington and Oregon (Lehmkuhl et al. 1991). Turner and Ruscher (1988) and Mladenoff et al. (1993) found that the perimeter-area fractal is a good indicator of landscape disturbance, indicating the conversion of complex wild landscapes to the simple geometry of human-dominated landscapes. Lehmkuhl et al. (1994), however, found that fractal indices are poorly correlated with logging disturbance measured as the area of clearcut in the landscape. Others have questioned the accuracy of the fractal and other indices of landscape pattern (Cale and Hobbs 1994; Groom and Schumaker 1996; Schumaker 1996).

There are more caveats for using landscape pattern metrics. Not all metrics may be useful in describing the attributes of all landscapes or in answering specific questions about landscapes relative to the ecology or management of a particular plant or animal community. Spatial pattern analysis software packages, such as FRAGSTATS (McGarigal and Marks 1993), are easy to use and generate much data on landscape metrics. Sorting through those results can be confusing and wasteful unless careful thought is given to the questions to be answered and to determining which metrics can answer those questions. Moreover, attention should be paid to the differences in scale and dimension of maps when comparing indices of pattern between locations. Turner (1989, 1990), Lehmkuhl and Raphael (1993), and O'Neill et al. (1996) found that estimates of some pattern indices for a simple landscape may vary depending on the scale of the map units (grain) or map dimensions (extent). O'Neill et al. (1996) provide guidelines for map grain (resolution) and area to avoid bias in calculating indices.

HISTORIC REFERENCE CONDITIONS

Swanson et al. (1994), Morgan et al. (1994), and Parsons et al. (1999) described the application of the range of natural variability as a reference against which to compare the current condition of landscapes. They stressed that the use of natural variability as a reference condition is not an attempt to turn managed landscapes into wilderness areas or return them to a single preexisting condition. Instead, it is an approach to meet ecological or social objectives by managing existing conditions in a landscape within historical natural range of conditions that may be sustainable for that ecosystem.

Caraher et al. (1992) applied the range of natural variability concept to the Blue Mountains landscape in eastern Oregon. They estimated the range of natural variability for several ecosystem components (i.e., distribution of several stages, stream shrub cover, streambank stability). They found that several ecosystem elements are currently outside the range of natural variability because of activities such as fire exclusion, tree planting, and fish restocking. Lehmkuhl et al. (1994) and Hessburg et al. (1999) compared the historical and current condition of several landscape variables and indices for forests of the interior Columbia River Basin. Gaines and Harrod (1994) and Gaines (2001) presented a process that could be used during watershed analysis to assess the condition of wildlife habitats and identify potential restoration projects. They mapped the current and projected historic distribution of habitats using GIS and, by comparing the differences in these maps, identified areas on which to focus habitat restoration. More examples and a complete discussion of the uses and limitations of historical data can be found in articles featured in Parsons et al. (1999).

Community-Ecosystem Monitoring

A community comprises the populations of some or all species coexisting at a site. An ecosystem includes the abiotic aspects of the environment and the biotic community. Monitoring at this level is important to the maintenance of ecosystem functions and integrity that have been identified as a main theme of ecosystem management (Haynes et al. 1996; Marcot et al. 1994). Land managers may be interested in monitoring communities or ecosystems to determine if current management strategies meet legal and social obligations to sustain the health, diversity, and productivity of ecosystems (Haynes et al. 1996).

Monitoring questions that could be asked include: How have management activities or natural disturbances affected species diversity in a particular community? What is the function of a species in the community or ecosystem? Where are the areas of high species richness, endemism, or rarity, and how well are they protected?

A common way of assessing biodiversity is by measuring the number and relative abundance of species in a community or ecosystem, often referred to as *species diversity*. Species diversity is a function of the number of species present (richness) and the evenness or equitability (relative abundance) of each (Hurlbert 1971). Hurlbert (1971) pointed out that although species diversity and species richness are often positively correlated, situations do exist in which increases in species diversity are accompanied by decreases in species richness. Care should be taken therefore, when only species richness (counts of the number of species) is used to evaluate biodiversity. On the other hand, species diversity indices also should be carefully used because it may be hard to interpret differences in species composition at different sites. For example, two sites may have similar indices of diversity but have entirely different species composition. One site may be primarily exotic species, whereas the other may be mainly native or endemic species.

DIVERSITY INDICES

Various indices and models have been developed to measure diversity within a community (Magurran 1988). In general, three main categories of measures are used to assess species diversity: (1) species richness indices, which measure the number of species in a sampling unit; (2) species abundance models, which have been developed to describe the distribution of species abundances; and (3) indices that are based on the proportional abundances of species such as the Shannon and Simpson indices (Magurran 1988).

Selecting the appropriate index in which to assess community-level diversity can be difficult, since there are many to choose from and each has advantages and disadvantages. Magurran (1988) suggested that important criteria for selection include the ability to discriminate between sites, dependence on sample size, the component of diversity being measured, and whether the index is widely used and understood. The most commonly used indices include the log series, species richness, Shannon index, and the Simpson index (Magurran 1988).

To quantify community-level biodiversity, an inventory or sample of the species present and their relative abundance must be completed. Various methods have been used to inventory plant, wildlife, and fish species. Summaries of the techniques can be found in Cooperrider et al. (1986), Barbour et al. (1987), and Kent and Coker (1992).

Many examples of the application of indices to assess diversity of species within a community are in the literature. For example, Thiollay (1992) used species richness, the Shannon index, an evenness index, and the Simpson index to assess the influence of selective logging on bird species diversity. He used a point-count method in which nine hundred and thirty-seven 0.25-hectare sample quadrants were inventoried for twenty minutes each. He discovered an overall 27 to 33 percent decrease of species richness, frequency, and abundance following logging, with a less-marked decline of species diversity and evenness.

Cushing and Gaines (1989) used the species richness, the Shannon index, and an equitability or evenness index to compare diversity of aquatic insects in three cold-desert streams and to evaluate the effects of winter spates on diversity. They collected quantitative macroinvertebrate samples at monthly intervals from each stream. They discovered that species diversity increased with increasing stream size and substratum diversity but declined following winter spates.

FUNCTIONAL GROUPS OR GUILDS

Some investigators have taken a different viewpoint by lumping species into functional groups or guilds. Many approaches for grouping species based on habitat or behavioral similarities and their potential problems have been discussed in the literature (Landres et al. 1988; Morrison et al. 1992; Noss and Cooperrider 1994). Walker (1992) suggests an approach in which species are grouped into guilds based on their function in the ecosystem, and then the relative importance of each guild is considered based on how a change in their abundance affects ecosystem and community processes. Gaines et al. (1989) grouped aquatic insects into trophic levels and functional groups and then used density and biomass as indicators of numerical and biomass dominance to assess community diversity in three cold-desert streams. They found that detritivores are the dominant (both numerically and as biomass) function in their study streams and that collecting (gathering and filtering) is the major feeding strategy used. From this information, they inferred that detritus in the form of fine particulate organic matter is the major food source for benthic invertebrates in small, headwater cold-desert streams. Thus, managing for streamside vegetation is important in maintaining the function of aquatic insect communities.

Birds are commonly grouped into foraging guilds (Mannan et al. 1984; Manuwal et al. 1987). For example, Mannan et al. (1984) monitored the effects of herbicides on avian foraging guilds. They identified a potential danger in examining only the summed response of all guild members in that intraguild responses were inconsistent, because a large increase of one or two species could mask the decline or absence of others (Mannan et al. 1984). Monitoring of functional groups or guilds may be most meaningful if combined with population-species-level monitoring.

RAPID ASSESSMENT TECHNIQUES

Another approach using groups of invertebrates and nonflowering plants as surrogates of biodiversity was presented by Oliver and Beattie (1993).

They estimated species richness of spiders, ants, polychaetes, and mosses, and divided them into recognizable taxonomic units (RTUs). These RTUs are taxa that are readily separated by morphological differences that are obvious to individuals with less training than professional taxonomists. They found that by using RTUs, there is little difference between classifications made by a biodiversity technician and those made by a taxonomy specialist. This could result in a considerable savings of time and money to complete the inventories needed to assess this one facet of biological diversity. This technique could be applied at different time periods to monitor the trends in the numbers of individuals within RTUs.

GAP ANALYSIS

Scott et al. (1993) presented a method, applied at the landscape or regional scale, to assess biological diversity at the community level. This approach, called *gap analysis*, provides a framework in which to obtain an overview of the distribution and conservation status of several components of biodiversity. The approach involves mapping, digitizing, and ground-truthing vegetation and species distribution data; digitizing biodiversity management areas and landownership maps; adding location data on all species and high-interest habitats such as wetlands and streams; and mapping, delineating, and ranking areas of high community diversity and species richness. These data are then used to identify "gaps" in protection of vegetation types and species-rich areas and provide land managers with information needed to make informed decisions about reserve selection and design, land management policy, and other conservation actions.

Population-Species Monitoring

A population is defined as all individuals of one species occupying a defined area and usually isolated to some degree from other similar groups. A species is generally defined as a group of organisms formally recognized as distinct from other groups.

Monitoring at this level may have the most relevance to meeting the species or population viability objectives of regional ecosystem initiatives. For example, land managers may decide to monitor a species or population in order to measure trends. This would be important in determining if management strategies maintain population viability.

Monitoring questions at the species or population level could include, What is the trend in the species or population? How is species or population abundance affected by land-management activities or natural disturbances? What is the probability of population or species persistence over a period of time in a specified area?

Most monitoring of biodiversity has occurred at the population-species level. Deciding which species or population to monitor has received considerable discussion, and no single approach is without pitfalls. Noss (1990) suggests five categories of species that may be selected for monitoring: (1) ecological indicators—species that signal the effects of perturbations on a number of other species with similar habitat requirements; (2) keystones—species on which the diversity of a large part of a community depends; (3) umbrellas—species with large area requirements, which if conserved would also conserve many other species; (4) flagships—popular, charismatic species; and (5) vulnerables—species that are rare, genetically impoverished, or for some reason prone to extinction. Species may also be selected to monitor for bioindicators and it may be appropriate to use species of invertebrates, fungi, lichens, and amphibians, as well as vertebrates and vascular plants (Marcot et al. 1997, 1998).

Monitoring a species or population may include counting of individuals but often involves the monitoring of habitat that is used by or is important to a species (Cooperrider et al. 1986; Noss 1990). Noss (1990) pointed out that monitoring habitat variables does not alleviate the need to monitor populations because the presence of habitat is no guarantee that the species is present. Conversely, monitoring only population variables could be misleading because some individuals may occur in areas of marginal habitat (Van Horne 1983). The most reliable approach would include monitoring both habitat and population variables (Noss 1990).

Various methods are available to monitor and inventory populations and species, and their habitats. Cooperrider et al. (1986) provide a comprehensive reference of the available methods for monitoring and inventorying fish and wildlife habitat and populations, Wilson et al. (1996) provide an overview of standard methods to measure and monitor mammals, and Heyer et al. (1994) summarize standard methods for amphibians. Several methods are available for plants but are not discussed in detail here. Where appropriate, we mention some common methods and provide references where more-detailed discussions of these techniques can be found.

ABUNDANCE INDICES

An index is usually a count statistic that is obtained in the field and carries information about a population (Wilson et al. 1996). An index is usually used when individuals of the species in question are difficult to observe and count, or capture and tag, or when a formal abundance estimate is too expensive or time-consuming. Abundance indices are divided into direct indices and indirect indices. Direct indices are based on direct observation of animals, either visually or through capture or harvest (Seber 1982). These include incomplete counts such as those obtained from aerial surveys, capture indices based on the number of individual animals captured per unit of time or effort, and harvest indices based on the number of animals harvested over a specific period of time or effort (Wilson et al. 1996). Indirect indices are based on evidence of an animals presence (Seber 1982): for example, track counts, scent station surveys, auditory indices (based on counts of animal sounds), structure surveys (numbers of nests, lodges, food caches, etc.), scat and other sign counts, and home-range size estimates (use of home range as an index of population density) (Wilson et al. 1996).

Gaines et al. (1995) provide an example of an indirect index using auditory responses of coyotes (*Canis latrans*) and gray wolves (*Canis lupus*) to simulated howling. They conducted a total of 2,137 howling sessions resulting in 215 responses by canids and an overall response rate-howling session (RR-S) of 10 percent. Two responses had the vocal characteristics of wolves for a RR-S of 0.1 percent, and 213 responses were of coyotes for a RR-S of 9.9 percent. It remains unknown at this time how well RR-S relates to actual population size or density, and thus its use as an abundance index should be approached cautiously.

Point counts, based on landbird vocalizations, are a commonly used method to index avian species populations (Ralph et al. 1993; Huff et al. 2001). Huff et al. (2001) present a methodology to monitor the trends in avian species numbers over time at a regional scale using point counts as an index to avian populations. Ralph et al. (1993) provides a comprehensive handbook of methods that can be used to monitor landbirds.

The advantages of using abundance indices are that they are less expensive and less time-consuming than more formal population estimation techniques. If appropriate indices are available, and if their relationships to abundance are known to be invariant over time and survey conditions, then there is no need to estimate an absolute density (Wilson et al. 1996). Such ideal conditions, however, almost never exist and often present potentially misleading conclusions regarding population size or density.

POPULATION ESTIMATES

Estimates of population size and density usually require significant investments of time and money. They are divided into complete counts, or a census, and estimates when complete counts are not possible (Wilson et al. 1996). In a complete count, the population size can be determined in a survey area from the number of individuals counted, with no correction for sampling observation probability (e.g., observability, sightability, visibility, and detectability). In most situations, counts of organisms in an area are incomplete and represent unknown portions of the total population. Thus, it is important to determine the probability of detecting an individual in a given survey area in order to estimate the population size or density.

Miller et al. (1997) provide an example of the use of radiotelemetry and replicated mark-resight techniques to estimate brown bear (*Ursus arctos*) and black bear (*Ursus americanus*) densities. They estimated a mean probability of sighting a marked individual by experienced pilot-observer teams to be 0.323 for brown bears and 0.321 for black bears. Brown-bear densities ranged from 10 to 551 bears per 1,000 square kilometers of habitat from fifteen study areas. Black-bear density ranged from 89 to 290 bears per 1,000 square kilometers from three study areas. These estimates were used to track changes in bear abundance that may result from environmental disturbances or human impacts, to estimate the total number of bears in Alaska, and as one component of determining sustainable harvest quotas.

Madsen et al. (1999) developed an effectiveness monitoring plan for the marbled murrelet (*Brachyrhamphus marmoratus*) for the Northwest Forest Plan. They incorporated population monitoring using line transects employed by surveyors from ships that traverse the marine habitats of murrelets. This monitoring plan provides a species-population level example of monitoring conducted specifically to determine if regional-scale objectives are being met.

POPULATION VIABILITY ANALYSIS

Population viability analysis (PVA) estimates the conditions necessary for a population to persist for a given period of time in a given place (Soulé 1987). The PVA allows for a prediction of the possible trend of a population and can provide insights into why a population may be decreasing. One of the disadvantages of a quantitative PVA is that a rigorous data set is needed to complete the analysis. These data are often expensive to obtain and require several years or decades of study. Because of the need to have such a rigorous and extensive data set, PVA may be too cumbersome to use to monitor biodiversity. When such data are available, however, PVA may be useful in establishing baseline population information to predict how management actions might affect viability. Recently, a new approach has been used to develop a qualitative PVA relying on the professional judgment of scientists familiar with a species or group of species (Thomas et al. 1993; Lehmkuhl et al. 1997). This qualitative approach has the advantage of requiring a less-rigorous data set that still meets the essential criterion of a PVA, in order to provide an estimate of the likelihood that a population will persist over a given time period.

Beier (1993) conducted a quantitative PVA for cougars (*Felis concolor*), which exist at low density and require large areas. Beier (1993) used population parameters that included mean litter size, juvenile survival rates, adult survival rates, and carrying capacity in his PVA model. These data were obtained from an extensive study of a cougar population in the Santa Ana Mountains of southern California. Beier also factored into his PVA model different sizes of habitat areas, levels of immigration, and some degree of environmental stochasticity. He found that there was a high likelihood of extinction if the amount of habitat that is currently available is reduced and if an important movement corridor is further degraded by land-development activities.

There are several studies that use PVA for rare plant species, and each takes a slightly different approach. We suggest reviewing these studies (e.g., Burgman and Lamont 1992; Harper 1977; Koopowitz et al. 1994; Manasse 1992; Menges 1990; Possingham et al. 1992) and selecting an approach that best fits the specific situation, while keeping in mind the following two points. First, when carrying out such analysis for plants, consider the mode of reproduction of the species under study. Plants that are *autogamous* (self-fertilizing) or *agamospermic* (seed production without fertilization) may persist indefinitely regardless of numbers, assuming that their fixed genomes provide the ability to exist in a given environment. Second, small populations of plants that are obligate out-crossers may be subject to loss of genetic diversity, such as from genetic drift. Given (1994) stated that 64 percent of the neutral genetic variation of a population will be lost through genetic drift in one hundred generations (see Lesica and Allendorf 1992, however). It follows that the rate of seedling recruitment into the adult population is an important aspect in determining how long small populations will persist.

A qualitative approach allows for an assessment of many species at broad spatial scales. Thomas et al. (1993) used this approach to PVA to assess different forest management alternatives in the Forest Ecosystem Management Assessment for the Northwest Forest Plan, as did Lehmkuhl et al. (1997) for the Interior Columbia Basin Ecosystem Management Project. However, although a qualitative approach allows for a rapid viability assessment without the commitment of resources to gain extensive life history data, it is not without some drawbacks. Qualitative assessments could be used as interim PVAs until resources become available to obtain the data necessary to test the many assumptions that interim PVAs are based on (Shaffer 1990).

Genetic Monitoring

Genetic diversity refers to the breadth of genetic variation within and among individual populations and species. Frankel (1970) postulated that genetic variation is essential for the long-term survival of endangered species, especially those that occur in fast-changing or harsh environments. Genetic diversity is a necessary prerequisite for future adaptive change or evolution, and, presumably, populations and species that lack genetic variation are at greater risk of extinction (Hamrick et al. 1991; Schaal et al. 1991). Land managers may decide to monitor genetic diversity to determine if management strategies are providing for species viability.

Examples of monitoring questions that could be asked include: What is the genetic diversity within a population or among populations and does it change over time? How has habitat fragmentation affected the genetic structure of a population or species?

Determining which population or species should be monitored for genetic diversity should occur when specific monitoring questions are developed. Most of the time, however, the resources available to conduct genetic diversity studies will be the limiting factor in their application. We therefore recommend the following criteria in the selection of populations or species for this level of monitoring: (1) species or populations that are limited in their numbers and distribution (e.g., endangered, threatened, and candidate species), (2) populations that are naturally fragmented or have become fragmented as a result of human activities and the likelihood of genetic interchange among component populations is low, (3) populations that are on the edge of a species range, and (4) species that naturally occur at low densities but may have wide distribution (e.g., large carnivores).

Lande and Barrowclough (1987) recommend that long-term population management programs should involve some form of direct monitoring of genetic variation. Three major types of characters have been used to estimate the level of genetic diversity: morphological, allozyme, and DNA sequences (Schaal et al. 1991).

MORPHOLOGICAL VARIATION

The measurement of morphological variation is the most easily obtained indicator of genetic diversity. Morphological measurements often can be obtained in the field or from field specimens and thus not require laboratory studies. Another advantage is that morphological characters may be ecologically adaptive (Schaal et al. 1991), meaning they are good indicators of genetic variation, local differentiation, or ecotypes. This method is often the most realistic when the biochemical analysis discussed below is impractical. Perhaps the greatest disadvantage is the assumption that morphological variation is a reliable indicator of underlying genetic variation. This assumption can be difficult to validate unless it is done in conjunction with allozyme or DNA analysis.

Several examples of the use of morphological measurements are available in the published literature. Rausch (1963) compared skull measurements of grizzly bears and brown bears from twenty-six different regions and found geographic variation among populations. Erickson (1945) found a hierarchical distribution of leaf-shape variation within and among groups of *Clematis fremontii reihlii* in the limestone glades of the Missouri Ozarks. Taylor et al. (1994) used needle and cone characteristics to determine patterns of similarity of spruce (*Picea* spp.) in western North America.

Analysis of morphological variation is also commonly presented in papers dealing with biosystematics. For example, Anderson and Taylor (1983) analyzed morphological variation within a large (50-hectare) population of mixed species of *Castilleja* to determine patterns of relationships. They measured ninety-eight characters of one hundred selected plants and subjected the data to various taximetric analysis, including cluster, discriminant, and principal components.

ALLOZYME VARIATION

Allozyme electrophoresis has been the most common method used to assess genetic diversity. Allozyme analysis provides an estimate of gene and genotypic frequencies within populations (Schaal et al. 1991). Such data can be used to measure population subdivision (Weir and Cockerham 1984), genetic diversity (Nei 1973), gene flow (Slatkin and Barton 1989), genetic structure of species, and comparisons among species (Taylor 1991). In general, allozyme variation does seem to be a good indicator of the overall level of variation within a genome (Hamrick 1989). However, because this analysis includes only a single class of genes, those encoding soluble enzymes, these genes may not always be representative of the variation within the genome (Schaal et al. 1991).

Analysis of allozyme variation has been used to describe the genetic structure of populations of bull trout (*Salvelinus confluentus*) in the Columbia and Klamath River drainages (Leary et al. 1993). Leary et al. (1993) discovered that the two populations had little genetic variation within the populations but significant genetic differences between populations, thereby indicating substantial genetic divergence. In fact, these populations would qualify as separate "species" under the Endangered Species Act according to criteria established for anadromous salmonid species (Leary et al. 1993).

Stangel et al. (1992) used allozyme variation analysis to survey genetic variation and examine population structure among twenty-six populations of the endangered red-cockaded woodpecker (*Picoides borealis*). They found a large among-population component of genetic variation when compared to other bird species (Stangel et al. 1992). Genetic variation was reduced in some small populations, but these small populations were important as reservoirs of unique genetic combinations (Stangel et al. 1992).

VARIATION IN DNA

The development of recombinant DNA technologies allows the direct measurement of genetic variation, which differs from indirect methods such as estimation of variation from a phenotype. There are, however, some significant drawbacks to DNA analysis. For example, the laboratory techniques are complicated, time-consuming, and costly. In addition, because only a small segment of the genome is analyzed at one time, there is a potential danger of misinterpreting conclusions about genetic variation in one type of sequence to the entire genome (Schaal et al. 1991).

Restriction-site analysis of mitochondrial DNA has been used to deduce population structure (Avise et al. 1987). Populations within species often have unique mitochondrial DNA genotypes that reflect the distance between populations or the presence of geographic barriers to genetic interchange (Wayne et al. 1992).

There are many examples of the application of DNA technology to the study of wildlife, fish, and plant species. Leary et al. (1993) used mitochondrial DNA restriction fragments in combination with allozyme analysis to study genetic variation in populations of bull trout. Haig et al. (2001) used random amplified polymorphic DNA (RAPD) to examine genetic variation hierarchically among local breeding areas, subregional groups, regional groups, and subspecies of spotted owls (*Strix occidentalis*). They used this information to support the designation of conservation units for spotted owls. A repeat evaluation of these studies (Leary et al. 1993 or Haig et al. 2001) at a future date could be used to evaluate whether genetic diversity is changing among the populations and if they are becoming more or less distinct.

Conclusion

In this chapter, we presented an approach to monitoring biological diversity. We suggest a three-phase approach that includes the identification of monitoring questions; identification of monitoring methods; and the analysis, interpretation, and integration of the data into management strategies. The identification of monitoring questions and methods could be accomplished through an interdisciplinary approach, building on baseline information provided in various ecoregional assessments. We have provided several examples of methods available for land managers to use in monitoring biodiversity that allow the quantification and interpretation of monitoring data.

Because monitoring biodiversity is a potentially large effort, we suggest it is better to monitor a few elements well than to spread resources too thin. In addition, monitoring should focus on elements that have relevance to key management issues. In order to monitor biodiversity at a variety of temporal and spatial scales, an interagency effort that is based on cooperation and collaboration will be needed. Collaboration between researchers and managers will be especially important. Managers with skills in the quantitative sciences will be needed in order to quantify and interpret monitoring information. Monitoring biological diversity is important for land managers to determine if management strategies meet legal mandates, to achieve goals and objectives identified in ecoregional initiatives, and to implement an adaptive approach to ecosystem management.

ACKNOWLEDGMENTS

We thank the Blue Mountains Natural Resources Institute for the initial impetus to produce a synthesis document to address this important issue. We also thank Bruce Marcot, Susan Piper, Lynn Starr, and Peter Morrison for their thorough review and thoughtful suggestions to improve this manuscript.

LITERATURE CITED

Allen, T. F. H., and T. B. Star. 1982. *Hierarchy: Perspectives for ecological complexity.* Chicago: University of Chicago Press.

Anderson, A. V. R., and R. J. Taylor. 1983. Patterns of morphological variation in a population of mixed species of *Castilleja* (Scrophulariaceae). *Systematic Botany* 83(3): 225–232.

Andren, H. 1994. Effects of habitat fragmentation on birds and mammals in landscapes with different proportions of suitable habitat: A review. *Oikos* 71:355–366.

Avise, J. C., J. Arnold, J., R. M. Ball, E. Bermingham, T. Lamb, J. E. Neigel, C. A. Reeb, and N. C. Saunders. 1987. Intraspecific phylogeography: The mitochondrial DNA bridge between population genetics and systematics. *Annual Review of Ecology and Systematics* 18:489–522.

Barbour, M. G., J. H. Burk, and W. D. Pitts. 1987. *Terrestrial plant ecology.* Menlo Park, Calif.: Benjamin/Cummings Publishing.

Beier, P. 1993. Determining minimum habitat areas and habitat corridors for cougars. *Conservation Biology* 7(1):94–108.

Brussard, P. F., D. D. Murphy, and R. F Noss. 1992. Strategy and tactics for conserving biological diversity in the United States. *Conservation Biology* 6(2):157–159.

Burgman, M. A., and B. B. Lamont. 1992. A stochastic model for the viability of *Banksia cuneata* populations: Environmental, demographic and genetic effects. *Journal of Applied Ecology* 29:719–727.

Burrough, P.A. 1986. *Principles of geographic information systems for land resources assessment.* Oxford, U.K: Clarendon Press.

Cale, P. G., and R. J. Hobbs. 1994. Landscape heterogeneity indices: Problems of scale and applicability, with particular reference to animal habitat description. *Pacific Conservation Biology* 1:183–193.

Caraher, D. L., J. Henshaw, F. Hall, W. H. Knapp, B. P. McCammon, and J. Nesbitt. 1992. *Restoring ecosystems in the Blue Mountains: A report to the regional forester and the forest supervisors of the Blue Mountain forests.* U.S. Department of Agriculture, Forest Service, Pacific Northwest Region, Portland, Ore.

Cooperrider, A. Y., R. J. Boyd, and H. R. Stuart, eds. 1986. *Inventory and monitoring of wildlife habitat.* U.S. Department of the Interior, Bureau of Land Management, Service Center, Denver, Colo.

Cushing, C. E., and W. L. Gaines. 1989. Thoughts on the recolonization of endorheic cold desert spring-streams. *Journal of the North American Benthological Society* 8(3):277–287.

Erickson, R. O. 1945. The *Clematis fremontii* var. riehlii population in the Ozarks. *Annals of the Missouri Botanical Garden* 32:413–460.

ESA (Endangered Species Act). 1973. U.S. Laws, Statutes, etc.; Public Law 93-205. Act of 28 December 1973 16 U.S.C. 1531–1536, 1538–1540.

Everett, R., C. Oliver, J. Saveland, P. Hessburg, N. Diaz, and L. Irwin. 1994. Adaptive ecosystem management. Pp. 340–354 in *Ecosystem management: Principles and applications.* Vol. 2. General Technical Report PNW-GTR-318. U.S. Department of Agriculture, Forest Service, Pacific Northwest Region, Portland, Ore.

Forman, R. T. T., and M. Godron. 1986. *Landscape ecology.* New York: John Wiley and Sons.

Frankel, O. H. 1970. Variation, the essence of life. Sir William Macleay Memorial Lecture. *Proceedings of the Linneaus Society* 95:158–169.

Gaines, W. L. 2001. Disturbance ecology, land allocations, and wildlife management. Pp. 29–34 in *Proceedings of the management of fire maintained ecosystems workshop,* British Columbia Forest Service, Whistler, British Columbia.

Gaines, W. L., C. E. Cushing, and S. D. Smith. 1989. Trophic relations and functional group composition of benthic insects in three cold desert streams. *Southwestern Naturalist* 34(4):478–482.

Gaines, W. L., and R. J. Harrod. 1994. Abstract. Watershed analysis for wildlife species. *Northwest Science* 68(2).

Gaines, W. L., G. K. Neale, and R. H. Naney. 1995. Response of coyotes and gray wolves to simulated howling in north-central Washington. *Northwest Science* 69(3):217–222.

Gast, W. R., Jr., D. W. Scott, C. Schmitt, D. Clemens, S. Howes, and C. G. Johnson Jr. 1991. *Blue Mountains forest health report: New perspectives in forest health.* U.S. Department of Agriculture, Forest Service, Pacific Northwest Region, Portland, Ore.

Given, D. R. 1994. *Principles and practice of plant conservation.* Portland, Ore.: Timber Press.

Groom, M. J., and N. Schumaker. 1996. Evaluating landscape change—pattern of worldwide deforestation and local fragmentation. Pp. 24–44 in *Biotic interactions and global change,* edited by P. M. Kareiva, J. G. Kingsolver, and R. B. Huey. Sunderland, Mass.: Sinauer Associates.

Grumbine, R. E. 1992. *Ghost bears: Exploring the biodiversity crisis.* Washington, D.C.: Island Press.

Haig, S. M., R. S. Wagner, E. D. Forsman, and T. D. Mullins. 2001. Geographic variation and genetic structure in spotted owls. *Conservation Genetics* 2(1):25–40.

Hamrick, J. L. 1989. Isozymes and analysis of genetic structure of plant populations. Pp. 335–348 in *Isozomes in plant biology*, edited by C. M. Schonewald-Cox, S. M. Chambers, B. Macbryde, and W. L. Thomas. Menlo Park, Calif.: Benjamin-Cummings.

Hamrick, J. L., M. J. W. Godt, D. A. Murawski, and M. D. Loveless. 1991. Correlations between species traits and allozyme diversity: Implications for conservation biology. Pp. 75–86 in *Genetics and conservation of rare plants*, edited by D. H. Falk and K. E. Holsinger. New York: Oxford University Press.

Harper, J. L. 1977. *Population biology of plants*. New York: Academic Press.

Harrod, R. J., W. L. Gaines, R. J. Taylor, R. Everett, T. Lillybridge, and J. D. McIver. 1996. Biodiversity in the Blue Mountains. Pp. 81–105 in *Search for a solution: Sustaining the land, people, and economy of the Blue Mountains*, edited by R. G. Jaindl and T. M. Quigley. Washington, D.C.: American Forests.

Haynes, R. W., R. T. Graham, and T. M. Quigley, eds. 1996. *A framework for ecosystem management in the interior Columbia basin and portions of the Klamath and Great Basins*. General Technical Report PNW-GTR-374. U.S. Department of Agriculture, Forest Service, Pacific Northwest Research Station, Portland, Ore.

Hessburg, P. F., B. G. Smith, S. D. Kreiter, C. A. Miller, R. B. Salter, C. H. McNicoll, W. J. Hann, and T. M. Quigley. 1999. *Historical and current forest and range landscapes in the interior Columbia River basin and portions of the Klamath and Great Basins*. Part 1. *Linking vegetation patterns and landscape vulnerability to potential insect and pathogen disturbances*. General Technical Report PNW-GTR-458. U.S. Department of Agriculture, Forest Service, Pacific Northwest Research Station, Portland, Ore.

Heyer, W. R., M. A. Donnelly, R. W. McDiarmid, L. L. Hayek, and M. S. Foster. 1994. *Measuring and monitoring biological diversity: Standard methods for amphibians*. Washington, D.C.: Smithsonian Institution Press.

Huff, M. H., K. A. Bettinger, H. L. Ferguson, M. J. Brown, and B. Altman. 2001. *A habitat-based point-count protocol for terrestrial birds, emphasizing Washington and Oregon*. General Technical Report PNW-GTR-501. U.S. Department of Agriculture, Forest Service, Pacific Northwest Research Station, Portland, Ore.

Hurlbert, S. H. 1971. The nonconcept of species diversity: A critique and alternative parameters. *Ecology* 52(4):577–586.

Kent, M., and P. Coker. 1992. *Vegetation description and analysis: A practical approach*. London: CRC Press.

King, A. W. 1997. Hierarchy theory: A guide to system structure for wildlife biologists. Pp. 185–214 in *Wildlife and landscape ecology: Effects of pattern and scale*, edited by J. A. Bissonette. New York: Springer-Verlag.

Koopowitz, H., A. D. Thornhill, and M. Andersen. 1994. A general stochastic model for the prediction of biodiversity losses based on habitat conversion. *Conservation Biology* 8(2):425–438.

Lande, R., and G. F. Barrowclough. 1987. Effective population size, genetic variation, and their use in population management. Pp. 87–124 in *Viable populations for conservation*, edited by M. E. Soulé. New York: Cambridge University Press.

Landres, P. B., J. Verner, and J. W. Thomas. 1988. Ecological uses of vertebrate indicator species: A critique. *Conservation Biology* 2(4):316–329.

Leary, R. F., F. W. Allendorf, and S. H. Forbes. 1993. Conservation genetics of bull trout in the Columbia and Klamath River drainages. *Conservation Biology* 7(4):856–865.

Lehmkuhl, John F., P. F. Hessburg, R. L. Everett, M. H. Huff, and R. D. Ottmar. 1994. *Historical and current forest landscapes of eastern Oregon and Washington*. Part 1. *Vegetation pattern and insect and disease hazards*. General Technical Report PNW-GTR-328. U.S. Department of Agriculture, Forest Service, Pacific Northwest Research Station, Portland, Ore.

Lehmkuhl, J. F., and M. G. Raphael. 1993. Habitat pattern around northern spotted owl locations on the Olympic Peninsula, Washington. *Journal of Wildlife Management* 57:302–315.

Lehmkuhl, J. F., M. G. Raphael, R. S. Holthausen, J. R. Hickenbottom, R. H. Naney, and J. S. Shelly. 1997. Effects of planning alternatives on terrestrial species in the interior Columbia River basin. Pp. 537–730 in *Evaluation of EIS alternatives by the Science Integration Team*, edited by T. M. Quigley, K. M. Lee, and S. J. Arbelbide. General Technical Report PNW-GTR-406. U.S. Department of Agriculture, Forest Service, Pacific Northwest Research Station, Portland, Ore.

Lehmkuhl, J. F., L. F. Ruggiero, and P. A. Hall. 1991. Landscape-scale patterns of forest fragmentation and wildlife richness and abundance in the southern Washington Cascade Range. Pp. 425–442 in *Wildlife and vegetation of unmanaged Douglas-fir forests*, edited by L. F. Ruggiero, K. B. Aubry, A. B. Carey, M. H. Huff, technical coordinators. General Technical Report PNW-GTR-285. U.S. Department of Agriculture, Forest Service, Pacific Northwest Research Station, Portland, Ore.

Lesica, P., and F. W. Allendorf. 1992. Are small populations of plants worth preserving? *Conservation Biology* 6(1):135–139.

Madsen, S., D. Evens, T. Hamer, P. Henson, S. Miller, S. K. Nelson, D. Roby, and M. Stapanian. 1999. *Marbled murrelet effectiveness monitoring plan for the Northwest Forest Plan*. General Technical Report PNW-GTR-439. U.S. Department of Agriculture, Forest Service, Pacific Northwest Research Station, Portland, Ore.

Magurran, A. E. 1988. *Ecological diversity and its measurement*. Princeton, N.J.: Princeton University Press.

Manasse, R. S. 1992. Ecological risks of transgenic plants: Effects of spatial dispersion on gene flow. *Ecological Applications*. 2(4):431–438.

Mannan, W. R., M. L. Morrison, and E. C. Meslow. 1984. Comment: The use of guilds in forest bird management. *Wildlife Society Bulletin* 12:426–430.

Manuwal, D. A., M. H. Huff, M. R. Bauer, C. B. Chappell, and K. Hegstad. 1987. Summer birds of the upper subalpine zone of Mount Adams, Mount Rainier, and Mount St. Helens, Washington. *Northwest Science* 61(2):82–92.

Marcot, B. G., M. A. Castellano, J. A. Christy, L. K. Croft, J. F. Lehmkuhl, R. H. Naney, R. E. Rostentreter, R. E. Sandquist, and E. Zieroth. 1997. Terrestrial ecology assessment. Pp. 497–1713 in *An assessment of ecosystem components in the interior Columbia Basin and portions of the Klamath and Great Basins*, edited by T. M. Quigley, S. J. Arbelbide, and S. F. McCool. General Technical Report PNW-GTR-405. U.S. Department of Agriculture, Forest Service, Pacific Northwest Research Station, Portland, Ore.

Marcot, B. G., L. K. Croft, J. F. Lehmkuhl, R. H. Naney, C. G. Niwa, W. R. Owen, and R. E. Sandquist. 1998. *Macroecology, paleoecology, and ecological integrity of terrestrial species and communities of the interior Columbia River basin and portions of the Klamath and Great Basins*. General Technical Report PNW-GTR-410. U.S. Department of Agriculture, Forest Service, Pacific Northwest Research Station, Portland, Ore.

Marcot, B. G., M. J. Wisdom, H. W. Li, and G. C. Castillo. 1994. *Managing for featured, threatened, endangered, and sensitive species and unique habitats for ecosystem sustainability*. General Technical Report PNW-GTR-329. U.S. Department of Agriculture, Forest Service, Pacific Northwest Research Station, Portland, Ore.

McGarigal, K., and B. J. Marks. 1993. *FRAGSTATS: Spatial pattern analysis program for quantifying landscape structure*. General Technical Report PNW-GTR-351. U.S. Department of Agriculture, Forest Service, Pacific Northwest Research Station, Portland, Ore.

Menges, E. S. 1990. Population viability analysis for an endangered plant. *Conservation Biology* 4:52–62.

Metzger, J. P., and E. Muller. 1996. Characterizing the complexity of landscape boundaries by remote sensing. *Landscape Ecology* 11:65–77.

Meyer, J. S., L. L. Irwin, and M. S. Boyce. 1998. *Influence of habitat abundance and fragmentation on northern spotted owls in western Oregon*. Wildlife Monograph No. 139.

Miller, S. D., G. C. White, R. A. Sellers, H. V. Reynolds, J. W. Schoen, K. Titus, V. G. Barnes, R. B. Smith, R. R. Nelson, W. B. Bullard, and C. C. Schwartz. 1997. *Brown and black bear density estimation in Alaska using radiotelemetry and replicated mark-resight techniques*. Wildlife Monograph No. 133.

Milne, B. T. 1988. Measuring the fractal geometry of landscapes. *Applied Mathematics and Computation* 27:67–79.

Mladenoff, D. J., M.A . White, J. Pastor, and T. R. Crow. 1993. Comparing spatial pattern in unaltered old forest and disturbed forest landscapes. *Ecological Applications* 3:294–306.

Morgan, P., G. H. Aplet, J. B. Haufler, H. C. Humphries, M. M. Moore, and W. D. Wilson. 1994. Historic range of variability: A useful tool for evaluating ecosystem change. *Journal of Sustainable Forestry* 2:87–111.

Morrison, M. L., B. G. Marcot, and W. R. Mannan. 1992. *Wildlife habitat relationships: Concepts and applications*. Madison: University of Wisconsin Press.

Nei, M. 1973. Analysis of genetic diversity in subdivided populations. *Proceedings of the National Academy of Sciences* 70:3321–3323.

NFMA (National Forest Management Act). 1976. U.S. Laws, Statutes, etc.; Public Law 94-588. Act of Ctt. 22, 1976. U.S.C. 1600 (1976).

Niemela, J. 2000. Biodiversity monitoring for decision-making. *Annales-Zoologici-Fennici* 37:307–317.

Noss, R. F. 1983. A regional landscape approach to maintain diversity. *BioScience* 33(11):700–706.

———. 1990. Indicators for monitoring biodiversity: A hierarchical approach. *Conservation Biology* 4(4):355–364.

———. 1991. From endangered species to biodiversity. In *Balancing on the brink of extinction: The Endangered Species Act and lessons for the future*, edited by K. Kohm. Washington, D.C.: Island Press.

Noss, R. F., and A. Y. Cooperrider. 1994. *Saving nature's legacy: Protecting and restoring biodiversity*. Washington, D.C.: Island Press.

Noss, R. F., and L. D. Harris. 1986. Nodes, networks, and MUMs: Preserving diversity at all scales. *Environmental Management* 10:299–309.

Nyberg, J. B. 1998. Statistics and the practice of adaptive management. Pp. 1–8 in *Statistical methods for adaptive management studies*, edited by V. Sit and B. Taylor. British Columbia Ministry of Forests Research Program, land management handbook, Number 42. Victoria, B.C.: British Columbia Ministry of Forests.

O'Neill, R. V., J. R. Krummel, R. H. Gardner, G. Sugihara, B. Jackson, D. L. DeAngelis B. T. Milne, M. G. Turner, B. Zygmunt, S. W. Christensen, V. H. Dale, and R. L. Graham. 1988. Indices of landscape pattern. *Landscape Ecology* 1:153–166.

O'Neill, R. V., C. T. Hunsaker, S. P. Timmins, B. L. Jackson, K. B. Jones, K. H. Ritters, and J. D. Wickham. 1996. Scale problems in reporting landscape pattern at the regional scale. *Landscape Ecology* 11:169–180.

O'Neill, R. V., K. H. Ritters, J. D. Wickham, and K. B. Jones. 1999. Landscape pattern metrics and regional assessment. *Ecosystem Health* 5:225–233.

Oliver, I., and A. J. Beattie. 1993. A possible method for the rapid assessment of biodiversity. *Conservation Biology* 7(3):562–568.

Parsons, D. J., T. W. Swetman, and N. L.Christensen. 1999. Uses and limitations of historical variability concepts in managing ecosystems. *Ecological Applications* 9(4):1177–1178.

Possingham, H. P., J. Davies, I. R. Noble, and T. W. Norton. 1992. A metapopulation simulation model for assessing the likelihood of plant and animal extinctions. Pp. 428–433 in *Proceedings of the 9th biennial conference on modeling and simulation* [meeting dates unknown], [location of meeting unknown]. Brisbane, Australia: Simulation Society of Australia, The Printing Office.

Ralph, C. J., G. Geupel, P. Pyle, T. E. Martin, and D. F. DeSante. 1993. *Handbook of field methods for monitoring landbirds.* General Technical Report PSW-GTR-144. U.S. Department of Agriculture, Forest Service, Pacific Southwest Research Station, Albany, Calif.

Rausch, R. L. 1963. Geographic variation in size in North American brown bears, *Ursus arctos*, as indicated by condylobasal length. *Canadian Journal of Zoology* 41:33–45.

Ringold, P. L., B. Mulder, J. Alegria, R. L. Czaplewski, T. Tolle, and K. Burnett. 1999. Establishing a regional monitoring strategy: The Pacific Northwest Forest Plan. *Environmental Management* 23:179–192.

Ripple, W. J., G. A. Bradshaw, and T. A. Spies. 1991. Measuring forest fragmentation in the Cascade Range of Oregon. *Biological Conservation* 57:73–88.

Romme, W. H., and D. H. Knight. 1982. Landscape diversity: The concept applied to Yellowstone Park. *BioScience* 32(8):664–670.

Schaal, B. A., W. J. Leverich, and S. H. Rogstad. 1991. A comparison of methods for assessing genetic variation in plant conservation biology. Pp. 123–134 in *Genetics and conservation of rare plants*, edited by D. A. Falk and K. E. Holsinger. New York: Oxford University Press.

Schumaker, N. H. 1996. Using landscape indices to predict habitat connectivity. *Ecology* 77:1210–1225.

Scott, J. M., F. Davis, B. Csuti, R. Noss, B. Butterfield, C. Groves, H. Anderson, S. Caicco, F. D'Erchia, T. C. Edwards Jr., J. Ulliman, and R. G. Wright. 1993. *Gap analysis: A geographic approach to protection of biological diversity.* Wildlife Monograph No. 123.

Seber, G. A. F. 1982. *The estimation of animal abundance and related parameters.* 2d. ed. New York: MacMillan.

Shaffer, M. L. 1990. Population viability analysis. *Conservation Biology* 4(1):39–40.

Slatkin, M., and N. H. Barton. 1989. A comparison of three indirect methods for estimating average levels of gene flow. *Evolution* 43:1349–1368.

Soulé, M. E., ed. 1987. *Viable populations for conservation.* New York: Cambridge University Press.

Stangel, P. W., M. R. Lennartz, and M. H. Smith. 1992. Genetic variation and population structure of red-cockaded woodpeckers. *Conservation Biology* 6(2):283–292.

Swanson, F. J., J. A. Jones, D. O. Wallin, and J. H. Cissel. 1994. Natural variability— implications for ecosystem management. Pp. 80–94 in *Ecosystem management: Principles and applications.* Vol. 2. General Technical Report PNW-GTR-318. U.S. Department of Agriculture, Forest Service, Pacific Northwest Research Station, Portland, Ore.

Taylor, R. J. 1991. The origin of *Lamium hybridum*: A case study in the search for the parents of hybrid species. *Northwest Science* 65(3):116–124.

Taylor, R. J., T. F. Patterson, and R. J. Harrod. 1994. Systematics of Mexican spruce revisited. *Systematic Botany* 19(1):47–59.

Thiollay, J. M. 1992. Influence of selective logging on bird species diversity in a Guianan rain forest. *Conservation Biology* 6(1):47–63.

Thomas, J. W., M. G. Raphael, R. G. Anthony, E. D. Forsman, A. G. Gunderson, R. S. Holthausen, B. G. Marcot, G. H. Reeves, J. R. Sedell, and D. M. Solis. 1993. *The report of the scientific analysis team. Viability assessments and management considerations for species associated with late-successional old growth forests of the Pacific Northwest.* U.S. Department of Agriculture, Forest Service, Portland, Ore.

Turner, M. G. 1989. Landscape ecology: The effect of pattern on process. *Annual Review of Ecology and Systematics* 20:171–197.

———. 1990. Spatial and temporal analysis of landscape pattern. *Landscape Ecology* 4:21–30.

Turner, M. G., R. Costanza, and F. H. Sklar. 1989. Methods to evaluate the performance of spatial simulation models. *Ecological Modeling* 48:1–18.

Turner, M. G., and C. L. Ruscher. 1988. Changes in landscape patterns in Georgia, USA. *Landscape Ecology* 1:241–251.

USDA (U.S. Department of Agriculture) and USDI (U.S. Department of Interior). 1994. Record of decision for amendments to Forest Service and Bureau of Land Management planning documents within the range of the northern spotted owl. Portland, Oregon.

U.S. Forest Service (USFS). 1997. *Wenatchee National Forest late-successional reserve assessment.* Pacific Northwest Region, Wenatchee, Wash.

Van Horne, B. 1983. Density as a misleading indicator of habitat quality. *Journal of Wildlife Management* 47:893–901.

Volland, L. A. 1980. Diversity as a land management issue. Portland, Ore.: USDA Forest Service, Pacific Northwest Region, unpublished report.

Walker, B. H. 1992. Biological diversity and ecological redundancy. *Conservation Biology* 6(1):18–23.

Wayne, R. K., N. Lehman, M. W. Allard, and R. L. Honeycutt. 1992. Mitochondrial DNA variability of the gray wolf: Genetic consequences of population decline and habitat fragmentation. *Conservation Biology* 6(4):559–569.

Weir, B. S., and C. C. Cockerham. 1984. Estimating F-statistics for the analysis of population structure. *Evolution* 38:1358–1370.

Wilcox, B. A. 1984. In situ conservation of genetic resources: Determinants of minimum area requirements. Pp. 639–647 in *National parks, conservation and development: The role of protected areas in sustaining society,* edited by J. A. McNeely and K. R. Miller. Washington, D.C.: Smithsonian Institution Press.

Wilson, D. E., R. F Cole, J. D. Nichols, R. Rudran, and M. S. Foster, eds. 1996. *Measuring and monitoring biological diversity: Standard methods for mammals.* Washington, D.C.: Smithsonian Institution Press.

Wilson, E. O. 1992. *The diversity of life.* Cambridge: Harvard University Press, Belknap Press.

With, K. A., and T. O. Crist. 1995. Critical thresholds in species' responses to landscape structure. *Ecology* 76:2446–2459.

Woodley, S. 1993. Monitoring and measuring ecosystem integrity in Canadian national parks. Pp. 155–176 in *Ecological integrity and the management of ecosystems,* edited by S. Woodley, J. Kay, and G. Francis. Delray Beach, Fla.: St. Lucie Press.

PART V

Summary and Synthesis

CHAPTER 15

Monitoring, Assessment, and Ecoregional Initiatives: A Synthesis

Joel C. Trexler and David E. Busch

Monitoring regional ecosystems occurs at large spatial and temporal scales, creating unique challenges and opportunities that are different from those encountered with monitoring projects in smaller, more homogeneous areas. Scale is significant to ecoregional initiatives because of the complexity of large natural systems and the function of large organizations needed to manage these types of areas. Implementing ecoregional initiatives may involve dozens (or even hundreds) of management actions of various magnitudes most of which require monitoring and assessment in their own right. We have attempted to address both ecological and institutional dynamics in the preceding chapters. In this final chapter, we draw together, and expand upon, many of the important points made by the authors of this text in their discussion of the challenges and opportunities presented by the development and implementation of monitoring systems for ecoregional initiatives.

Challenges for Ecologists

Decision makers look to ecologists to retrospectively identify the effects of their actions on ecosystems. Increasingly, managers also call on ecologists to anticipate the effects of alternative management scenarios to assist in choosing the best plan among several options (NSTC 1997). The former expectation requires monitoring studies that identify trends in ecosystem state, ideally as quickly as possible after a management plan is implemented. The latter expectation requires knowledge of ecosystem function that can be used to develop predictive models of ecosystem condition resulting from anthropogenic manipulation of environmental drivers. Both expectations represent major challenges for ecological science (Duncan and Kalton 1987; Woodward et al. 1999).

Monitoring ecosystem initiatives is distinct from other types of environmental impact assessment (e.g., Schmitt and Osenberg 1996) because of the lack of meaningful "reference conditions" for whole ecosystems. There is growing appreciation that the position of habitats in a landscape affects their local dynamics. Although it may be possible to find habitats within an ecosystem that are represented at multiple sites and which can be used as local references, there are no replicate ecosystems that can serve as references at the landscape scale (Hargrove and Pickering 1992). Surprises can arise in management at the ecosystem scale, not only because expectations are based on analysis at seemingly similar sites in different landscape contexts, but also because management effects can be modified by extreme climatic events that occur infrequently (Holling 1986). The interaction of local environmental context and environmental variability with long-term, often subtle, climatic trends produces the diversity of ecosystem dynamics that commonly defy simple regional generalities (Stoddard et al. 1998). Thus, monitoring at a regional scale provides unique challenges both for application and interpretation of ecological principles.

The temporal scale over which ecosystems respond to management actions creates challenges for ecoregional monitoring. Interaction among ecosystem components may be nonlinear and indirect, possibly displaying thresholds, and may not be manifested for long periods after a management action (Holling 1986). It is believed that the time scale of ecosystem response is proportional to the size and habitat diversity of the ecosystem (Harris 1980; Allen and Hoekstra 1992; Powell 1995). Thus, the full implications of management at the ecoregion scale may emerge only over long periods. In addition, short-term responses may be misleading when applied to the long-term prospects for ecoregional initiatives, while management-induced change may be difficult to separate from natural variability over short-term scales (Jassby 1998). Complex indirect ecological effects are not well addressed by an effectiveness-monitoring program alone without coupling it to validation monitoring or research. Because research is often funded separately, monitoring data from management situations can be difficult to interpret given the multiple natural and anthropogenic influences simultaneously driving environmental change (Stow et al. 1998). Thus, targeted studies and experimental research leading to a mechanistic understanding of ecosystem function must be paired with monitoring programs to yield the understanding needed to anticipate the impact of management alternatives on future ecosystem state.

To address these challenges, data gathering for monitoring must be conducted with a geographic coverage that is consistent with the scale of environmental driving factors. The placement of study sites must match the spatial scale of environmental gradients, which cannot be known a priori. Unfortunately, it may be impossible to synthesize findings from geographically isolated monitoring sites, termed sentinel sites (Duncan and Kalton 1987; Gibbons and Munkittrick 1994), even when a relatively large number of such sites are studied (Jassby 1998). Fixed, intensively studied sentinel sites may be useful when placed at targeted "hot spots" for management, but a framework of spatially extensive sampling is critical to address regional patterns (Stoddard et al. 1998; Urquhart et al. 1998).

Monitoring for the initiatives highlighted in this volume builds upon a "bottom-up" view of ecosystem control, emphasizing the role of physical factors in shaping ecosystem process. However, "top-down" biotic processes are also important for ecosystem management and may be another source of surprise outcomes of management. For example, the effect of management on piscivorous fishes in aquatic ecosystems with trophic cascades could have important implications for ecosystem function (Carpenter and Kitchell 1993; DeAngelis et al. 1997). Monitoring biotic community processes may require a level of temporal resolution and frequency of sampling that is difficult to apply at a landscape scale. The typical solution, adoption of a relatively small number of sentinel sites for intensive study where impacts are anticipated, should be integrated with a spatially resolved monitoring design, preferably incorporating a stratified random component. Ultimately, ecosystem-scale analyses will be required to integrate local effects at the regional level.

Challenges for Decision Makers

The large spatial scale of ecoregional initiatives creates unique challenges for managers because of the special demands for integrating personnel and financial resources that such initiatives demand. It is worth noting the potential for regional-scale monitoring programs to be more cost efficient than a collection of smaller project-, species-, or habitat-specific monitoring programs. While considerable economy of scale may be obtained in large efforts, the resources provided are seldom great enough to monitor all important dimensions of an ecoregion adequately, particularly over extended time periods. One reason for this is that large projects generally incorporate many goals that are challenging to prioritize and evaluate in a common framework so that the number of important indicators can exceed the resources available to track them. Also, ecoregional-scale programs can generate their own bureaucracies with institutional inertia and possible competition for support that must be overcome. Often, a multi-organizational framework must be developed with the strength to set the programmatic agenda and allocation of responsibility and resources.

The long time frame over which the implementation and effects of large-scale projects occur places high demands on database management (Palmer and Mulder 1999; Palmer, chap. 8). Ecoregional initiatives require standardization of techniques across broad landscapes, potentially long time spans, and among disparate working groups. This level of coordination can be exceptionally difficult to realize. Compounding the problems of standardization is the tendency for technical specialists to develop more performance measures than can be adequately assessed. This tendency derives from the spatial complexity and heterogeneity of large ecoregions and has the potential to lead to an incoherent mass of information that fails to assess initiative goals. Choosing from among the myriad possible parameters to measure those that are most sensitive and informative relative to project goals may be the ultimate determinant of a successful monitoring program. Making such choices judiciously requires some ability to anticipate the future direction of a project, along with long-term sustained support at a level adequate to keep key monitoring elements in place.

Ecoregional Monitoring within Broader Frameworks

One of the great challenges in framing ecoregional monitoring programs is establishing their relevance to other local and national priorities and initiatives. Although there are examples of successful continental-scale monitoring programs for species or groups of species and for a variety of classes of environmental parameters, few programs actually characterize the status and trends of multiple, whole ecosystems. This is due to the difficulties inherent in integrating information across areas and resource types (National Research Council 1995). Thus, the fit of monitoring for most regional ecosystem initiatives, including the examples used here, within proposed national or international monitoring frameworks is questionable (Bricker and Ruggiero 1998). For example, the specialized nature of the environment in the south Florida and the lower Colorado River ecoregions may contribute to a lack of fit with national and international monitoring programs even though their uniqueness makes them the target of international attention. Aspects of Northwest Forest Plan monitoring are thought to be consistent with comprehensive national and international evaluations of natural resources (Mulder et al. 1999). However, explicit linkages to national programs such as Forest Health Monitoring or Forest Inventory and Analysis (Morrison and Marcotte 1995) have not been fully detailed.

Broad-scale monitoring frameworks often lack acceptance due to diverse opinions about needs for monitoring information. Some view such frameworks as solely addressing programmatic themes. Others feel that monitoring frameworks should integrate analytical themes, while still others view such frameworks in terms of taxonomy or geographic scale. It is seldom recognized that these visions can be complementary and integrated into a single, comprehensive framework. Accordingly, we propose a conceptual system that incorporates programmatic, analytical, geographic, and taxonomic themes into a common monitoring framework (fig. 15-1). This framework borrows from one proposed for integrating the nation's environmental monitoring and research networks and programs (NSTC 1997). We have modified that original framework to more

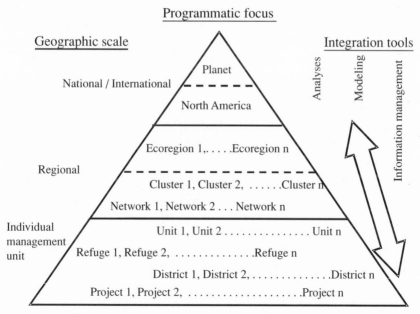

Figure 15-1. Ecoregion-scale monitoring integrated in local and national frameworks. The means for such integration are envisioned as involving database structure and analysis, ecological modeling, and consideration of scale across levels of ecological complexity.

explicitly describe the structural characteristics enabling it to be relevant at a variety of spatial scales. Perhaps most important, this conceptual framework attempts to depict how the nation's biotic resources are monitored through projects focused on issues ranging in scope from the local to the global.

Some monitoring programs are dependent on networks of specialists or sampling facilities providing status and trend information from their own locality. Certain national and international monitoring initiatives are based on aggregating pertinent data without explicit attempts to unite independent monitoring projects focused on ecosystem condition at the local level. Employing the depicted monitoring framework helps visualize potential pathways for integrating status and trend information conducted at different scales, using analytical or modeling techniques to span levels of ecological hierarchy.

In most of the examples in this book, there is a focus on monitoring systems within a given ecoregion (fig. 15-1). Although not all of the nation's ecoregions have operable systems to detect status and trend, the suite of regional-scale monitoring projects is becoming more representative of the nation's key ecoregions (Busch and Trexler, chap. 1). Regional monitoring programs could further the depiction of national status and trends if comprehensive standards of data management and analytical tools were brought to bear using the methods of integration suggested by our proposed conceptual framework.

Often, the monitoring focus is on a national park, wildlife refuge, or other type of reserve, a multiple-use forest or rangeland district, a water project, or analogous land- or water-management unit (fig. 15-1). At a minimum, the proposed framework provides a means of visualizing such localized projects in regional and national contexts. Beyond this, the framework encourages the organization of monitoring projects along lines of programmatic and geographic scale. One promising example using this type of hierarchical context is the National Park Service's Long-term Ecological Monitoring Program, which is making a transition from the development of monitoring protocols on a prototypical unit basis (i.e., "park by park") and toward the development of monitoring programs for networks of units from recognized ecoregions (National Park Service 2001).

Where land and water management units are geographically extensive, clusters of refuges, parks, districts, projects, and the like may help define ecoregions from which to derive status and trend information. Under the monitoring framework (fig. 15-1), ecoregions can likewise be aggregated to determine national status and trends. By conducting analyses that cut across land- or water-management unit boundaries and zones of responsibility, the framework enhances our ability to describe status and trends associated with ecosystems and biota at regional and national scales. Thus, the full potential of ecoregional analyses is only likely to be realized by working in the context of multi-jurisdictional clusters including various partner organizations' lands and waters.

In a similar fashion, the proposed framework (fig. 15-1) contributes to the development of a contextual basis for integrating determinations about the population status of biota, as well as trends occurring across taxa and geographic scales. The framework provides a way to view monitoring of single species or isolated processes in community or ecosystem contexts extending, at least hypothetically, to regional or national levels. The framework can also serve as the starting point for integrated analyses of databases and increased development and use of ecological models, which aid our understanding of biota and systems across geographic scales and monitoring foci. In this manner, such a framework has the potential to extend our integration of monitoring themes across all levels of ecological hierarchy.

The Connection between Monitoring and Adaptive Management

Monitoring is a critical component of any iterative, adaptive process guiding an ecoregional initiative because it provides information to managers about the state of the initiative's implementation. Ideally, this includes both information on ecological status, by way of effectiveness monitoring, and the state of managers' actions, by way of implementation monitoring. Clearly, no systematic response to the implementation of an ecoregional initiative can be carried out without an infusion of information from these types of monitoring, ideally through a formal and intentional process. Just as engineers must maintain a continual flow of communication with their work crews on the progress and problems of construction, environmental managers must seek data on the state of the ecological systems they manage. Monitoring provides the information needed for managers to act in a timely fashion based on ecological responses to environmental management as issues arise. However, the complexity of the temporal and spatial scales over which ecological systems respond to management actions makes the identification of leading indicators difficult. Rapid responses may not be indicative of long-term patterns, and short-term patterns may not be indicative of ecological changes unfolding over longer time scales. The inherent nonlinearity of ecological processes, including threshold effects, can make prediction difficult. Such complexities make adaptive management of natural systems a challenging endeavor.

Tracking ecosystem status and trends may generate much of the data needed to understand the mechanisms underlying ecological change. Monitoring is generally the basis for a system that supports ecoregional programs of analysis, modeling, interpretation, and ultimately decision making. Comprehensive monitoring systems create the potential for feedback loops between the monitoring plan, ecological understanding, and improvement in the monitoring design, in an adaptive monitoring process (Holling 1978; Walters 1986). Despite this, there is a growing literature on the failure of projects to attain the ideal of adaptive management, often as a result of institutional inertia (Halbert 1993; Gunderson et al. 1995; Walters 1997). How can this inertia be overcome? Increasingly, ecosystem management is appreciated as a large-scale experiment (Walters and Green 1997) because of the large uncertainty inherent in the enterprise. One of our best examples of such management "experiments" is the recent test "floods" created in the Colorado River (Patten et al. 2001; Stevens and Gold, chap. 4). In some cases, ecosystem experiments may be viewed as contradictory to the legal mandates of managers (e.g., the National Environmental Policy Act or Endangered Species Act), which by necessity require a cautious, methodical approach to the management of public resources. Ultimately, full implementation of adaptive management may require revisiting the legal mandates underlying natural resources management in general.

Challenges to establishing ecoregional monitoring programs include vague objectives, inconsistent or absent methods to assess indicators, and a lack of existing knowledge about system variability at relevant temporal or spatial scales. An iterative process of refining monitoring schemes as projects are implemented can solve these problems, including experimental testing of monitoring programs (Ringold et al., chap. 3). Monitoring systems must be designed to anticipate the development of improved methods and sampling designs, and should include mechanisms to develop links between the growing body of scientific knowledge and increasingly sophisticated resource management needs. An adaptive approach to monitoring may also provide a mechanism to overcome challenges that arise when regional initiatives are formed from previously independent management programs, with independently developed monitoring plans. Operating in such a dynamic environment can be challenging for those who must work with goals that change or are vague. This requires organizational flexibility and staff adaptability. Throughout the process of optimizing monitoring program design, long-term security and compatibility of data gathered should be maintained as a top priority. Multi-level review is one mechanism to engage stakeholders throughout the process, including review and comments by independent panels of scientific or management experts.

Setting Targets and Goals

The establishment of goals and targets for ecoregional initiatives is a critical step, often determining the potential for success (Ogden et al., chap. 5). Monitoring has a critical role in determining targets for restoration through the establishment of baseline conditions, providing data to determine the mechanisms that lead to success and setting limits on what can be evaluated and achieved through an initiative's implementation. In many cases in North America, the explicit or implicit ecoregional initiative goal is to move toward conditions that are more characteristic of those found prior to an area's settlement by Europeans. There are limits to the information that can be provided to establish "natural" or "historical" conditions; even definitions of these states are controversial (e.g., Angermeier 2000; Povilitis 2001). With the realization that natural systems are dynamic, ecoregional initiative targets are often directed toward restoring performance measures to within a "historic range of variability." However, identifying the historic range of variability may be particularly challenging because of the tendency to focus on extreme events. Callicott et al. (1999) suggest that conservation goals may be grouped into two schools, *compositionalism* and *functionalism*. Compositionalists emphasize biological diversity, biological integrity, and ecological restoration, while functionalists emphasize ecological services, ecological stability and sustainability, and ecosystem rehabilitation and management. Callicott et al. (1999) argue that these concepts are complementary in relevance for areas intended as reserves (compositionalism) and human-inhabited and exploited areas (functionalism). In this framework, areas intended for reserves are appropriately targeted to fulfill some degree of "naturalness" while human-inhabited areas would be targeted to achieve sustainability. In practice, ecoregional initiatives may encompass both types of goal because of their large geographic extents. Thus, it may be necessary for monitoring programs to incorporate different types of standards for different portions of ecoregions.

In spite of its limitations, ecological integrity is a critical concept in setting goals and identifying targets for monitoring ecoregional initiatives. Definitions of integrity are often vague, and there is no universal definition of a "healthy" ecosystem (Callicott et al. 1999; Karr and Chu 1999). However, Karr (1981) set the goal of establishing a variety of indicators that include function-based, structure-based, and composition-based metrics. Because single indicators cannot assess function, structure, and composition, defining quantifiable indicators of desired attributes is a major challenge for establishing a useful monitoring design. Appropriate indicators should include those that link ecological scales, as well as societal values, to the integrity concept even though ultimate end points are often derived from legal mandates. The use of conceptual frameworks incorporating ecosystem function, structure, and composition, can be the key to devising monitoring programs that address ecoregional goals and targets in a comprehensive manner.

The Process of Monitoring

A major challenge faced by managers setting up monitoring plans for ecoregional initiatives comes from the need to integrate monitoring across the range of scales that such initiatives encompass. It is vital that information from different management areas and different disciplines be consistent and comparable in order to facilitate analyses. A high level of attention to data management is needed for successful implementation of ecoregional initiatives. Managers must work with scientists for complementarity across groups that implement a monitoring program. To accomplish this, project-wide data standards and information-management protocols need to anticipate long-term use of information from monitoring. As Palmer points out in chapter 8, standardized data-quality procedures are ultimately a requirement for natural resource management organizations charged with monitoring at this scale.

Reconciling the need for local monitoring with the need to understand ecoregional status and trends is another major challenge for managers. Ecoregional initiatives are often carried out as a series of local projects, which can make demands for localized monitoring support that compete with those required for assessment of regional-scale dynamics. Large-scale regional monitoring typically requires a longer time horizon than does small-scale local project monitoring because the ecological processes anticipated at the landscape scale may develop over longer time frames and cannot be expected to be a simple sum of the trajectories of each local area. Thus, ecoregional monitoring programs require well-coordinated, multi-scalar components. One key to successfully conducting regional scale monitoring in an integrated fashion with local monitoring can come through collaboration among organizations. Coordinated regional reporting and decision-making mechanisms must be established, including the development of incentives for personnel to work in multidisciplinary, interorganizational settings to achieve common goals.

Finally, a history of past management practices can be used to make predictions about the impact of new ecoregional initiative plans. Historical data, including the use of old records and paleoecological reconstruction, are often critical for setting targets, even though historical data were frequently collected in an ad hoc fashion. Similarly, comparisons across space may be used as a surrogate for time in some cases to predict the effects of management actions. Fourqurean and Rutten (chap. 10) explain how disturbance events can also provide important learning opportunities if monitoring programs are structured to function through, and acquire information from, extreme events. A plan to capitalize on disturbance as an adaptive management learning source may favor some redundancy in data gathering to minimize the loss of data under such events.

Selecting Indicators

Indicators must be chosen to encompass the multiple scales of ecoregional initiatives. Although core sets of indicators should almost always be employed at all scales, different parameters may be required at each level of scale (both spatial and temporal). Some important attributes of ecosystem function may vary at scales not readily measurable in the field. Ecoregional initiative managers may have to evaluate trade-offs in attaining the goals of monitoring status versus monitoring of trends. In chapter 11, Hemstrom describes the integration of remotely-sensed and plot-based data used to achieve a goal of monitoring ecoregional vegetation change. Where available, satellite imagery and aerial surveys are powerful tools that can permit standardized depictions of resource status across scales and over broad areas (Vande Castle 1995; Bradshaw 1998). However, many key aspects of ecoregional status, such as the density and distribution of keystone animal species, may not be assessed remotely. Further, it may not be possible to match the demands of managers for information at "policy-relevant" scales (especially temporal scales) with the limitations on data gathering that are constrained by ecological scale (Rykiel 1998).

One challenge that arises in planning ecoregional monitoring is the choice of appropriate methods for monitoring at the regional scale given the expectation that efforts be sustained for many years as an initiative unfolds. The demands for data synthesis and comparability require that methods be as simple as possible to maximize potential for sustained consistency in data over the course of a multi-decadal project. On the other hand, indicators should include a variety of ecosystem attributes, from species, community, biogeochemical, and process domains. Consideration should be given to developing a mix of indicators ranging between those that are response-oriented and those that are predictive or stress-oriented (Noon, chap. 2). Focal species monitoring emphasizes monitoring of species that play a disproportionate functional role in an ecosystem by transferring matter or energy, structuring the environment, creating opportunities for other species, or regulating other species (classic keystone species). Frederick and Ogden (chap. 12) describe how information on individual species or species groups may provide links across habitats and system scales, which can be especially desirable where organisms move across areas within or outside a regional initiative. Areas outside the ecoregion that are unrelated to the initiative may influence species, especially those organisms that move beyond the ecoregion within their life cycle (e.g., Browder 1985; Walters et al. 1992; Frederick et al. 1996). Accounting for deleterious or beneficial effects on such species from environmental change outside the region is challenging, but the failure to do so can create misleading impressions of success or failure (Walters 1997).

In chapter 14, Gaines and others provide a comprehensive view of the types of indicators that can be considered for monitoring biodiversity. To evaluate ecosystem status and trends, a spectrum of indicators should usually be monitored because no single class of indicators will be highly sensitive across the range of both spatial and temporal scales relevant to a regional initiative. The motivation for choice of indicators ranges from their policy relevance to their level of inherent variability and statistical power to detect change (Osenberg et al. 1994). For example, estimates of a species' abundance may be relatively imprecise, but such estimates are often closely tied to the key targets of restoration (e.g., decline of a listed or charismatic species). Life history attributes, such as demographic performance measures, may be more informative and accurately tracked than abundance. Structural factors, such as habitat connectivity, or physical parameters such as nutrient pollution, may also be important indicators of restoration progress when it is impractical to monitor populations or communities. Contaminants, species or genetic diversity, and the distribution of invasive species are all attributes that may be intimately linked to ecoregional initiative objectives. Importantly, if restoration targets are biotic, the clearest indicators for such targets are not necessarily biotic. Physical factors can be proxies for biotic targets, providing that indicators can be accurately measured once linkages between the biotic and physical factors are well established. While it is nearly always desirable to incorporate some form of monitoring of biological targets when they are the motivation for management, including proxies may provide high information return for the monitoring resources expended.

Achieving Balance in Regional Monitoring Programs

A number of chapters in this book emphasize the need for monitoring to be comprehensive during the early phases of implementation. This is because there is invariably a need for information to test assumptions about causal relationships linking the indicators monitored with ecoregion initiative targets or performance measures. In many cases, such relationships must be made clear if monitoring programs are to become more "habitat-based" and thus more efficient. Quantitative model development can play a complementary role, enhancing system understanding and monitoring efficiency, when used in a coordinated fashion with monitoring and research (DeAngelis et al. 1998; Sklar et al. 2001).

The monitoring approach selected will have effects on the cost of implementation (fig. 15-2). When the number of indicators is high or the monitoring protocols are procedurally complex, costs will tend to be high. Characteristically, this type of monitoring is based on sampling the effects of program implementation. By necessity, monitoring programs tend to be retrospective, or effects-oriented, at the outset because information is either scarce or unusable as long as relationships about the linkages between ecosystem stressors and their effects remain as assumptions. Even when elegant ecosystem conceptual models are developed, the relationships that they describe cannot necessarily be validated using quantitative models and hypothesis testing research. Thus, a conservative approach often means that a number of indicators will be monitored, some of

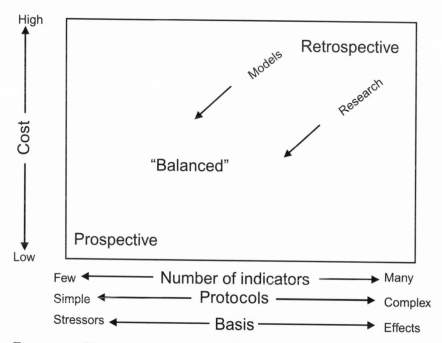

Figure 15-2. Monitoring approach: the effects of indicator number, biotic complexity, and monitoring basis on costs for ecoregional monitoring programs.

which may prove less useful than initially hoped at providing inferences about ecoregional initiative targets.

Over time, investment in effects-based monitoring can yield a payoff in lower cost provided that the transition toward the sampling of fewer or simpler indicators can be made. However, reducing the number of indicators sampled also means running the risk that the remaining indicators provide an oversimplified description of the ecosystem (fig. 15-2). Such a transition, therefore, not only requires the development of a robust understanding of cause and effect, but also requires supporting models and research that clarify the linkages between them. While some contend that the transition to solely stressor-based monitoring can be successfully achieved, actual examples appear to be rare. We therefore advocate programs that incorporate the efficiency and lower cost of prospective monitoring when investigating causes but that retain elements of effects-based monitoring to validate models, test assumptions, and support important, science-based monitoring research. This type of "balanced" approach is more likely to ensure the continued interest and involvement of scientists in ecoregional monitoring programs.

Role of Modeling and Information Management

In chapter 9, Sauer and others point out that deficiencies in the collection and management of data can impede effective trend analysis. Throughout this book, modeling and information management are emphasized as being important to the successful attainment of monitoring—and, by extension, ecoregional initiative—goals. Foresight about how information is collected and used is essential to getting the most from what are sometimes substantial investments in data acquisition.

Analytical Models

Although not required for formulating ecosystem conceptual frameworks, analytical models can facilitate the formalization of conceptual models and make the underlying assumptions of conceptual models explicit. As DeAngelis and others point out in chapter 6, all monitoring is ultimately indirect and requires "models" to interpret. Use of modeling allows for better interpretation of indicators, even permitting corrections for the inevitable challenges of changing monitoring technology and sampling intensity. Moreover, modeling permits integration of data gathered at different scales, permits resolution of spatial complexity, and may assist in identification of performance measures that lead to the development of efficient indicators. Ultimately, models can be used to evaluate causal relationships, permitting projection to the future for comparing management scenarios and evaluating success. Monitoring and modeling should develop in tandem in an iterative process that improves both enterprises.

Although analytical models can contribute to the monitoring enterprise, they are not a panacea. A major challenge to such models is their complexity, which presents challenges in both the reality of the output and the time required to build and run the models. Highly parameterized models can produce output that is an artifact of the model structure but which is exceedingly hard to identify as such (Levin 1992). Extensive exploration of the parameter space for such complex models may be impossible because of time limitations. The experience in south Florida may be instructive, where the time frame over which managers and policy makers required answers from models has at times hampered the development and testing of complex simulation tools. This has provided an impetus for the development of less-complex landscape models that can be run quickly and with less technical expertise than simulation models (Curnutt et al. 2000). As complex as some management models become, their inability to incorporate all drivers means it may not be possible to resolve the relative impact of altering a subset of possible environmental drivers through a particular management action. Different models, incorporating various subsets of drivers, may correctly reveal that some or all are important. However, choosing the cogent model among several for predicting management outcomes may not be possible, a situation Walters (1997) has termed a "battle of models." Thus, models can help target investment in empirical analysis of environmental drivers and increase management efficiency, but they cannot replace empirical evaluation altogether.

Managing Information

Given the reality of fiscal and personnel constraints, there are few opportunities to build comprehensive ecoregion-scale monitoring networks from the ground up. Many of the most valuable programs for monitoring ecosystem initiatives have come from the institutionalization of what were originally long-term, broad-scale research projects. In some cases, a rich legacy of data has accumulated and its value has been recognized as the needs for trend analysis became more obvious. However, the ability to complete such analyses did not come without forethought and a cooperative spirit on the part of those responsible for collecting, storing, and analyzing information. Adherence to sound information management increasingly allows ecologists to use data collected by other specialists to address questions at broad spatial, temporal, and thematic scales (Michener et al. 1997). It is clear that data must be managed effectively if there is to be hope for development of functional ecosystem monitoring systems at a regional scale. Information management is a crosscutting issue supporting all aspects of adaptive management from the collection of monitoring data, to reporting, to initiating change in management approaches (Palmer and Mulder 1999). Some basic tenets should be adhered to as systems of information support for ecoregion monitoring are designed. Monitoring information management systems should address the information assembly and transfer inherent in adaptive management and adaptive monitoring (fig. 15-3). Monitoring information management systems should foster data defensibility, access, use, and permanence. Information management needs are both geospatial and tabular (i.e., nonspatial biological, physical, chemical, social, and economic attributes must be addressed). A structured approach to system development will result in coherent targets and better achievement of goals.

Exemplary design processes for monitoring information systems should utilize steps similar to these:

1. Analyze current information.
2. Define new requirements.
3. Design a new information management system.
4. Develop the system.
5. Implement the system.
6. Evaluate and maintain the system.

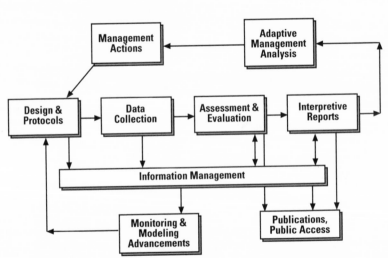

Figure 15-3. Overview of monitoring functions, including pathways for (1) protocol design, data collection, and analysis; (2) information management and adaptive monitoring; and (3) adaptive management, including the decision-making process.

An information infrastructure must be in place to support adaptive management with findings and recommendations from monitoring programs in a time frame that matches project development. For many ecosystem initiatives, it is conceivable that the data required to make an information management framework viable will need to be assembled from the monitoring systems of several different organizations. Also, there is likely to be an added responsibility to share monitoring information and analyses with ecosystem initiative partners, including the public. The challenges and pitfalls facing the designers of information management systems for regional monitoring networks are many and diverse, and incentives for data management are not always clear at the outset of program design. Data management is often not included in project funding, yet costs for this aspect of adaptive management can reach 35–40 percent of a viable monitoring program's budget (Davis in press). Because the reporting and use of metadata standards is not yet common, duplication and inconsistency in data collection and management often leads to lost opportunities. Past experiences indicate that full-time data managers are needed for large-scale initiatives. Resource specialists are not necessarily the best information managers and vice versa; team approaches are generally required to make monitoring information systems functional. Years of valuable legacy data are frequently at the greatest risk of loss. This happens when there is no provision for quality assurance, standardization of protocols, or common record formats for monitoring data, and access to monitoring information becomes practically limited to those that collect it.

Conclusion

Perhaps the ultimate challenge for ecoregional initiatives is combining information on ecological conditions and processes with what are judged to be socially and financially acceptable states. As covered in chapter 7 by Reynolds and Reeves, expert systems provide a formal method for combining information from disparate data sources. While encouraging the use of quantitative monitoring information from probabilistic sampling designs, this approach permits use of imprecise or incomplete systems of data and makes use of logical constructs to develop relationships from potentially abstract concepts relating to ecosystem condition. We feel that one of the most critical uses of monitoring information is for communication about uncertainty at key steps as a large ecoregional initiative progresses. Once managers are aware of the uncertainty associated with initiative targets, they can act to reduce uncertainty or otherwise minimize its associated risks.

As we pointed out in our introductory chapter, several large ecoregional initiatives are now well underway. Sharing lessons among similar efforts should be an important priority of those involved in order to avoid reinventing the proverbial wheel. We have attempted in this book to contribute to such a transfer of ideas, partly by presenting several examples that showcase what has been developed and implemented to date. Beyond this, our greater goal has been to identify elements of ecoregional monitoring systems that can, and ultimately must, be integrated to produce a functional and complete program. Our hope is that future initiatives will add to this knowledge so that ecosystem monitoring can become a well-accepted process that rewards society for its investment of financial and institutional resources with the ability to manage natural systems efficiently and effectively.

LITERATURE CITED

Allen, T. F. H., and T. W. Hoekstra. 1992. *Toward a unified ecology*. New York: Columbia University Press.

Angermeier, P. L. 2000. The natural imperative for biological conservation. *Conservation Biology* 14:373–381.

Bradshaw, G. A. 1998. Defining ecologically relevant change in the process of scaling up: Implications for monitoring at the "landscape" level. Pp. 227–249 in *Ecological scale: Theory and applications*, edited by D. L. Peterson and V. T. Parker . New York: Columbia University Press.

Bricker, O. P., and M. A. Ruggiero. 1998. Toward a national program for monitoring environmental resources. *Ecological Applications* 8:326–329.

Browder, J. 1985. Relationship between pink shrimp production on the Tortugas and water flow patterns in the Florida Everglades. *Bulletin of Marine Science* 37:839–856.

Callicott, J. B., L. B. Crowder, and K. Mumford. 1999. Current normative concepts in conservation. *Conservation Biology* 13:22–35.

Carpenter, S. R., and J. F. Kitchell. 1993. *The trophic cascade in lakes*. Cambridge, U.K.: Cambridge University Press.

Curnutt, J. L., J. Comiskey, M. P. Nott, and L. J. Gross. 2000. Landscape-based spatially explicit species index models for Everglades restoration. *Ecological Applications* 10:1849–1860.

Davis, G. E. In press. Vital signs monitoring program design, implementation, and applications: A case study from Channel Islands National Park, California. In *Measuring and monitoring biodiversity for conservation science and adaptive management*, edited by J. K. Reaser and F. Dallmeier. Washington, D.C.: Smithsonian Institution Press.

DeAngelis, D. L., L. J. Gross, M. A. Huston, W. F. Wolff, D. M. Fleming, E. J. Comiskey, and S. Sylvester. 1998. Landscape modeling for Everglades ecosystem restoration. *Ecosystems* 1:64–75.

DeAngelis, D. L., W. F. Loftus, J. C. Trexler, and R. E. Ulanowicz. 1997. Modeling fish dynamics in a hydrologically pulsed ecosystem. *Journal of Aquatic Ecosystem Stress and Recovery* 6:1–13.

Duncan, G. J., and G. Kalton. 1987. Issues of design and analysis of surveys across time. *International Statistical Review* 55:97–117.

Frederick, P. C., K. L. Bildstein, B. Fleury, and J. C. Ogden. 1996. Conservation of large, nomadic populations of white ibises (*Eudocimus albus*) in the United States. *Conservation Biology* 10:203–216.

Gibbons, W. N., and K. R. Munkittrick. 1994. A sentinel monitoring framework for identifying fish population responses to industrial discharges. *Journal of Aquatic Ecosystem
Health* 3:227–237.

Gunderson, L .H., C. S. Holling, and S. S. Light. 1995. *Barriers and bridges to the renewal
of ecosystems and institutions.* New York: Columbia University Press.

Halbert, C. L. 1993. How adaptive is adaptive management? Implementing adaptive management in Washington State and British Columbia. *Reviews in Fish Biology and
Fisheries* 1:261–283.

Hargrove, W. W., and J. Pickering. 1992. Pseudoreplication: A sine quo non for regional
ecology. *Landscape Ecology* 6:251–258.

Harris, G. P. 1980. Temporal and spatial scales in phytoplankton ecology. Mechanisms,
methods, models, and management. *Canadian Journal of Fisheries and Aquatic Sciences*
37:877–900.

Holling, C. S. 1986. The resilience of terrestrial ecosystems: Local surprise and global
change. Pp. 292–317 in *Sustainable development of the biosphere*, edited by W. C. Clark
and R. E. Munn. Cambridge, U.K.: Cambridge University Press.

———, ed. 1978. *Adaptive environmental assessment and management.* New York: John
Wiley and Sons.

Jassby, A. D. 1998. Interannual variability at three inland water sites: Implications for sentinel ecosystems. *Ecological Applications* 8:277–287.

Karr, J. R. 1981. Assessment of biotic integrity using fish communities. *Fisheries* 6:21–27.

Karr, J. R., and E. W. Chu. 1999. *Restoring life in running waters.* Washington, D.C.:
Island Press.

Levin, S. A. 1992. The problem of pattern and scale in ecology. *Ecology* 73:1943–1976.

Michener, W. K., J. W. Brunt, J. J. Helly, T. B. Kirchner, and S. G. Stafford. 1997.
Nongeospatial metadata for the ecological sciences. *Ecological Applications* 7:330–342.

Morrison, M. L., and B. G. Marcotte. 1995. An evaluation of resource inventory and monitoring program used in national forest planning. *Environmental Management*
19:147–156.

Mulder, B. S., B. R. Noon, T. A. Spies, M. G. Raphael, C. J. Palmer, A. R. Olsen, G. H.
Reeves, and H. H. Welsh. 1999. The strategy and design of the effectiveness monitoring program for the Northwest Forest Plan. General Technical Report GTR-PNW-
437, U.S. Department of Agriculture, Forest Service, Portland, Ore.

National Park Service. 2001. *The national resource challenge.* [Online at
www.nature.nps.gov/challenge/nrc.htm]

National Research Council. 1995. *Review of EPA's Environmental Monitoring and
Assessment Program: Overall evaluation.* Washington, D.C.: National Academy Press.

NSTC (National Science and Technology Council). 1997. *Integrating the nation's environ-
mental monitoring and research networks and programs: A proposed framework.* The
Environmental Monitoring Team, Committee on Environment and Natural Resources,
National Science and Technology Council, Washington, D.C.

Osenberg C. W., R. J. Schmitt, S. J. Holbrook, K. E. Abusaba, A. R. Flegal. 1994.
Detection of environmental impacts: Natural variability, effect size, and power analysis.
Ecological Applications 4:16–30.

Palmer, C. J., and B. S. Mulder. 1999. Components of the effectiveness monitoring pro-
gram. Pp. 69–97 in *The strategy and design of the Effectiveness Monitoring Program for the
Northwest Forest Plan*, edited by B. S. Mulder et al. General Technical Report PNW-
GTR-437, U.S. Department of Agriculture, Forest Service, Pacific Northwest Station,
Portland, Ore.

Patten D. T., D. A. Harpman, M. I. Voita, and T. J. Randle. 2001. A managed flood on the
Colorado River: Background, objectives, design, and implementation. *Ecological
Applications* 11:635–643.

Povilitis, T. 2001. Toward a robust natural imperative for conservation. *Conservation Biology* 15:533–535.

Powell, T. M. 1995. Physical and biological scales of variability in lakes, estuaries and the coastal ocean. Pp. 119–138 in *Ecological time series*, edited by T. M. Powell and J. H. Steele. New York: Chapman and Hall.

Rykiel, E. J., Jr. 1998. Relationships of scale to policy and decision making. Pp. 485–497 in *Ecological scale: Theory and applications*, edited by D. L. Peterson and V. T. Parker. New York: Columbia University Press.

Schmitt, R. J., and C. W. Osenberg, eds. 1996. *Detecting ecological impacts: Concepts and applications in coastal habitats*. San Diego: Academic Press.

Sklar, F. H., H. C. Fitz, Y. Wu, R. Van Zee, and C. McVoy. 2001. The design of ecological landscape models for Everglades restoration. *Ecological Economics* 37:379–401.

Stoddard, J. L., C. T. Driscoll, J. S. Kahl, and J. H. Kellogg. 1998. Can site-specific trends be extrapolated to a region? An acidification example for the northeast. *Ecological Applications* 8:288–299.

Stow, C. A., S. R. Carpenter, K. E. Webster, and T. M. Frost. 1998. Long-term environmental monitoring: Some perspectives from lakes. *Ecological Applications* 8:269–276.

Urquhart, N. S., S. G. Paulsen, and D. P. Larsen. 1998. Monitoring for policy-relevant regional trends over time. *Ecological Appliations* 8:246–257.

Vande Castle, J. 1995. Integration of spatial analysis in long-term ecological studies. Pp. 48–53 in *Ecological time series*, edited by T. M. Powell and J. H. Steele. New York: Chapman and Hall.

Walters, C. 1997. Challenges in adaptive management of riparian and coastal ecosystems. *Conservation Ecology* 1(2):1. [Online at www.consecol.org/vol1/iss2/art1/index.htm]

Walters, C. J. 1986. *Adaptive management of renewable resources*. New York: McMillan.

Walters, C. J., and R. Green. 1997. Valuation of experimental management options for ecological systems. *Journal of Wildlife Management* 61:987–1006.

Walters, C. J., L. Gunderson, and C. S. Holling. 1992. Experimental policies for water management in the Everglades. *Ecological Applications* 2:189–202.

Woodward, A., K. J. Jenkins, and E. G. Schreiner. 1999. The role of ecological theory in long-term ecological monitoring: Report on a workshop. *Natural Areas Journal* 19:223–233.

Glossary

adaptive ecosystem management. A process for testing hypotheses through management experiments in natural systems, collecting and interpreting new information, and making changes based on monitoring information to improve the management of ecosystems. An ongoing process conducted during the implementation of an ecosystem initiative for monitoring how well ecosystem management achieves its desired targets and for using monitoring assessments as a basis for making improvements in the design and operation of management plans.

adaptive monitoring. Similar to and integrated with the concept of adaptive management, the idea that monitoring programs should themselves be adaptable because of changes in ecosystem management goals, better knowledge about targets, or more efficient means for collecting monitoring information. The challenge for those designing and operating long-term monitoring programs is to make such programs adaptable while preserving the validity of the information collected through the use of effective data management and the use of core indicators to make status and trend detection valid through periods of change.

assessment. The process whereby monitoring data are interpreted in the context of questions and issues pertaining to ecosystem management targets. Assessment also includes the development of statistical relationships from monitoring data, other forms of data analyses and interpretation, and modeling linked to the objectives of the initiative.

attribute. A component of a natural system or conceptual model of such a system. A subset of the potential biological elements or components of natural systems, which is representative of overall ecological conditions or important human values. Attributes can be used to select indicators to monitor the known or hypothesized effects of ecosystem stressors.

baseline. Condition of a system at time zero to which system changes may be referenced and compared. Choosing a point in time as a baseline is generally a subjective and arbitrary process resulting in choices for baselines from among temporal phases such as (1) upon the initiation of a new management regime, (2) prior to European settlement, or (3) upon the acquisition of the first adequate or complete set of monitoring data.

compliance monitoring. Tracking and verifying implementation of regulations or management plans; synonymous with *implementation monitoring*.

conceptual ecological model. A graphical representation of a set of relationships among factors important to the function of an ecosystem or its subsystems. Conceptual models generally link stressors, which may be natural or representative of anthropogenic perturbation, with ecosystem attributes, which are often the source for monitoring indicators.

core indicators. A subset of the overall suite of potential indicators that is represented at all times and places in a monitoring data collection program. Core indicators are important for comparing subregions within ecoregional monitoring areas and to the applicability of monitoring information across projects conducted at different spatial and temporal scales.

critical pathway. A component of a conceptual model. The known or hypothesized linkage(s) connecting stressor(s), ecological effects, and attributes, which appears to explain a high proportion of the ecological response shown in the model (i.e., the ecologically most important response pathways resulting from a stressor).

ecoregion. Ecosystems of regional geographic extent.

ecoregional initiative. Programs that seek to alter land and water management to conserve and restore species and environments, as well as social and economic systems, across regional landscapes.

effectiveness monitoring. Evaluating system status and trends resulting from the implementation of a management plan; evaluating whether the management plan achieves the desired outcomes or predicted targets.

effects-oriented monitoring. Monitoring indicators of ecosystem response to management actions using response-oriented indicators.

health. Term used to indicate ecosystem status relative to targets or ideals. Because targets and ideals are usually subjective, this term is often ambiguous.

implementation monitoring. Also known as *compliance monitoring,* this form of status and trend detection helps evaluate how closely management plan prescriptions were followed.

indicator. A parameter that characterizes an important aspect of a natural system's condition and which may be measured to provide direct or indirect information on the state of an ecological resource.

inventory. A comprehensive determination of a system's status at a single point in time. An initial inventory of a system's status is often used as the baseline condition.

monitoring. The systematic process of collecting and storing data related to particular natural and human systems at specific locations and times. Determination of a system's status at various points in time yields information on trends, which is fundamental to the potential for monitoring to detect system change (also termed *status and trend detection*).

performance measure. The specific measures for ecosystem initiative targets. Collectively, performance measures provide an index of ecosystem condition and the overall progress of a management plan toward achieving its targets.

predictive monitoring. Monitoring an indicator that is predictive of ecosystem response in the future (a leading indicator) and is linked to a critical stressor.

prospective monitoring. An approach to monitoring that involves the determination of the status and trend of stressors that have demonstrated linkages to ecosystem effects. Also known as *predictive monitoring* because the cause and effect linkages are well-enough established to make reliable predictions about ecosystem effects in response to stressors. A program of retrospective monitoring, research, and modeling is often envisioned as maturing into one where much of the monitoring is prospective in nature.

protocol. The standardized methodology employed to collect data about a monitoring indicator.

response-oriented indicator. An indicator that responds to ecosystem state change and reflects response to stressors. *See* effects-oriented monitoring.

retrospective monitoring. An approach to determining ecosystem status and trend based on an evaluation of effects that can be linked via assumptions to causes. Retrospective monitoring is applied where information permits assumptions to be made about cause-and-effect linkages, but the accumulation of monitoring information, research hypothesis testing, or modeling does not support a prospective approach. *See* prospective monitoring.

status. The condition of an indicator, or suite of indicators, demonstrating a system's condition at one point in time.

stressor. A component of a conceptual model. A physical, chemical, or biological entity or process that induces effects on individuals, populations, communities, or ecosystems. Stressors include natural processes and structures, including forms of disturbance (e.g., naturally caused fire, native pathogens, or climate-driven hydrological change), and may also represent anthropogenic alterations of natural processes or structures (e.g., invasion of exotic species, stream diversion or impoundment, or vegetation manipulation).

stress-oriented indicator. An indicator that is predictive of ecosystem response in the future (a leading indicator) linked to a critical stressor. *See* predictive monitoring.

target. The desired endpoint of an ecosystem management strategy or plan, often dynamic or expressed in terms bounded by a range of variability.

trend. The change in a system's status between two or more points in time. Monitoring of indicators' status at multiple points in time is fundamental to the ability to detect trends.

validation monitoring. Monitoring that is specifically directed at the testing of assumptions about a system's cause-and-effect relationships, specifically at cause-and-effect relationships existing between management activities and monitoring indicators or the resources being managed.

Contributors

Jim Alegria, U.S. Bureau of Land Management, P.O. Box 2965, Portland, OR 97208

Sarah Bellmund, Biscayne Bay National Park, 40001 State Road 9336, Homestead, FL 33043

Kelly Burnett, USDA Forest Service, Pacific Northwest Research Station, 3200 SW Jefferson Way, Corvallis, OR 97331

Laura A. Brandt, South Florida Water Management District, 3301 Gun Club Road, West Palm Beach, Florida 33416-4680

David E. Busch, USGS, Forest and Rangeland Ecosystem Science Center, Regional Ecosystem Office, P.O. Box 3623, Portland, OR 97208

John H. Chick, Illinois Natural History Survey, Great Rivers Field Station, 450 Montclair Ave., Brighton, IL 62012

E. Jane Comiskey, Department of Ecology and Evolutionary Biology, University of Tennessee, Knoxville, TN 37996

Raymond L. Czaplewski, USDA Forest Service, Rocky Mountain Forest and Range Experiment Station, 240 W. Prospect, Fort Collins, CO 80526-2098

Steven M. Davis, South Florida Water Management District, 3301 Gun Club Road, West Palm Beach, FL 33416-4680

Donald L. DeAngelis, USGS Florida Caribbean Science Center, Department of Biology, University of Miami, Coral Gables, FL 33124

James W. Fourqurean, Department of Biological Sciences and Southeast Environmental Research Center, Florida International University, Miami, FL 33199

Peter Frederick, Department of Wildlife Ecology and Conservation, University of Florida, P.O. Box 110430, Gainesville, FL 32611-0430

William L. Gaines, USGS Grand Canyon Monitoring and Research Center, current affiliation: Okanogan and Wenatchee National Forests, 215 Melody Lane, Wenatchee, WA 98801

Barry D. Gold, U.S. Geological Survey, Conservation and Science Office, The David and Lucille Packard Foundation, 300 Second St., Los Altos, CA 94022

Louis J. Gross, Department of Ecology and Evolutionary Biology, University of Tennessee, Knoxville, TN 37996

Lance H. Gunderson, Department of Environmental Studies, Emory University, Atlanta, GA 30322

Richy J. Harrod, USDA Forest Service, Okanogan and Wenatchee National Forests, 215 Melody Lane, Wenatchee, WA 98801

Miles A. Hemstrom, USDA Forest Service, Pacific Northwest Research Station, Portland Forestry Science Laboratory, 620 SW Main St., Portland, OR 97208

John F. Lehmkuhl, USDA Forest Service, Pacific Northwest Research Station, Forestry Sciences Laboratory, 1133 N. Western Avenue, Wenatchee, WA 98801

William A. Link, USGS Patuxent Wildlife Research Center, 11510 American Holly Drive, Laurel, MD 20708

William F. Loftus, USGS Florida Caribbean Science Center, Everglades National Park Field Station, 40001 State Road 9336, Homestead, FL 33034-6733

Wolf M. Mooij, Centre for Limnology, Netherlands Institute of Ecology, Rijksstraatweg 6, 3631, AC Nieuwersluis, The Netherlands

Barry Mulder, U.S. Fish and Wildlife Service, 911 NE 11th Avenue, Portland, OR 97232

James D. Nichols, USGS Patuxent Wildlife Research Center, 11510 American Holly Drive, Laurel, MD 20708

Barry R. Noon, Department of Fisheries and Wildlife Biology, Colorado State University, Fort Collins, CO 80523

M. Philip Nott, The Institute for Bird Populations, P.O. Box 1346, Point Reyes Station, CA 94956-1346

John C. Ogden, South Florida Water Management District, 3301 Gun Club Road, West Palm Beach, FK 33416-4680

Craig J. Palmer, Harry Reid Center for Environmental Studies, University of Nevada, Las Vegas, 4505 Maryland Parkway, Las Vegas, NV 89154-4009

Gordon H. Reeves, USDA Forest Service, Pacific Northwest Research Station, 3200 SW Jefferson Way, Corvallis, OR 97331

Keith M. Reynolds, USDA Forest Service, Pacific Northwest Research Station, 3200 SW Jefferson Way, Corvallis, OR 97331

Paul L. Ringold, Western Ecology Division, National Health and Environmental Effects Research Laboratory, Office of Research and Development, Environmental Protection Agency, 200 SW 35th Street, Corvallis, OR 97333

Leanne M. Rutten, Department of Biological Sciences and Southeast Environmental Research Center, Florida International University, Miami, FL 33199

John R. Sauer, USGS Patuxent Wildlife Research Center, 11510 American Holly Drive, Laurel, MD 20708

Lawrence E. Stevens, Grand Canyon Wildlands Council, P.O. Box 1315, Flagstaff, AZ 86002

Tim Tolle, USDA Forest Service, Region 6, P.O. Box 3890, Portland, OR 97208

Joel C. Trexler, Department of Biological Sciences, Florida International University, Miami, FL 33199

Index

Abundance: Braun-Blanquet scale, 264–265, 270; ecosystem, 308–310; indices, 389–390; as performance measure, 359; relative, 359, 360, 361, 385

Across Trophic Level System Simulation (ATLSS), 363

Adaptive ecosystem management (AEM): Colorado River ecosystem, 102–127, 113, 122–124; definitions, xii, 229, 425; improvement through feedback, 126, 150–151, 412; link with monitoring, 10, 31, 411–412; overview, 101–102, 190

Adaptive Management Working Group (AMWG), 104, 109, 111, 112–114, 122

Adaptive monitoring: Aquatic Conservation Strategy, 86, 88, 89; for biodiversity, 379; Colorado River ecosystem, 111–127; definition, 425; design, 81–90; disadvantages, 88, 90; examples, 85–88; Florida Everglades, 139, 150–160; goal-setting, 81, 112–114; three barriers to, 82–85

Administrative aspects. *See also* Personnel: "champions," 102; Colorado River ecosystem, 105–110, 123–124; institutional inertia, 412; multijurisdictional areas, 136, 280–281, 411; tribal, 105, 315

Aerial surveys, 233, 282, 326, 327, 328, 390, 415

Age structure, 271, 272, 277, 280

Algae: blooms, 261; calcareous green, 274, 276; competition, 261; periphyton, 152, 153, 154, 156, 160; phytoplankton C:N:P ratio, 263

Alligator, 137, 152, 153, 155, 157–158, 175, 358

Allozyme variation, 394–395

Analytic hierarchy process (AHP), 207

Anthropogenic perturbation. *See* Stressors, human

Aquatic Conservation Strategy (ACS), 76–78, 86, 88, 89, 190–191, 199

Aquatic/Riparian Effectiveness Monitoring Program (AREMP), 205–206

Aquatic systems: conceptual model, 7, 114–115, 260–264; dissolved oxygen, 325; interaction with terrestrial systems, 110–111; marine. *See* Seagrass beds; productivity and food webs, 159–160

ArcView, 196, 201

Assumptions. *See* Validation monitoring

Attribute, definition, 28, 425–426

Background data. *See* Baseline; Reference conditions

Baseline: definition, 2, 4, 246;

Baseline: (*continued*)
development, 85–86; late-seral old-growth forest, 312–313; pre-Glen Canyon Dam data, 110; as restoration target, 353–354; "shifting baseline syndrome," 354
Bass, largemouth, 358, 365
Bayesian methods, 62, 195, 237, 238, 242
Beach habitat building flow (BHBF), 108
Bear: black, 391; brown, 391, 394; grizzly, 394
Belostomatidae, 357
Benchmark distribution, 57, 58
Benchmark value, 30–31, 56, 339, 353
Bioaccumulation, 321, 335, 366
Biodiversity: in conceptual models, 41, 42, 47, 52, 300; definition, 377; ecoregional, 20, 377–396; effects of disturbance, 111; forest, 15, 38; indicators, 171, 387, 389; legal mandates, 55, 378; in Northwest Forest Plan, 79, 300; role of focal species, 37; survey for initial status, 378
Biodiversity hierarchy theory, 377–378
Biodiversity monitoring, 20; community-ecosystem scale, 377, 385–388; landscape scale, 380, 381–385; population-species scale, 380, 388–395; three-phase approach, 379–381
Bird monitoring surveys. *See also* Breeding Bird Survey; Christmas Bird Count; Seabirds at Sea: assessment, 338–341; examples and data, 231–234, 329–337; Florida Everglades, 175, 321–344; limitations, 322; waterfowl, 233–234
Birds: banding or marking, 338–339; Chesapeake Bay, 342; endangered species, 108, 111, 120, 176; estimation of population change from survey data, 227–249; as flagship species, 175, 323–325; grassland, 243–244; migration, 241, 336, 337; nesting. *See* Nesting location; Nest initiation; Pacific Northwest forests, 301; rookeries in south Florida, 277, 367–368; vocalization point counts, 390; wading, Florida Everglades: attributes that led to success, 342–344; movement to other ecosystems,

335–337, 338–339; nesting pattern, 180–182, 367–368; ridge-and-slough model, 153, 155–156, 158, 160; role as indicators, 175, 323–325; surveys, 321–344
Braun-Blanquet abundance scale, 264–265, 270
Breeding Bird Survey (BBS), 228, 232, 234, 236, 238–239, 245, 322
Breeding site fidelity, 323

California, northwestern. *See* Northwest Forest Plan
Canadian Wildlife Service, 233–234
Canals. *See* Flood protection canals in Florida
Carbon: C:N:P ratio, 263–264, 265, 271, 273, 274, 275; sequestration, 52
Castilleja, 394
Catch-per-unit-effort (CPUE), 365, 390
Cattails, 138, 152, 155, 156, 159
Causation: link between stressors to indicators, 50, 51, 64, 142, 321; and magnitude of stressor, 40; in predictive monitoring, 34–35; use of monitoring and research, 31–32, 142
Celtis reticulata, 116
Central and South Florida Project (C&SF Project), 137, 138
Charismatic fauna, 20, 342–343, 353
Christmas Bird Count (CBC), 228, 233, 234, 236, 237, 239–241
Chub: bonytail, 107, 119; humpback, 106, 119
Chubsuckers, 157, 365
Cichlid, Mayan, 366
Ciconiiformes, 322
Clean Water Act (CWA), 30
Clematis, 394
Climate, 258–259, 298, 322. *See also* Hurricanes
Colorado River, 14, 103, 104, 106, 107, 108, 109, 111
Colorado River Basin Salinity Control Project, 13
Colorado River Compact, 106
Colorado River ecosystem. *See also* Grand Canyon: administrative context, 105–110; characteristics and change through time, 110–111; literature

review, 117–121; map, 104; monitoring, 102–127. *See also* Grand Canyon Monitoring and Research Center; overview, 103–104; restoration, 11

Colorado River Storage Project, 106

COMDYN program, 245, 248

Communication: dispute resolution techniques, 62; with the public, 65, 144, 204; with resource managers, 6, 144; with stakeholders, 9, 102

Community-ecosystem scale, 47, 171, 241–245, 377, 380, 385–388

Community structure analysis, 386–387

Compliance monitoring, 30, 38, 426

Compositionalism, 413

Comprehensive Everglades Monitoring and Assessment Plan, 148–150

Comprehensive Everglades Restoration Plan (CERP): applied science strategy, 141–143, 161; conceptual models, 143–148, 162; eight lessons learned, 160–162; modeling examples, 175–184; overview, 136, 138–139, 355–357; performance measures, 358–367

Conceptual models: Colorado River ecosystem, 110, 114–115; definition, 426; development, 41–42, 46–53; Florida Everglades, 143–148, 162; Northwest Forest Plan, 3, 16, 299–301, 318; south Florida seagrass beds, 260–264

Connectivity: definition, 293; between forest reserves, 79, 81; indices, 383; landscape, 52, 381; of Northwest forests, 291, 293, 309, 311–312, 317

Contaminants, 322, 328–329, 334–335, 359, 366–367, 416

Coordination of efforts: in biodiversity monitoring, 396; interagency monitoring, 76, 83, 84, 93, 94, 280–281; multijurisdictional aspects, 136, 280–281, 411; team approach for research, 149; use of knowledge-based systems, 205–206

Coral reefs, 258, 270, 276

Cost. *See* Financial aspects

Cottonwood, 11

Coyote, 390

Crayfish, 155, 157, 160, 323, 327, 356, 357, 358

Credibility, 215, 291

Crocodile, American, 175

Cultural aspects, 116, 120

Dams, 39. *See also* Glen Canyon Dam

Data. *See also* Information archiving; Information management; Information transfer: in adaptive management, 16, 17, 19; analysis. *See* Knowledge-based systems; Statistics; annual summaries, 214; archiving, 219, 419; assurance, 9; comparability, 83, 84, 88, 93, 415; duplicated information, 82; essential characteristics, 218; gaps in: advantages of knowledge-based systems, 203–204; Colorado River ecosystem,124, 110; Florida Everglades, for fish, 363; NetWeaver computation of influence, 198, 201; Pacific Northwest forests, 84, 85; inconsistent or changing collection methods, 83–84; information flow, 213–214; management systems, 214–216, 218–220, 222–223, 419–421; metadata, 215, 222; permanence, 211–212, 215, 218, 408, 420; quantification of imprecise information. *See* Fuzzy logic; survey, 90, 227–249; visualization, 268–269

Database development and management, 8, 9, 408

Data entropy, 215, 222

Decision-making: analytic hierarchy process, 207; challenges in, 407–408; indicator changes which trigger, 33; and information flow, 213–214; linked with monitoring, 5, 8, 46, 61–64, 87, 142, 161, 291, 411–412; regional capacity, 93, 416–417; use of both monitoring and research findings, 5, 87; use of marginal return concept, 86; use of statistical testing, 44

Decision theory, 46, 63

Decomposition, 42, 292, 293

Detectability, 230–231, 249, 391

Die-offs, 137, 259

Dioxins, 334

Disease, 299, 324, 329

Disturbance events: anthropogenic. *See* Stressors, human; effects on biodiversity, 111; large-scale habitat

Disturbance events: (*continued*)
loss, 79; in monitoring design, 283;
natural. *See* Climate; Fire; Hurricanes;
overview, 29, 39, 40; succession
following, 29; in temporal scale
decisions, 92
Disturbance indices, 382–383
Diversity indices, 386–387
DNA analysis, 394–395
Documentation, 122–124, 212, 214, 219
Dominance indices, 382, 383
Douglas-fir, 300, 301, 314
Dove, mourning, 238–239
Dragonfly naiads, 357
Driving factors, 12–14, 33, 154, 407

Eagle, bald, 120
Ecological diversity, 292, 308–310
Ecological fitness, 37
Ecological integrity, 37–38, 413
Ecological resilience, xiii, 193
Ecological risk assessment, 141
Ecological succession. *See* Succession
Economic aspects. *See* Financial aspects
Ecoregion, 2, 426
Ecoregional initiatives: challenges for
decision makers, 407–408; challenges
for ecologists, 405–407; definitions, 2,
12–14; drivers for, 12–14;
implementation, 405, 416–417. *See also*
Goal-setting; Indicators, selection;
Targets; major U.S., 3; monitoring
biodiversity for, 377–396; monitoring
within broader frameworks, 408–411;
outline of topics, 18–21; role of
information management, 419–421;
role of modeling, 418–419
Ecosystem abundance, 308–310
Ecosystem aspects: conceptual models.
See Conceptual models; function as
determinant for monitoring, 14–16;
guilds, 387; "health," 6, 30, 37, 138,
167, 193, 413–414, 427; lack of
reference conditions, 406; modeling
underlying variables, 176–177;
monitoring as early warning system,
31; predator-prey relationships, 52,
159–160, 179, 324–325, 331, 367;
variability, 41–42, 64, 172, 370–371,
384–385

Ecosystem-community scale. *See*
Community-ecosystem scale
Ecosystem management, 189
Ecosystem Management Decision
Support (EDMS), 195–202, 204, 207
Ecosystem science, 4–5, 10, 32, 141–143,
161, 405–406
Edge effects, 303, 305, 382, 384
Education. *See* Information transfer;
Outreach programs
Eel, Asian swamp, 369
Effectiveness monitoring, 5, 167,
205–206, 291, 298, 302, 317, 426
Effects-oriented monitoring. *See*
Retrospective monitoring
Egret, 322; great, 334, 335; little, 324;
snowy, 324, 335
Endangered species: charismatic fauna,
20; Colorado River ecosystem, 106,
107, 108, 111, 119–120; Florida
Everglades, 144, 176; qualification and
allozyme variation, 394; as species-
based indicator, 52, 389
Endangered Species Act (ESA), 11, 13,
53, 103, 106, 378, 412
Endpoints, 6, 38, 141, 383
Engineer species, 52, 54. *See also* Focal
species
Environmental impact statements (EIS):
Glen Canyon Dam, 103, 107, 108,
125; Northwest Forest Plan, 290, 297
Environmental Monitoring and
Assessment Program (EMAP), 32, 49
Environmental surveillance. *See*
Monitoring
Eustronglyides, 324
Eutrophication: Florida Everglades, 141,
152, 154, 155, 159; Florida seagrass
beds, 258, 261–264, 277; model,
261–264
Evenness index, 386, 387
Everglades. *See* Florida Greater
Everglades restoration
Everglades Landscape Model (ELM),
363
Everglades National Park, 12, 140, 157,
178, 281, 325–327, 355–357
EVERKITE model, 179, 180
Exotic species, 386, 416; Colorado River
ecosystem, 11, 14, 111, 124; Florida

Everglades, 14, 153, 366, 369; Pacific Northwest forests, 299

Expert systems. *See* Knowledge-based systems

Extinction, 30, 247, 248, 392, 393

Failure, reasons for, 32–34, 43, 102, 412

Falcon, peregrine, 120

False negatives. *See* Type II errors

False positives. *See* Type I errors

Field sampling methods: airboat electrofishing, 365; an adaptive approach, 86, 87; birds, 230–231, 249; Braun-Blanquet assessment, 264–265, 270; choosing where to measure, 266–268, 277, 283; design considerations, 33, 59–61; financial aspects, 90, 91; location redundancy, 283; radiotelemetry, 184, 391; sample frame, 231; simplicity of methods, 283; spatial scale issues, 90, 91–92; transects, 326, 327, 328, 391

Financial aspects: biodiversity monitoring, 379; data management, 222, 421; economy of scale, 91, 407–408; fiscal cycles and ecosystem response time, 261; implementation, 91, 407–408, 417; indicator selection, 51; long-term budget with adaptive approach, 88; sampling strategy selection, 90, 91, 407–408

Fir, Pacific silver, 290

Fire: Colorado River ecosystem, 14; Florida Everglades, 14, 153, 154, 155, 156, 159; Pacific Northwest forests, 292, 298–299, 307, 309, 315; prescribed, 298; as stressor, 39, 352; suppression, 309

Fish: allozyme variation, 394; anadromous, 12, 40, 56, 199, 342, 394; Colorado River, 12, 106, 107, 108, 111, 119; Columbia River, 342, 394; exotic species, 11, 14, 111, 366, 369; Florida Everglades, 153, 157, 333, 341, 351–371; habitat suitability, 199, 204; large, monitoring communities, 365–66; mercury in, 152, 157, 335, 358, 366–367, 368; monitoring surveys, 40, 341, 356; in Northwest Forest Plan, 40; nursery habitat, 124;

parasites, 111; water temperature as indicator for, 56

Flagfish, 361

Flagship species. *See* Focal species

Flood protection canals in Florida, 137, 138, 152–153, 258–259, 369

Floods: in Aquatic Conservation Strategy, 78; Colorado River, 103, 111, 123–124; Florida Everglades, 152, 154; and nest abandonment, 324

Florida Bay. *See* Seagrass beds

Florida Everglades: bird surveys, 321–344; fish. *See* Fish, Florida Everglades; historical record, 140, 329, 339, 341–342, 351–371; history of monitoring, 354–357; hydrological patterns, 176–177, 325, 329–331; hypotheses to explain changes, 154–156; invertebrates, 153, 155, 157, 356, 358; maps, 326, 332, 355; as migration site, 337; modeling examples, 175–184; nine ecological subregions, 147; pre-drainage condition, 140, 151, 154, 325; ridge-and-slough conceptual model, 151–160; salinity, 138, 155, 333; "true" and "greater," 136–137

Florida Fish and Wildlife Conservation Commission, 366

Florida Greater Everglades restoration: adaptive monitoring plan, 148–160; applied science strategy, 135–136, 141–143, 161; conceptual model and performance measures, 143–148, 162; eight lessons learned, 160–162; endpoints, 13; goal-setting, 351–371; legislation, 13, 38–139; overview, 11–12, 138–139; scientific strategy. *See* Comprehensive Everglades Restoration Plan; uncertainties, 139–140

Florida International University, 357, 358

Florida Keys National Marine Sanctuary, 147, 258, 266, 268. *See also* Seagrass beds

Florida Water Conservation Areas, 12, 158, 325–327, 332, 341

Flycatcher, southwestern willow, 108, 111, 120

Focal species: definition, 37; Florida Everglades, 175; Northwest Forest Plan, 15, 40, 54, 290; overview of

Focal species: (*continued*)
  concept, 53–55; selection, 55, 389; as
  species-based indicator, 52, 53, 415–416
Food resources: and foraging success, 325,
  333; and hydrological patterns,
  329–331; and reproductive success,
  323–324, 328, 337, 358
Food web: Florida Everglades, 159–160,
  323–325, 334; Northwest forests, 298;
  role of birds, 321, 323, 334; as "top-
  down" biotic process, 407
Forest Ecosystem Management
  Assessment Team (FEMAT), 190, 191,
  290, 297, 312, 392
Forest Health Monitoring, 409
Forest Inventory and Analysis (FIA),
  220, 409
Forest management, 39, 74, 79, 297, 298,
  301, 305, 306
FORPLAN, 204
Fractal indices, 382, 383, 384
FRAGSTATS software, 384
Frog: hylid, 160; northern leopard, 111,
  119; pig, 157, 358; ranid, 160
Fuzzy logic, 193–195, 197, 198, 199,
  200–203, 204

Gap analysis, 388
Gar, Florida, 365
Generalized additive models, 238
Generalized linear models, 234, 236, 238
Genetic drift, 392
Genetic monitoring, 393–395
Genetic scale, 47, 171, 377, 380
Geographic information systems (GIS),
  174, 176, 195, 382
Geomorphology, 52, 103, 111
Gibbs Sampling, 242
Glen Canyon: historical background,
  106–109; monitoring for adaptive
  management, 102–105, 109–127
Glen Canyon Dam. *See also* Colorado
  River ecosystem: Environmental
  Impact Statement, 103, 107, 108, 125;
  map, 104; Record of Decision, 105,
  125; USFWS decisions, 107, 108
Glen Canyon Environmental Studies
  (GCES), 13, 107, 108, 123, 124–125,
  127
GLIM software, 234, 237, 239, 240

Global positioning systems (GPS), 170
Global scale, 21, 410
Glossary, 425–428. *See also* Terminologies
Goals: agency *vs.* interagency, 94;
  challenges of adaptive approach, 88;
  compositionalism *vs.* functionalism,
  413; conversion to measurable targets,
  135, 142, 168; federal hierarchy, 76, 81;
  refinement from policy to monitoring
  activities, 82; in survey data, 229–230,
  249
Goal-setting, 9, 10, 36–39, 81, 112–114,
  351, 370, 413–414
Grain. *See* Spatial grain; Temporal grain
Grand Canyon. *See also* Colorado River
  ecosystem: historical background,
  106–109; monitoring for adaptive
  management, 102–127; water-release
  programs, 17; World Heritage Site, 103
Grand Canyon Monitoring and Research
  Center (GCMRC), 105, 108, 109, 110,
  111–126
Grand Canyon National Park, 11, 103, 105
Grand Canyon Protection Act, 105, 108,
  125
Grasses, terrestrial, 40, 176, 243
Groundwater, 14, 159, 258
Guilds, 387

Habitat connectivity. *See* Connectivity
Habitat degradation and loss: Colorado
  River ecosystem, 103–104, 108,
  110–111; Florida Everglades, 137–138,
  152; large-scale disturbances, 79;
  Pacific Northwest forests, 79, 109;
  species' sensitivity to, 53; as stressor,
  39; threshold effect on extinction, 30
Habitat fragmentation, 299, 383, 384
Habitat-species relationships. *See*
  Landscape models
Habitat-use patterns, 322
Heavy metals, 329, 368
Hemlock, western, 290, 301, 317
Heron, 322; great white, 324
Historical background: Colorado River
  ecosystem, 105–110; Florida
  Everglades, 137–138, 351–371; south
  Florida seagrass beds, 282
Historical record: for benchmark value of
  indicators, 30–31, 339, 414–415;

Colorado River ecosystem, 124–125; effects of inconsistent methods, 83; Florida Everglades, 140, 329, 339, 341–342, 369; late-seral old-growth forest, 314; limitations in, 413; of management actions, 92–93; need for information management, 211–212; passerine birds, 322; use in biodiversity monitoring, 384–385

Hurricanes, 259, 275–276, 279–280, 355, 356

Hydrological patterns, 37; Florida Everglades, 176–177, 325, 329–334, 341–342, 360; south Florida, 257–259

Hydrological restoration. *See* Florida Greater Everglades restoration

Hypotheses: concerning the Florida Everglades, 151, 154–156, 158, 160, 162; in coordination of science with monitoring, 4, 146, 148; in indicator monitoring, 44, 61–62; null, 44, 61, 248; ranking for levels of uncertainty, 146, 158; testing, 6, 62, 151, 162

Ibis, 322; white, 155, 160, 323, 333, 336, 337, 358

Implementation monitoring, 5, 80, 167, 427

Indicators: allozyme variation, 394; benchmark value, 30–31, 339; of biodiversity, 171, 387, 389; birds, 177, 321–325, 335; condition, 28, 65, 86; core, 426; definition, 51–53; of ecological sustainability, 51–53, 389; expected values and trends, 43–44, 55–59; Florida panther, 183; frequency distribution of values, 57–59; of initiative performance, 6; late-seral old-growth forest, 49, 302–305; for legislative implementation and compliance, 30; links to stressors, 50; the need for, 169–171; for predation rate, 173–174; proxy, 6; seagrass C:N:P ratio, 262, 279; selection, 40, 41, 42–43, 50–51, 64, 415–416; snail kite, 179; spatial grain of, 77; as surrogates for endpoints, 38, 170; types, 52–53, 171, 429; use of a "cluster," 170; values that trigger response, 44–45, 58; wood stork, 181

Indicator species, 53, 54. *See also* Focal species

Information. *See* Data

Information archiving, 124–125, 419

Information management: approaches, 221; budget considerations, 421; Colorado River AEM, 124–125; concepts, 218–220; and quality assurance, 211–224; role in ecoregional initiatives, 419–421

Information transfer, 124–125, 213–214, 419, 420

Infrastructure, 88, 420–421

Initiatives. *See* Ecoregional initiatives

Insect outbreaks, 299

Interagency monitoring, 76, 83, 84, 93, 94. *See also* Multijurisdictional aspects

Interagency monitoring design group (MDG), 80

International scale, 1, 21, 408–409, 410

Internet, 125, 215, 268

Invasive species. *See* Exotic species

Inventory: definition, 2, 427; local species completeness, 248; of south Florida seagrass beds, 269; of U.S. forests, 220

Invertebrates, Florida Everglades, 153, 155, 157, 356, 358

Isolation index, 383

Jackknife estimator, 245

Kestrel, American, 239, 240

Keystone species, 52, 54, 389, 415. *See also* Focal species

Killifish: bluefin, 360, 361; least, 361; marsh, 361; pike, 366

Kite, snail, 175, 178–180, 331, 335, 358

Knowledge-based systems: analysis products, 199–202; definition, 192; evaluation within larger contexts, 204; extending to multiple spatial scales, 205; as framework for collaboration, 205–206; landscape analysis with EMDS system, 195–198; to monitoring and evaluation, 192–195; Nestucca Basin, 198–202; repeatability and adaptability, 206

Lake Havasu, 11

Lake Mead, 11, 104, 106

Lake Okeechobee, 137, 138, 355. *See also* Florida Greater Everglades restoration
Lake Powell, 11, 103, 104, 107
Landsat Thematic Mapper (TM), 298, 299, 301, 315
Landscape analysis, 195–198, 204, 205, 382–384
Landscape heterogeneity, 52, 381
Landscape models, 15, 86, 87
Landscape scale: in biodiversity monitoring, 380, 381–385; in bird survey data, 229; in conceptual model, 47; Florida Everglades, 159; indicators, 171; international efforts, 1; lack of replicate ecosystems, 406; late-seral old-growth forest indicators, 297, 298, 302–303, 317–318
Land-use aspects, 47, 296
Late-seral old-growth forest (LSOG): definitions, 292, 293–296; fragmentation analysis, 383; indicators, 49, 302–305; monitoring, 39, 289–318; in Northwest Forest Plan, 79; Oregon Coast Range example, 314–317; use of inventory data, 220; vegetation mapping, 294–296, 312–313
Late-successional reserve assessments (LSR), 379
Lessons learned, 160–162, 221–224, 283, 317–318
Limpkin, 358
Lizard, zebra-tailed, 111
Locally weighted least squares (LOESS), 234, 238
Long-Term Ecological Monitoring Program, 410

Management. *See* Natural resource management
Man and the Biosphere, Human Dominated Systems program, 141
Manatee, West Indian, 335
Mangrove swamps, 155, 331, 357
Maps, use in biodiversity monitoring, 384, 385
Markov-Chain Monte Carlo methods (MCMC), 228, 237, 242, 243–244
Marl prairies, 138, 154, 155, 359, 368–369
Marshes. *See* Florida Greater Everglades restoration

Mercury: bioconcentration in birds, 329, 334, 335; in fish, 152, 157, 335, 358, 366–367, 368
Mesquite, 11, 116
Methods: biodiversity monitoring, 380–381; changing, 83–84, 412; inconsistent, 83, 84; nonexistant, 82; sampling. *See* Field sampling methods; standardization issues, 83–84, 93, 339
Methylmercury. *See* Mercury
Metrics, landscape, 382–384
Mexico, 106, 371. *See also* Colorado River ecosystem
Migration, 241, 336, 337, 416
Models: abiotic conditions, 178–180; analysis and interpretation, 182–184; bioenergetics, 174; calibration, 363; capture-recapture, 145–146; conceptual. *See* Conceptual models; ecosystem components in, 41–42, 46–53, 146, 148, 176–177; eutrophication, 261–264; feedback processes, 48, 172; Florida Everglades, 175–184; generalized additive, 238; habitat conditions, 55; hierarchical, 242–245, 249; landscape scale. *See* Landscape models; log-linear, 234, 235; monitoring of ecosystems, 7, 167–186. *See also* Monitoring, conceptual framework; "natural system," 351–352, 359, 363, 364, 367–370; the need for, 169–171, 184–186; Pacific Northwest forests, 305; population, 177–182; predictive, 305; ridge-and-slough, 152–153; role in ecoregional initiatives, 418–419; sensitivity analyses, 185; simulation, 354, 365, 419
Mollies, sailfin, 361
Monitoring: achieving balance, 416–417; analysis of results, 182–184, 199–202, 204, 268–269; an outline of, 18–21; biodiversity. *See* Biodiversity monitoring; Colorado River ecosystem, 111–127; compliance. *See* Compliance monitoring; conceptual aspects, 16–18; conceptual framework: Colorado River ecosystem, 110, 114–115; Florida Greater Everglades restoration, 143–148, 162; Northwest Forest Plan,

3, 16, 299–301; overview, 5–8, 14, 16. *See also* Goals; Objective-setting; south Florida seagrass beds, 260–264; conceptual issues. *See also* Terminologies; a catalyst for change. *See* Monitoring, link with decision-making; expected values and trends, 55–59; key components, 33, 36–46, 168, 169–173; the legacy, 32–34; model development, 46–53; overview, 27–30, 64–65, 167–168; predictive or retrospective, 31, 34–36, 172–173; what can monitoring tell us, 31–32; why is monitoring necessary, 30–31; continuum of research and, 4–5, 10, 32; coordination of efforts. *See* Interagency monitoring; definitions, 2, 27, 167, 427; design of programs. *See* Monitoring design; as a discipline, 8–9; early warning system, 31; ecoregional within broader frameworks, 408–411; ecosystem function as determinant for, 14–16; effectiveness, 5, 167, 205–206, 291, 298, 302, 317, 426; failure, 32–34, 43, 102, 412; focal species. *See* Focal species; genetic, 393–395; implementation, 5, 80, 167, 427; importance, 1–21; late-seral old-growth forest, 314–317; link with decision-making, 5, 8, 46, 61–64, 87, 142, 161, 291, 411–412; Northwest Forest Plan. *See under* Northwest Forest Plan; objective-setting, 9, 82, 168; predictive. *See* Predictive monitoring; program evaluation, 10; retrospective, 34–36, 417–418, 428; spatial scales, 90–92; statistics, 59–61; surveys. *See* Survey data; temporal scales, 92–93; three-phased approach, 5–6; validation, 5, 80, 175, 297, 406, 428

Monitoring design: accommodating multiple scales, 60, 90–93, 167–186; administrative aspects. *See* Infrastructure; Personnel; as an adaptive process, 81–90, 283; eight steps to, 15–16, 65; information management issues, 213–224; interagency group, 80–81; Northwest Forest Plan, 290; seagrass communities of south Florida, 260–269, 283; three essential elements, 33, 169–173

Monte Carlo methods, 174, 228, 237, 242

Morphological variation, 393–394

Mosquitofish, 335, 358, 360, 361

Multiagency activities. *See* Interagency monitoring

Multijurisdictional aspects, 136, 280–281, 411

Murrelet, marbeled, 12, 15, 40, 54, 298, 391

Muskrat, 111

National Academy of Sciences, 107, 108, 123, 127

National Audubon Society, 325, 326. *See also* Christmas Bird Count

National Biological Service, 356

National Environmental Policy Act (NEPA), 13, 105, 106, 412

National Forest Management Act (NFMA), 30, 38, 378

National Marine Fisheries Service (NMFS), 74, 199

National parks, 30

National Park Service (NPS), 38, 75, 105, 106, 378, 410

National Park Service Organic Act, 106

National Research Council (NRC), 107, 123, 127

National scale, 105, 106, 408–410

Native Americans, 116

Natural resource management: data management goals, 214–216; linking ecoregional initiatives to, 9, 381; multijurisdictional, 136, 280–281, 411; policy in adaptive management, 16, 17; ties with monitoring, 10, 55–56, 142, 381; use of fuzzy logic, 193–194

Natural resource managers, 6, 85, 405–406

Natural system model (NSM), 351–352, 359, 363, 364, 367–370

Nature reserves, 410, 413

Nesting location, 322, 331–333, 337

Nesting population size, 322, 330–331

Nest initiation, 180–182, 333–334

Nests, alligator, 152

Nestucca Basin, Oregon, 198–202

NetWeaver, 194, 196, 197, 201–202, 206
Nitrogen: C:N:P ratio, 263, 265, 271,
    273, 274, 275; fixation, 292–293
North American Bird Conservation
    Initiative, 229, 232
North Coast Watershed Assessment
    Project, 206
Northwest Forest Plan: abundance
    outcomes and thresholds, 309–310;
    aquatic objectives. See Aquatic
    conservation strategy;
    Aquatic/Riparian Effectiveness
    Monitoring Program, 205–206; area,
    11, 290; biodiversity monitoring, 379;
    conceptual model, 7, 16; ecological
    objectives, 77–79; information flow,
    213–214; land ownership, 75, 315;
    landscape pattern metrics, 382–384;
    late-seral old-growth forest
    monitoring, 39, 49, 220, 289–318;
    maps, 75, 291; monitoring design,
    80–93; monitoring goals, hierarchy of
    federal, 76, 81; national and
    international implications, 409; Oregon
    Coast Range example, 314–317;
    overview, 11, 74–80, 190–191;
    planning units, 75, 84, 90, 93;
    presidential conference, 74; Record of
    Decision, 13–14, 190; spatial scales,
    90–92; standards and guidelines, 91;
    temporal scales, 92–93; Terrestrial
    Objectives, 79; use of forest inventory
    data, 220
Nutrient cycling, 37, 153
Nutrient enrichment. See Eutrophication

Objective-setting, 9, 82, 168
Observability, 230–231, 249, 391
Old-growth forest, 15, 39, 74, 296. See
    also Late-seral old-growth forest;
    Northwest Forest Plan
Oregon Coast Range. See Northwest
    Forest Plan
Oregon Department of Forestry, 84
Organochlorines, 329
Otter, 155; Colorado River, 111
Outcomes, predicted, 10, 31, 62–63,
    308–312
Outreach programs, 102, 125
Owl, northern spotted: DNA analysis,

395; importance of scale in monitoring,
    11, 74, 378; map of range and federal
    land ownership, 75; as priority and
    focal species, 15, 40, 54
Oxygen, dissolved, 325

Pacific Northwest. See Northwest Forest
    Plan
Panther, Florida, 12, 175, 182, 183, 184,
    335, 392
PANTRACK model, 183
Patchiness, 52, 83, 381, 382, 383
Peer review, 102, 116, 122
Performance measures: as assessment
    tools, 145; definitions, 167, 427;
    Florida Greater Everglades restoration,
    142, 143–150, 156–158, 161, 353–354,
    358–367; restoration, 359–365;
    selection, 353, 354, 408; use of term, 6,
    353; workable number for sustained
    monitoring, 149
Periphyton, 152, 153, 154, 156, 160
Personnel: data managers, 421; demands
    upon, 88, 93; incentives, 94;
    management and supervision, 94, 396;
    quality assurance training, 222; senior
    scientists, 162; volunteers, 232
Perturbation. See Disturbance events;
    Stressors
Pesticides, 329, 335
Phosphorous: C:N:P ratio, 263–264, 265,
    271, 273, 274, 275; eutrophication,
    141, 152, 154, 155, 159; as limiting
    factor, 365
Pikeminnow, Colorado, 106, 119
Poecilia latipinna, 361
Poisson distribution, 234, 236–237, 249
Poisson regression, 238, 362
Policy: in adaptive management, 16, 17;
    links with monitoring, 10, 81, 82, 88,
    89, 142, 161, 317; policymakers in
    adaptive monitoring design, 85;
    "policy-relevant" scales, 415
Polychlorinated biphenyls (PCB), 334
Population: definitions, 336, 388;
    delineation, and bird migration, 336;
    distribution surveys. See Aerial surveys;
    estimation of change in, 227–229,
    234–249, 390–391; local extinctions,
    247, 248; source and sink, 336;

trajectories, 235, 236, 239; variation factors other than stressor, 321
Population modeling, 177–178
Population-species scale, 47, 171, 377, 380, 388–395
Population variability analysis (PVA), 391–392
Prawns, freshwater, 356
Predator-prey relationships, 52, 159–160, 179, 324–325, 331, 367
Predictive monitoring: Bayesian parameter estimation, 62–63; comparison with actual response, 145; definition, 427; the need for, 172–173; overview, 15, 31, 34–36, 64
Primary productivity: effects of eutrophication, 261–262; Florida Everglades, 152, 153, 160, 368; as indicator, 37, 40, 52; measurement methods, 265–266
Productivity, 159–160, 331–333. See also Primary productivity
Projection, 172, 173
Proposals, request for, 122
Proposition, 196, 203
Prospective monitoring, 417–418, 427
Protocols, 4, 122, 428
Provincial scale, 89, 297
Public lands, multi-use, 30
Public perception: charismatic fauna, 20, 342–343, 353; wading birds, 342–343

Quality assurance: alternative approaches, 221; conceptual bases, 213–220; definition of terms, 217; need for, 211–212; recommendations, 221–224
Quality assurance project plan, 217, 218, 222
Quality control, 9, 217, 353
Quality management, 217
Quality system, 217

Radiotelemetry, 184, 391
Random amplified polymorphic DNA analysis (RAPD), 395
Ranking: stressors, 65; trends, 242, 243, 244; uncertainty, 146
Rare species, 85–86, 87
Reclamation Act, 106
Recognizable taxonomic units (RTU), 388

RECOVER team, 149, 150, 151
Recreation, 116, 120–121, 144, 259, 298
Redfield ratio, 263–264
Reference conditions. See also Baseline: and historical record, 92–93; late-seral old-growth forests, 307–312; limitations in, 406, 413; restoration to the "natural system," 351–352, 413; use in biodiversity monitoring, 384–385
Reference sites, 352, 363, 371
Refuges, 410, 413
Regional scale: in Aquatic Conservation Strategy, 89; bioregional assessment, 9; and conservation of biodiversity, 378; initiatives. See Ecoregional initiatives; need to integrate programs, 93, 414; Northwest Forest Plan, 74, 76; vs. local scale, 91
Regulatory aspects, 12–14, 30, 38, 53, 378
Relative abundance, 359, 360, 361, 385
Remote sensing: late-seral old-growth forests, 294–296, 303, 305, 313, 314–315; seagrass beds, 281–282; use of indicators, 6; for vegetation change, 415
Reproductive success, 322, 327–328, 336
Research: "best available science" criteria, 138; Colorado River ecosystem, 116–121; competitive approach, 116; continuum of monitoring and, 4–5, 10, 32, 406–407; late-seral old-growth forest, 313–314; peer review, 102, 116, 122; prioritization strategy, 146, 150, 158; project and program review, 122–124; seagrass beds, 282; study site selection, 407; team approach, 149
Resources. See Natural resource management
Restoration: application of performance measures, 359–365; in Aquatic Conservation Strategy, 76, 77, 78; Colorado River ecosystem, 113, 114; definitions, 139, 140, 141; Florida Everglades. See Florida Greater Everglades restoration; to the "natural system," 351–352, 413; program evaluation, 10; science strategy for, 135–136, 141–143, 370
Retrospective monitoring, 34–36, 417–418, 428

Rhynchospora, 154
Ridge-and-slough habitat, 151–160, 357, 359, 360, 368–369
Riparian systems. *See also* Colorado River ecosystem; Streams: beach habitat, 108, 109; conceptual model, 7; floodplain processes, 7; plant communities, 14; primary productivity and nutrient enrichment, 261–264; recreational use, 116; suitability indicators, 6
Robustness, 246

Salinity: Colorado River, 14; Florida Everglades, 138, 155, 333; seagrass beds, 270
Saltcedar, 11, 124
Sampling. *See* Field sampling methods
SAS software, 237
Satellite imagery, 282, 415
Sawgrass, 138, 151, 154, 156, 159, 176, 369
Scale. *See also* Spatial scale; Temporal scale: accommodating multiple, 60, 90–93, 167–186, 205; economy of, 91, 407–408; importance in ecoregional initiatives, 405; selection, 47, 49–50
Seabirds at Sea, 322
Seagrass: age structure, 271, 272, 277, 280; C:N:P ratio, 263–264, 265, 271, 273, 274, 275, 279; salinity tolerance, 270
Seagrass beds: effects of coral reefs, 270–271; effects of hurricanes, 276, 279–280; Florida Everglades, 137; moving from status to trends, 276–283; recolonization, 276, 279–280; seasonal aspects, 269, 271–273, 275; south Florida, 257–259, 261–264
Sediment load, 39, 103–104
Sediment regime, in Aquatic Conservation Strategy, 78
Sensitive species, 52, 53
Sensitivity analyses, 185
Sentinel sites, 407
Sequestration, 52, 159
Shannon index, 386, 387
Shrimp: grass, 160, 356, 357; penaeid, 335
Signal, 29, 51, 64
SIMPDEL model, 182, 183, 184
Simpson index, 386

SIMSPAR model, 178
Siuslaw National Forest, 199, 314, 315
Snails: apple, 157, 178–179, 356, 358; Kanab ambersnail, 107, 108, 119
Societal aspects, 15, 46, 53, 116. *See also* Recreation
Soil accretion, 152, 155, 156
Soil erosion, 154, 368
South Florida Ecosystem Restoration Task Force, 13
South Florida Water Management District, 138, 356
Sparrow: Baird's, 244; Cape Sable seaside, 175, 176–177, 186; Henslow's, 243–244
Spatial aspects: addressed in initiatives, 10; choosing where to measure, 266–268, 277–278; role in ecosystem protection, 74, 76–80; species-level population analysis, 247–248; visualizing patterns in data, 247–248, 268–269
Spatial grain, 76, 77, 322
Spatially explicit individual-based model (SEIB), 174, 178, 179, 182
Spatially explicit species index model (SESI), 176
Spatial scale: accommodating multiple, 37, 90–92, 167–186, 205, 261, 277–278, 414; and biodiversity, 301; birds as indicators, 322, 342; ecological response to management action, 56, 57; in field survey sampling, 90; use of geographic strata, 237–238, 240; variability of environmental features, 84
Species co-occurrence, 248
Species dispersal, 79, 83–84, 247, 312
Species distribution, 81
Species diversity, 385–387. *See also* Biodiversity
Species groups, analysis of composite change, 241–245
Species-habitat relationships. *See* Landscape models
Species-population scale, 47, 171, 377, 380
Species richness, 245–248, 385–387
Species turnover, 247
Spikerush, 154, 357, 368
Spoonbill, 322
Stakeholders, 9, 102, 105, 109, 114, 125, 161

Standardization issues: data permanence, 211–212, 215, 218, 408, 420; with interagency monitoring, 280; methods, 83–84, 93; surveys, 249, 339

Stand-scale, 296, 297, 298, 303, 306, 314, 315

Stand structure, 301, 303, 313

Statistics: abundance indices, 389–390; biodiversity monitoring data analysis, 380, 381; decision theory, 46; diversity indices, 386–387; false positives or negatives. *See* Type I errors; Type II errors; frequency distribution of indicator values, 57–59; gap analysis, 388; landscape indices, 382–384; parameter estimation *vs.* hypothesis testing, 44, 62; population variability analysis, 391–392; power, 44, 60, 61; practical issues, 59–61; precision, 60, 61; sensitivity or bias of test, 44, 383; significance issues, 241; software packages, 234, 237, 239, 240, 384; time-series analysis, 234, 274, 276

Status, 2, 28, 378, 415, 428. *See also* Status and trend detection

Status and trend detection: estimation of population change, 227–229, 234–249; for indicator variables, 43–44, 55–60, 64; late-seral old-growth forest, 306–307; south Florida seagrass beds, 276–283; three-phased approach, 17; use in development of monitoring, 8, 410–411

Stochastic variation, 29. *See also* Disturbance events; Stressors

Stork, wood: as focal species, 175; food resources and breeding timing, 323; nesting timing, 333; nest initiation, 180–182; population estimation, 336; surveys, 325–326, 328

Streams. *See also* Aquatic Conservation Strategy; Riparian systems: adaptive monitoring design, 86, 88, 89; channel processes, 7; cold-desert, 387; indicators of in-stream condition, 77–78, 86; in-stream flow, 78

Stress-oriented monitoring. *See* Predictive monitoring

Stressors. *See also* Disturbance events; Fire: in adaptive management model,

16, 17; definitions, 29, 428; ecological resistance to, 37; examples, 39; human: in conceptual model, 48, 115; Florida Everglades, 137–138, 152–153, 161; Florida Keys, 258; historical conditions, 352; Pacific Northwest forests, 191, 298; in temporal-scale decisions, 92; link to indicators, 50, 304; natural, 7. *See also* Climate; Fire; Hurricanes; in predictive and retrospective monitoring, 34; ranking, 65

Success. *See* Performance measures

Succession, 17, 29, 299, 308, 313–314

Sucker, razorback, 119

"Survey and manage" species, 85–86, 87

Survey data: aerial and satellite, 282, 326–328; biodiversity, 378; birds of the Florida Everglades, 326–328, 329–337; estimation of change in population, 227–229, 234–249; indirect indices for, 390; sampling and spatial scale, 90; seagrass beds, 282

Sustainability: ecological, 51–53, 189, 206; Florida Everglades restoration goals, 141, 160–161; forest management, 74; Montreal Process criteria, 206

Systematic Reconnaissance Flight program (SRF), 327, 328, 339–340

Targets: conversion from goals and objectives, 135, 144–145, 413–414, 416; definitions, 6, 428; Florida Everglade fish restoration, 358–367

Target variables. *See* Performance measures

Technical Working Group (TWG), 105, 109, 114

Temporal aspects: addressed in initiatives, 10; data loss over time, 211–212, 215; "long term" in late-successional old-growth forests, 307–308; long time-series data, 342; time lags in ecosystem processes, 17, 56, 57, 261, 406

Temporal grain, 92, 322

Temporal scale: accommodation of multiple, 37, 92–93, 167–186; effects of stressor, 54; eutrophication on seagrass beds, 263, 282–283; variability of environmental features, 28, 84

Terminologies, 2, 4, 5, 217, 293–296, 425–428
Terrestrial Objectives of Northwest Forest Plan, 79
Threshold effects, 411
Thresholds: criteria for contaminants, 158; ecological, 30; of indicator values, 44, 45, 56, 64; Northwest Forest Plan, 309–310, 311
Time-series analysis, 234, 274, 276, 342
Topography, 52, 298
Trees: establishment, maturation and death, 292; of Glen Canyon, 116; inconsistent classification methods, 83, 84, 93; islands in the Florida Everglades, 137, 152, 154, 155, 157
Trend: definitions, 4, 28, 428; detection. See Status and trend detection; Florida Everglades, 342; Pacific Northwest forests, 306–307, 316; ranked, 242, 243, 244; seagrass beds, 276–283; trade-offs with monitoring status, 415
Trophic levels. See Food web
Trout: bull, 394; cutthroat, 56; rainbow, 115
Type I errors, 31, 32, 44, 81
Type II errors, 32, 44, 45, 59, 81

Umbrella species, 52, 54, 389. See also Focal species
Uncertainty: in conceptual model, 45, 47; in decision analysis, 62; in the monitoring signal, 56, 64; in natural system model, 370; ranking hypotheses for levels of, 146, 158; in restoration, 136, 139–140
U.S. Army Corps of Engineers (ACE), 138, 356
U.S. Bureau of Land Management (BLM), 74, 75, 290, 315, 316, 317. See also Northwest Forest Plan
U.S. Department of Agriculture (USDA), 13–14, 75
U.S. Department of Energy (DOE), 216
U.S. Department of the Interior (DOI), 13–14, 75, 105, 190
U.S. Environmental Protection Agency (EPA), 32, 49, 75, 141, 216, 358

U.S. Fish and Wildlife Service (USFWS), 105, 107, 108, 124, 126, 233–234
U.S. Forest Service (USFS). See also Northwest Forest Plan: Pacific Northwest land ownership, 75, 290, 315, 316, 317; resource management plans, 74; Sierra Nevada lands, 49
U.S. Geological Survey (USGS), 105, 175, 356, 357
Utricularia purporea, 156

Validation monitoring, 5, 80, 175, 297, 406, 428
Value of indicator, 28, 29, 44–45
Vegetation: and bird biodiversity, 301; Colorado River ecosystem, 116; community structure, 78; ecological role, 78; exotic species, 111, 124; Florida Everglades, 152, 159; GIS-based mapping, 176; maps of late-successional forests, 294–296, 312–313, 315–317; population variability analysis, 392; reproductive strategies, 392; south Florida. See Seagrass; Seagrass beds; structure, 52, 301, 303, 313; trees. See Forest management; Trees

Warbler, cerulean, 238
Washington. See Northwest Forest Plan
Water lily, 154
Water management: canals and Florida seagrass beds, 257–258; canals in the Florida Everglades, 137, 138, 152–153; Colorado River, 123–124; effects of climate, 258; Pacific Northwest forests, 78
Water quality: in Aquatic Conservation Strategy, 77; as endpoint, 383; Florida Everglades, 152; monitoring, 9; south Florida seagrass beds, 258, 275, 278–279; temperature as indicator, 56–57; turbidity, 111
Water Resources Development Act (WRDA), 138–139
Watersheds, Pacific Northwest, 191, 198–199, 200, 385

Watershed scale, 76, 77, 191
Water table, 78
Weighted average, 238, 242
Western Area Power Administration, 105, 124
Wetlands. *See* Florida Everglades; Florida Greater Everglades restoration
Whale, bowhead, 62

White House Office on Environmental Policy, 74, 76
Willow, Goodding, 11, 111
Wind, 275–276, 279, 299. *See also* Hurricanes
Wolf, gray, 390
Woodpecker, red-cockaded, 55, 395